ADVANCES IN

AGRONOMY

VOLUME 31

CONTRIBUTORS TO THIS VOLUME

J.-C. Fournier

Umesh C. Gupta

M. G. Hale

L. D. Moore

K. Németh

P. H. Nye

Ginette Simon-Sylvestre

R. R. Smith

N. L. Taylor

G. L. Terman

ADVANCES IN

AGRONOMY

Prepared in cooperation with the

AMERICAN SOCIETY OF AGRONOMY

VOLUME 31

Edited by N. C. BRADY

International Rice Research Institute
Manila, Philippines

ADVISORY BOARD

H. J. GORZ, CHAIRMAN

K. M. KING   R. B. GROSSMAN   T. M. STARLING

J. B. POWELL   J. W. BIGGAR

M. STELLY, EX OFFICIO,

ASA Headquarters

1979

ACADEMIC PRESS

A Subsidiary of Harcourt Brace Jovanovich, Publishers

New York   London   Toronto   Sydney   San Francisco

709051

ACADEMIC PRESS, INC.
111 Fifth Avenue, New York, New York 10003

United Kingdom Edition published by
ACADEMIC PRESS, INC. (LONDON) LTD.
24/28 Oval Road, London NW1 7DX

LIBRARY OF CONGRESS CATALOG CARD NUMBER: 50–5598

ISBN 0–12–000731–2

PRINTED IN THE UNITED STATES OF AMERICA

79 80 81 82   9 8 7 6 5 4 3 2 1

CONTENTS

v

THE UNIVERSITY

OF MANITOBA

THE AVAILABILITY OF NUTRIENTS IN THE SOIL AS DETERMINED BY
ELECTRO-ULTRAFILTRATION (EUF)

K. Németh

VOLATILIZATION LOSSES OF NITROGEN AS AMMONIA FROM SURFACE-APPLIED
FERTILIZERS, ORGANIC AMENDMENTS, AND CROP RESIDUES

G. L. Terman

DIFFUSION OF IONS AND UNCHARGED SOLUTES IN SOILS AND SOIL CLAYS

P. H. Nye

BORON NUTRITION OF CROPS

Umesh C. Gupta

CONTRIBUTORS

Numbers in parentheses indicate the pages on which the authors' contributions begin.

J.-C. FOURNIER (1), *National Institute of Agricultural Research, Department of Soil Science, Laboratory of Soil Microbiology, Dijon Cedex, France*

UMESH C. GUPTA (273), *Research Branch, Research Station, P.O. Box 1210, Agriculture Canada, Charlottetown, Prince Edward Island, Canada C1A 7M8*

M. G. HALE (93), *Department of Plant Pathology and Physiology, Virginia Polytechnic Institute and State University, Blacksburg, Virginia 24061*

L. D. MOORE (93), *Department of Plant Pathology and Physiology, Virginia Polytechnic Institute and State University, Blacksburg, Virginia 24061*

K. NÉMETH (155), *Büntehof Agricultural Research Station, Bünteweg 8, 3000 Hannover 71, Federal Republic of Germany*

P. H. NYE (225), *Soil Science Laboratory, Department of Agricultural and Forest Sciences, University of Oxford, Oxford, England OX1 3PF*

GINETTE SIMON-SYLVESTRE (1), *National Institute of Agricultural Research, Department of Soil Science, Station of Soil Science, Versailles, France*

R. R. SMITH (125), *United States Department of Agriculture, Madison, Wisconsin 53716*

N. L. TAYLOR (125), *Department of Agronomy, University of Kentucky, Lexington, Kentucky 40546*

G. L. TERMAN* (189), *Soils and Fertilizer Research Branch, National Fertilizer Development Center, Tennessee Valley Authority, Muscle Shoals, Alabama 35660*

---

* Deceased.

PREFACE

Agronomy as a profession has a variety of meanings. But in all cases, it has a connotation that relates in some way to the production of crops. It is only natural that those scientists interested in crops and their culture find through agronomy a common interest with those interested in soils, the medium on which most crop plants grow.

Favorable weather in most parts of the world in 1977 and 1978 has permitted improved food production technologies to function. As a result, the yields of the primary food staples have increased and food stocks have risen to more acceptable levels. The technologies which have made possible this state of affairs are due in part to the work of soil and crop scientists and to their ability to communicate with each other internationally.

Contributions presented in this volume continue to give evidence of the international exchange of scientific information. The authors of the seven articles are from five countries, and these articles relate to problems of international significance. The first, summarizing research on the effects of pesticides on soil microorganisms, is in response to continued international concern with environmental quality. Likewise, a review of factors affecting the loss of ammonia from soils has implications for environmental quality but more importantly for the rising costs of energy required to produce nitrogen fertilizers.

Three articles are concerned with basic soil properties, their measurement, and factors affecting them. Root exudates is the subject of one, a follow-up of an earlier *Advances in Agronomy* article on this general subject. The other two articles are concerned with soil chemistry, one with the movement of ions in soils, and the other with a unique way of measuring the availability of essential elements in the soil.

A review article on boron nutrition brings scientists up to date on our knowledge of this important element. In addition, the genetics of one of the world's most important forage crops, red clover, should be of keen interest to crop and animal scientists alike.

Thanks must be extended to the ten authors who prepared articles for this volume. They have done a real service to their fellow soil and crop scientists.

N. C. BRADY

EFFECTS OF PESTICIDES ON THE SOIL MICROFLORA

Ginette Simon-Sylvestre* and J.-C. Fournier†

*I.N.R.A.,Department of Soil Science, Station of Soil Science, Versailles, France
and
†I.N.R.A., Department of Soil Science, Laboratory of Soil Microbiology,
Dijon Cedex, France

I. Introduction

The fertility of soil—that is, its capacity to produce more or less plentiful crops—depends not only on its physical constitution and its stock of nutrients, but also on the intensity of the biological processes that take place within it. Indeed, the activity of the microflora of the soil is generally favorable to vegetation—for instance, by the fixation of atmospheric nitrogen, the production

1

of nitrates, sulfates, and carbonic anhydride, the breakdown of animal and plant remains into compounds more easily available for the plants, and the removal from the soil of some of the diverse products that may be added to it, such as pesticides.

The soil appears to be a system biologically in equilibrium, but this equilibrium is a precarious one, and each disturbance of the environment presents the risk of modifying the activity of the microflora and consequently the soil's fertility. So the farmer, although he has no direct control over biological phenomena, may nevertheless try to influence them, either by his choice of farming methods or by application of various dressings (organic matter, lime, etc.). The increasing use of pesticides, although intended to protect the crops, may alter this equilibrium, by direct or indirect action, after short, average, or long periods of time, depending on whether the product acts quickly or persists longer in its initial state or in its metabolic forms. The study of these secondary effects of pesticides on the soil microflora indicates an interest that is all the more pronounced as the number of products offered to the farmers increases every year; treatments are often carried out at rather short intervals, they are sometimes superposed, and the chemical diversity of the products sold is very great. In 1972, the sale of pesticides reserved for agricultural use in France had already amounted to 1.4 billion francs—2% of the value of the products harvested; in the United States it reached 5%, and every five years since 1958 the consumption of pesticides has doubled. As long as the farmer must use these products in order to maintain the soil and its crops in a sanitary condition, there is an element of risk for the soil microflora; this risk cannot be disregarded. It seems justifiable to give our attention to the possible effects of pesticides on the soil microflora. The subject is broad, and it is difficult to establish its limits.

In addition to the technical difficulties of organizing a practical research program, both in establishing the experimental protocols and in performing the analyses, numerous other difficulties occur in the interpretation of the data. The small doses of pesticides used, as well as the often low solubility of these products, make their distribution in the soil very heterogeneous. The problems that arise in determining the value of the data derive from these defects in the conditions of experimentation. Methods of microbiological analysis often lack precision, and the interpretation of the data then becomes difficult; moreover, since the relationships between the various groups of microorganisms are not well known, it is not easy to establish such relationships among all the different effects recorded separately. The great variety of protocols used by research workers does not simplify the problem. As in all biological research, the experimental conditions can cause perceptible modifications in the biological processes. A final difficulty must be mentioned: In the association soil–plant–microflora to which a pesticide is applied, the distinction between the direct effects of the product on the microorganisms and the indirect effects that occur through the plant is sometimes difficult to establish and requires particular attention.

The following bibliographic review presents the data of research carried out during the last fifteen years, and more particularly during the last five years. The effects of pesticides on the microflora in general (count and activity) and on the microorganisms responsible for the major biological processes are considered in Sections III and IV. A few earlier publications have dealt with the subject, but with the increasing use of pesticides a readjustment and a broader base of information seem indispensable in determining the present state of knowledge. We do not mean to imply, however, that this review is exhaustive, so numerous are the papers in this field. In Sections V and VI, an innovation appears with the discussion of the incidence of pesticides in pathogenic microflora and in decomposing microflora. Until now, these subjects have been studied separately, most often by specialists, particularly in research concerning the pathogens. In addition, owing to the important part that technique plays in these studies of microflora in the presence of pesticides, we have presented first (Section II) a critical review of the various analytical methods and experimental protocols used.

II. Methodology Applied in the Study of the Effects of Pesticides on the Soil Microflora

A. CRITICAL SURVEY OF THE DIFFERENT METHODS USED

The microorganisms of the soil are extensive. According to Dommergues and Mangenot (1970), bacteria are the most numerous, varying in density between 10^6 and 10^9 bacteria per gram of soil. Actinomycetes and fungi are less dense by factors of 10 and 100, respectively. The biomass of this population is equally important. For the main groups, Pochon and De Barjac (1958) suggest the following figures (given in kilograms of living matter per hectare): 500 for bacteria, 700 for actinomycetes, and 1000–1500 for fungi. In comparison with these figures, the amounts of pesticide applied seem insignificant, generally about 1 ppm (= 1 mg/kg of soil), and rarely reaching 10 ppm. Moreover, the method of application of these products and their solubility (often very low) make their distribution into or on the soil heterogeneous, which leads to some difficulties in sampling, not only for chemical but also for microbiological analyses.

Apart from these general difficulties, the majority of workers studying the effects of pesticides on the soil microflora adopt the two groups of methods most often used by soil microbiologists: (a) Count methods for determining the total microflora and some special microorganisms. This is the direct measurement of the qualitative change appearing after pesticide treatments. (*b*) Indirect techniques for estimating the activity of the total microflora or of some microorganisms that play an important part in the biological cycles. We should also mention some particular methods such as the measurement of the biomass in the soils.

1. Composition of the Soil Microflora: Counts of the Microorganisms

Very rarely are counts of the microorganisms made directly with a microscope, in soil suspensions from experiments with pesticide treatments. Such counts have the advantage of being carried out under ecological conditions, but it is difficult to distinguish between living microorganisms and dead ones, for evidently the dead organisms are not counted in the characterization of the biological level of a soil. The use of fluorescence and of coloration techniques, however, makes it possible to reduce such errors to a certain extent.

In most cases, the count of the microorganisms is carried out after they have been grown on media favorable to their development. These are usually synthetic organic media, either liquid or solid (with gelose) or mineral (silicogel). The seeding is effected from inoculum suspensions, more or less diluted with water, into the medium or on the surface. After a period of incubation, the number of the microorganisms is estimated from the count of the colonies or from the estimation of the highest dilution that permits the growth of the microorganisms. In the latter case, probability tables, prepared by MacCrady (1915) and Swaroop (1951), indicate the number of microorganisms per gram of soil on the basis of the observed growths. This method of determination may be used on soil samples taken directly in the field from treated experimental plots, or in trials carried out *in vitro* in a laboratory, or in a greenhouse.

We have listed here some of the culture media used most often for counting the total microflora and the main groups of microorganisms.

For counting the total microflora, the soil aqueous extract seems to be preferred, whether it is liquid or solidified by gelose (Huge, 1970; Kaiser and Reber, 1970; Kaszubiak, 1970; Lozano-Calle, 1970; Catroux and Fournier, 1971; Karki *et al.*, 1973; Simon *et al.*, 1973; Simon-Sylvestre, 1974; Oleinikov *et al.*, 1975). Some workers add yeast and mineral salts to this medium (for example, Bunt and Rovira, 1955; Voets and Vandamme, 1970; Voets *et al.*, 1974). Other media used for the determination of the total microflora include that of Kaunat (1965) (mineral gelose medium with casein, peptone, and glucose) used by Kaiser and Reber, (1970); and Thoronton's medium, used by Sharma and Saxena (1974).

For fungi, Martin's medium (1950) with rose bengal is often used (Huge, 1970a; Kaszubiak, 1970; Voets and Vandamme, 1970; Tu, 1972; Camper *et al.*, 1973; Karki *et al.*, 1973; Simon *et al.*, 1973; Sharma and Saxena, 1974; Simon-Sylvestre, 1974). It contains streptomycin, which impedes the growth of bacteria without having the drawbacks of the acidified media (Houseworth and Tweedy, 1973; Focht and Josseph, 1974), in which certain fungi do not grow because of the excessive acidity, whereas other quickly growing species overgrow all the preparations. However, some fungi (*Pythiaceae,* in particular) are sensitive to streptomycin and cannot grow on media containing this antibiotic. Finally, we should mention the work of Kaiser and Reber (1970), who use a gelose medium with Maltea Moser and chloramphenicol.

Bacteria alone are often counted on nutritive gelose (Houseworth and Tweedy, 1973; Focht and Josseph, 1974) or on Czapek's gelose medium (Wainwright and Pugh, 1974). Spore-forming bacteria may be separated from the others by heating the soil for 10 minutes at 90 °C (Voets *et al.*, 19740 or by heating a soil suspension for 15 minutes at 75°C (Oleinikov *et al.*, 1975).

Actinomycetes are generally counted on media containing antibiotics: nystatine and actidione in the work of Camper *et al.* (1973), and albamycine and streptomycin in the studies of Focht and Josseph (1974). Sharma and Saxena (1974), on the other hand, use only a (nitrates + saccharose) medium.

The (gelose + sodium albuminate) medium is often chosen to evaluate bacteria and actinomycetes (Chandra, 1966; Tu, 1972; Camper, 1973). Van Faassen (1974), however, carries out the separation of bacteria and actinomycetes on a (soil extract + gelose) medium enriched in glycerol, asparagine, casein, glucose, and cycloheximide as antifungic.

In addition to providing estimates of the totals of the major groups of microorganisms, these methods may be used in research on the identification and evaluation of the percentages of various microorganisms. We must, however, place restrictions on the use of Martin's medium, which often affects the morphology of the colonies. This method of identification, which is long and dull, is suitable only for certain microorganisms, such as the pathogenic ones.

Some criticisms may be expressed concerning methods of dilution counts:

(1) Regardless of the technique used, the number of counted microorganisms is always much lower than that read by direct numeration, by a factor of 10–1000 on the average for the total numeration; very special organisms, such as anaerobic bacteria, autotrophic bacteria, and strictly cellulolytic organisms, do not grow because their feeding requirements are too specific.

(2) In spite of the standardization of the techniques of dilution, inoculation, and incubation, the precision of the measurements remains relative, for in addition to technical errors, there are the difficulties of sampling due to the unavoid-

able heterogeneity of the analyzed soils (Meynel and Meynel, 1965; Ricci, 1974). Whatever the circumstances, differences lower than a half power are not often significant.

(3) No distinction can be made between inactive microorganisms and those really active in the soil. In fact, the dilution methods often lead to the reduction or to the suppression of the phenomena of biostasis and synergism, which normally take place in the soil.

(4) The significance of the count of microorganisms sometimes has a restricted but far-reaching effect in the case of organisms with vegetative and reproductive forms, such as fungi. In fact, the mycelia may be cut into several pieces, the spores scattered and the count is not precise.

2. Measurement of Microflora Activity

An interesting method of testing the effects of pesticides on the soil microflora consists in measuring its biological activity. We may evaluate either the total activity or the activity of a particular group of microflora.

Measurement of the activity of the soil microflora provides ordinary indexes of the biological state of the soils and therefore of their fertility. We may determine the real activity of the microflora or its potential activity—that is, its ability to adapt to new ecological conditions, to the addition of various substances or substrates, or to the modification of any of the environmental factors.

There are two types of techniques for determining total activity: (a) Classical techniques, which allow the activity of the microflora proper to be characterized. The treated soil may then be used just as it is, as the basis of the analysis—the ecological conditions are then respected—or just for seeding of the media (grains of soil, suspension dilutions). (b) Other techniques, which determine the activity of the microflora by measurement of enzymatic activity.

a. *Classical Techniques. Measurement of biological activity.* The measurement of biological activity, described by Pochon and Tardieux (1962), consists in evaluating the rapidity with which the soil microorganisms grow in the more or less synthetic, liquid or solid media used for the counts of the microorganisms, and always in the presence of a specific substrate. The number of living microorganisms (see the tables of MacCrady) increases as the incubation progresses. The speed of the reaction is thus known, and also its limits.

These simple methods give a first approximation of the activity of the microflora, but most of the criticisms expressed with regard to the count methods by suspension dilutions apply here also.

These techniques are still used to study most of the major physiological groups of the soil microflora, for example, the microorganisms active in the nitrogen and carbon cycles: ammonifying bacteria (with tyrosine), nitrifying bacteria (with ammonium sulfate and sodium nitrite), denitrifying bacteria (with potassium nitrate), proteolytic bacteria (with serum), and amylolytic bacteria (with starch). They are mentioned in the reports of Huge (1970a), Kaiser and Reber (1970), Lozano-Calle (1970), Ritter *et al.* (1970), Catroux and Fournier (1971), and Simon *et al.* (1973).

Kinetic measurement of degradation of the substrates. All these kinetic measurements are made directly on the soil itself. In *incubation,* a substrate is added to the soil, which is incubated under favorable conditions of temperature and moisture. The products formed from the added substrate are chemically determined in the course of time. The activity curve of the microorganisms studied can be plotted from the data. This method is particularly suitable for studying ammonification and nitrification. The nitrogenous substrate added varies, depending on the workers. For nitrification, it is most often ammonium sulfate (Chandra and Bollen, 1961; Balicka and Sobieszczanski, 1969; Bardiya and Gaur, 1970; Dubey, 1969; Dubey and Rodriguez, 1970; Szember *et al.,* 1973; Wainwright and Pugh, 1973; Campbell and Mears, 1974; Focht and Josseph, 1974; Horowitz *et al.,* 1974a; Voets *et al.,* 1974; Abueva and Bagaev, 1975). Correa Salazar (1976) uses a variant of this method, in which soil suspensions take the place of soils. Sometimes the following substrates are used: monoammonium phosphate (Shaw and Robinson, 1960; Bartha *et al.,* 1967), ammonia (Chandra and Bollen, 1961), urea (Vlassak and Livens, 1975), asparagine (Szember *et al.,* 1973), cotton meal, (Eno, 1962), peanut oil cake (Akotkar and Deshmukh, 1974), or plant remains (Bliev, 1973).

In the case of organic substrates, the breakdown of the nitrogenous compounds allows measurement of the activity of the ammonifying as well as the nitrifying bacteria.

A variant of this technique, *percolation,* is also used for nitrification studies. Through a column filled with soil a solution of ammonium sulfate circulates in a closed circuit; subsequent chemical analyses give information about the biological evolution of this ammonium salt (Jaques, *et al.,* 1959; Torstensson, 1974). Urea may also be used (Namdeo and Dube, 1973). Technical difficulties make this method a difficult one.

Finally, measurement of the *mineralization* of the organic matter in the soil, under well-controlled conditions, can give good information on biological activity (Drouineau and Lefevre, 1949; Dommergues, 1960). However, not many papers on this technique mention research on the effects of pesticides.

Measurement of total soil respiration. Measurement of the total soil respiration (oxygen uptake and evolution of carbon dioxide) is often carried out according to Warburg's technique, directly on the soil, treated or not—just as it is (endogenous respiration) (Giardina *et al.,* 1970; Tu, 1975; Weeks and Hedrick, 1971), or enriched with nutrients to measure their effects on the metabolism (glucose: Bartha *et al.,* 1967; Tu, 1972; mannitol: Johnson and Colmer, 1958). This technique is suitable for short-term measurements and generally for small soil samples. For long-term measurements the technique of sweeping the soil samples with CO_2-free air is often used. In its passage through the soil, this air becomes heavy with CO_2 from the catabolic activities; it is then possible to trap and to titrate the CO_2 formed. Some workers use titrated solutions of baryta (Elkan and Moore, 1960; Agnihotri, 1971; Smith and Weeraratna, 1975) or of soda (Eno, 1962; Grossbard, 1971; Van Faassen, 1974).

Measurement of oxygen uptake is often carried out with electrolytic respirometers. The general principle of the operation of these apparatuses is the incubation of soil samples in closed cells and the periodic introduction of oxygen, by electrolysis, to compensate for the soil uptake (MacGarity et al., 1958; MacFayden, 1961).

The measurement of soil respiration is sometimes used in the study of particular groups of microflora. Thus, Anderson and Domsch (1973), using antifungic or antibiotic substances, were able to distinguish the relative importance of fungi and of bacteria in soil respiration. However, there seem to be no studies on the combined use of pesticides.

Measurements of radioactivity.   Labeled elements are helpful in making these determinations. Thus, the radiorespirometric technique represents a neat and practical method of measuring the rapidity of the mineralization of a radioactive substrate (labeled with ^{14}C). Mayaudon (1973) offers a true index of the biological activity of the soil, expressed in relation to the pmoles ($10^{-12} M$) of $^{14}CO_2$ evolved, in the course of the mineralization of *dl*-glutamic acid. The tests may be repeated on treated and control plots. Mayaudon (1973) studied the effects of the different pesticides used alone or in mixtures under sugar beet. Domsch et al. (1973), with a similar technique, followed the influence of benomyl on the breakdown of ^{14}C-labeled glucose.

Other radioactive substrates may be used for biological studies, such as those of Lauss and Danneberg (1975) on the decomposition of plant residues labeled with ^{15}N, in the presence of pyramin. Sometimes the pesticide itself is labeled, but its microbial degradation is studied more often than its effects on the soil microflora (Grossbard, 1970b, 1973).

Measurements in situ.   In order to work under more ecological conditions, some workers make their activity measurements directly *in situ,* either on the soil treated with pesticides under field conditions, or on the plants. Respiration of the soil, according to Dommergues's technique (1968), may be measured *in situ,* but not many papers have been published on pesticide research. The situation is similar with Billes' method (1971), which includes titration by electrolysis of CO_2 dissolved in a sodium chloride solution colored by phenol-phthalein.

Cellulolysis, in contrast, has attracted several workers: its study involves periodic checks on the behavior of strips of calico and of cellulose powder, both buried in the treated soil. The decrease in the strength of the calico strips, measured mechanically, and the loss of weight of the cellulose powder, enclosed in nylon bags, are dependent on the activity of cellulolytic organisms (Klyuchnikov *et al.*, 1964; Grossbard, 1974; Grossbard and Wingfield, 1975; Grossbard and Long, 1975, oral communication).

The method of "mold soil" also gives direct measurements of activity, but the data obtained are only qualitative. The soil serves as culture medium; it is increased with elective substrates chosen according to the research objectives. Microbial readings are carried out on lamellae, set on the soil. This method is described in an earlier work on DDT and hexachlorocyclohexane (Drouineau *et al.*, 1947).

In our final example, relating to symbiotic nitrogen fixation, plants are used for the determination of biological activity. Immediately after the rooting up of the leguminous plants, the root nodules are counted and examined under the microscope, and the leghemoglobin is determined. Goss and Shipton (1965) follow this technique on plants the seeds of which have been disinfected with an insecticide. Garcia and Jordan (1969) use the same method after spraying herbicides on a leguminous plant field.

b. Measurement of the Enzymatic Activity of Soils. We must also consider this technique, which is based on soil enzymology. In fact, a large portion of the enzymes present in the soil is of biological origin, both extra- and intracellular. Several workers have endeavored to evaluate this fraction after a pesticide treatment, hoping to find a correlation between enzymatic activity and soil fertility. Thus, some research workers have determined the presence of several enzymes such as tryptophanase, urease, phytase, invertase, and saccharase (Voets and Vandamme, 1970; Bliev, 1973; Zubets, 1973; Verstraete and Voets, 1974; Voets *et al.*, 1974; Karanth *et al.*, 1975; Lethbridge and Burns, 1976). Others have limited their studies to one determination only—for example, the determination of nitrogenase in studies on atmospheric nitrogen fixation (Eisenhardt, 1975; Neven *et al.*, 1975; Peeters *et al.*, 1975, or of dehydrogenase, which gives an estimate of the soil's respiratory activity (Ulasevich and Drach, 1971; Karki *et al.*, 1973; Karki and Kaiser, 1974; Van Faassen, 1974).

These enzymatic methods are attractive because of their simplicity and their good reproducibility, as compared with the classical techniques used in soil microbiology; yet interpretation of the data requires some prudence, in view of the fact that at present we have limited knowledge of the enzymes—their origins, the relationships between them, and the role they play in soil fertility.

3. Measurement of the Biomass

This kind of measurement, which has not received much attention, provides an estimate of the effects of pesticides on the soil microflora.

Several methods have been suggested for estimating the microbial biomass, including measurements of ATP (MacLeod *et al.*, 1969; Lee *et al.*, 1971;

Ausmus, 1973) and measurement of muramic acid (Millar and Casida, 1970). The biomass may also be calculated from the biovolume by multiplying the number of organisms in a soil sample by the volume of an organism of medium volume (Russell, 1973).

A particularly interesting method is proposed by Jenkinson and Powlson (1976). These authors deduce the biomass of the microorganisms from the measurement of the flush of CO_2 evolved after fumigation of the soil with CH_3Cl. They note that the CO_2 evolved comes from the carbon mineralization in the bodies of the microorganisms killed by the treatment; 50% of this carbon is mineralized during the ten days after the treatment by the surviving or reinoculated microorganisms recolonizing the medium.

4. Complementary Techniques

The preceding techniques for measuring activity are, in general, applied in research concerning the major biological cycles, the carbon and nitrogen cycles. They constitute the classical techniques, as opposed to less familiar methods, which are, nevertheless, also interesting, owing to the additional information they provide, especially with respect to the effects of pesticides on the soil microflora.

a. *Phosphorus and Sulfur Cycles.* Research relating specifically to the phosphorus and sulfur cycles includes studies on the mineralization of organic phosphorus (Tyunyaeva *et al.*, 1974, according to the technique of Menkina); studies on the oxidation of elementary sulfur (Tu, 1970, 1972, 1973) (laboratory experiments with incubation of powdered sulfur in the presence of pesticides); and studies on the mineralization of organic sulfur (Simon-Sylvestre and Chabannes, 1975) (monthly determinations of sulfates on soil samples from experimental plots treated under field conditions).

b. *Pure Cultures.* Assays have also been carried out with pure cultures of soil microorganisms. It is easy to follow the effects of a pesticide on these varied organisms, but the responses are individual. Furthermore, we cannot assume that these results would be the same for soil, where the interactions and competition between the microorganisms are numerous.

This kind of experiment requires pure cultures of different organisms, such as symbiotic nitrogen-fixing organisms (Brakel, 1963; Makawai and Ghaffar, 1970; Pajewska, 1972; Mendoza, 1973; Pantera, 1974; Suriawiria, 1974; Eisenhardt, 1975; nonsymbiotic nitrogen-fixing organisms (Pajewska, 1972; Szegi *et al.*, 1974; Peeters *et al.*, 1975); fungi (Lorinczi, 1974); Cellulolytic organisms (Lembeck and Colmer, 1967; Szegi, 1970); Cellulolytic fungi (Grossbard, 1974); nitrifying bacteria (Garretson and San Clemente, 1968); and bacteria (Kulinska and Romanov, 1970; Ujevic and Kovacikova, 1975). Moreover, with this type of experiment, we must determine the action of pesticides on antibiotic production by actinomycetes (Krezel and Leszczynska, 1970) and bacteria (Kosinkiewicz, 1970; giberellic substance production by bacteria (Sobieszczanski, 1970); free amino acid secretion by bacteria and actinomycetes (Balicka *et al.*, 1970); and pigment formation (Kulinska and Romanov, 1970). We should also consider the influence of pesticides on some antagonisms and on pathogenic microorganisms.

Finally, pure strains of microorganisms may be used as "indicators" of soil phenomena. Most of the tests that evaluate the toxicity of substances may be applied to the study of the effect of pesticides on the microflora. Thus, Breazeale and Camper (1972) studied the action of a line of herbicides on the growth rate of different microorganisms.

5. Conclusions

All these methods of biological analysis of soil have flaws, yet they contribute considerably to the analysis of the major biological phenomena of soil. However, no numerical values are obtained. All the data are comparative, even for the techniques that seem to be the most ecological; the conditions are often altered in order to magnify the phenomena, and the optimum or potential activity is estimated rather than the actual activity. We must add to this criticism the fact that the use of elective media allows us to study only the main groups of microorganisms; the interrelations between them in the soil are not considered, in order to simplify the study of the problem. Therefore, it is difficult to interpret the data obtained in the laboratory in terms of field conditions.

In short, we must emphasize comparative measurements and always work under the same conditions, with the same techniques, on soil samples freshly collected or stored at low temperatures. We should also prefer measurements of activity, which give better information about the behavior of the phenomena, to counts, which may lead to uncertain results.

B. EXPERIMENTAL PROTOCOLS FOR THE STUDY OF THE INFLUENCE OF PESTICIDES ON THE SOIL MICROFLORA

This section will present a fairly extensive, but not exhaustive, discussion of the characteristic conditions of experimentation described in recent studies of the interactions between pesticides and the soil microflora. The different factors of variation may be divided into three major groups, based on their association with the environmental conditions, the pesticide, or the research worker.

Our purpose is to draw attention to the great diversity that exists among the experimental protocols. As the studies concerned deal with biological processes, which are very sensitive to environmental conditions, this diversity is reflected in the variability of the effects studied.

Herbicides, which stimulated much of the research, have received the most attention. To simplify the discussion and to give the reader a general overview of the subject, the results are presented in tables that are very schematic and sometimes incomplete, for research workers often omit from their description of the experimental conditions those factors affecting the results of pesticide treatment that have already been proved.

1. Factors Associated with Environmental Conditions

Environmental conditions play an important part in the effect of pesticide treatments on the soil microflora and therefore deserve particular attention.

a. Noncontrolled Physical Factors. In field studies—that is, *in vivo*—under noncontrolled climatic conditions (Tables I and II), where the scientist can regulate neither the rain nor the temperature, the data obtained the first year in a specific location for a certain chemical are not necessarily repeated the next year (Davidson and Clay, 1972; Simon-Sylvestre, 1974). Our limited knowledge of the different climatic factors makes it difficult to demonstrate the part played by

TABLE I
Field Experiments

Pesticide	Soil type	Crop	Duration of the trial	Authors
Simazine, atrazine, 2,4-D	Sod podzol, peat podzol	Corn	1 year	Mashtakov *et al.* (1962)
Simazine, atrazine	Middle loam	Corn, lupine	1 year	Kozlova *et al.* (1964)
Simazine	Tchernozem	Corn	16 weeks	Kulinska (1967)
Simazine between rows	Tchernozem	Strawberry plants	2 years	Bakalivanov and Nikolova (1969)
Atrazine	Humus-bearing podzol	Corn	2 years	Ulasevich and Drach (1971)
Alachlor, propachlor	Degraded tchernozem	Corn	9 years	Husarova (1972)
Simazine, prometryn 2,4-D	Clayey illuvial podzol, slightly loamy	Barley, flax, winter rye	2 years	Kozyrev and Laptev (1972)
Sodium chlorate	Sandy loam		150 days	Karki *et al.* (1973)
Dalapon, paraquat	Grassland sward	Grass		Namdeo and Dube (1973)
Dinoseb	Leached podzol	Rotation	3 years	Szember *et al.* (1973)
Simazine, atrazine, cyanazine, 2,4-D, TCA	Clayey soil	Wheat	1 year	Deshmukh and Shrikhande (1974)
Atrazine	Loamy sand	Apple trees	25 years	Voets *et al.* (1974)

each of them, and the research worker can determine the effects of their action for a given year only.

Moreover, these investigations *in vivo* are performed most often in the presence of a plant that may vary in size and age from a small plant sprung up some weeks before, to a fruit tree several decades old. In experiments in plots, the size of the plot often depends on the plant being tested; the area may vary from a few square meters for vegetable crops (Ritter *et al.*, 1970; Kolesnikov *et al.*, 1973) to several hundred square meters for arable crops, especially in Russia (Yurkevich and Tolkachev, 1972; Fisyunov *et al.*, 1973). Some "rhizosphere" effects modify the actions of pesticides and render the problem more complicated. Thus, the species cultivated and the age of the plants are important variables that must be considered. The soil tested is another variable—a factor that is often underestimated by many microbiologists, in both *in vivo* and *in vitro* studies. This factor will be discussed later in this section.

b. Controlled Physical Factors. In the greenhouse. This part of our study is concerned with experiments in pots, which are intermediate between field trials and laboratory assays. The most important characteristics are given in Table III. The fact that such experiments are usually performed under controlled conditions of moisture and temperature brings them closer to experiments *in vitro*, whereas the frequent presence of the plants reproduces in a rather artificial manner the "rhizosphere" effects observed in the fields. The choice of the *plant tested,* as well as the *duration of the cultivation period,* is limited by the size of the pots. The quantity of soil may vary between 0.9 kg (Kaiser and Reber, 1970) and 20 kg (Catroux and Fournier, 1971; Steenbjerg *et al.*, 1972). The material used for the pots is sometimes specified, such as Mistcherlich pots (Lozano-Calle, 1970; Wanic and Kavecki, 1972; Kozaczenko and Sobieraj, 1973), or Wagner pots (Kozlova *et al.*, 1964; Malichenko, 1971). Yet, considerable information on the scheme of these experiments may be missing (the preservation of moisture, the problem of water flow, etc.).

Moreover, some microbiologists wait a longer period of time before beginning their experiments in order that the balance of the microflora may be recovered after the pots have been filled. Thus, Guillemat *et al.* (1960) delay their experiments for 15 days at 20°C; Smith (1972) delays for 2 weeks to bring the pots to field capacity; and Horowitz and Blumenfeld (1973) wait 4 months. This procedure seems reasonable, since the favorable conditions of moisture and temperature that are established often result in a waking of the microorganisms, which is immediately indicated by a noticeable increase in their number and in respiratory activity. On the other hand, Eno (1962) works immediately with air-dried soil, as do Allott (1969) and Houseworth and Tweedy (1973).

The *temperatures* used in greenhouses vary, depending on the workers, and even in the same room they are not always kept constant during the experiments.

TABLE II
Experiments *in Vivo* (Plots)

Pesticide	Soil type	Size of plot	Crop	Duration of the trial	Authors
2,4-D, simazine, monuron, dinoseb, petroleum oils	Sod podzol (clayey and loamy sand)	10 m²		2 years	Shklyar et al. (1961)
Simazine, atrazine	Sandy soil	50–56 m²		1 year	Klyuchnikov et al. (1964)
Methabenzthiazuron	Loamy sand, sandy loam	1 m²	Spring barley, winter barley	1 year	Huge (1970a)
Paraquat, simazine, diuron, linuron, chlortiamid, dichlobenil				5 years	Huge (1970b)
Prometryn, carbamate, 2,4-MCPB, dinoseb			Serradella	70 days	Kaszubiak (1970)
Methoprotryne	Alluvial soil		Wheat	1 year	Micev (1970)
Vapam, DD, chloropicrin	Calcareous clay	2 × 5 m	Tomato	2 years	Ritter et al. (1970)
MCPA, simazine, linuron, triallate	Sandy loam	4 × 22.5 m	± Corn, barley, carrot	7–8 years	Grossbard (1971)
2,4-D, 2,4-MCPA, SO$_4$H$_2$		1 × 10 m	Spring wheat	4 months	Bertrand and De Wolf (1972)
2,4-D, simazine	Tchernozem (slightly humic)	400 m²	Corn	1 year	Yurkevich and Tolkachev (1972)
TCA, EPTC, cycluron[a] chlorbufam	Light chestnut soil	60 m²	Sugar beet	5 years	Zharasov et al. (1972)

14

Herbicides/fungicides	Soil type	Plot size	Crop	Duration	Reference
Paraquat	Fine sand	4.8 × 7.5 m	Without crop	1 year	Camper et al. (1973)
Simazine, atrazine	Tchernozem (slightly humic)	350 m²	Corn, then winter wheat	2 years	Fisyunov et al. (1973)
Prometryn, triflualin, chloropropham	Podzol	1.5 × 3.40 m	Vegetables	4 years	Kolesnikov et al. (1973)
Simazine, atrazine, linuron	Tchernozem (slightly humic)	150 m²	Corn	5 years	Kudzin et al. (1973)
Simazine, atrazine, dalapon, TCA	Leached podzolic tchernozem	24 trees	Apple trees	13 years	Belohov (1974)
Simazine, diuron, amitrole, 2,2-DPA	Sandy clayey loam	53.5 m²	Apple trees	6 years	Campbell and Mears (1974)
Simazine, prometryn, diuron, fluometuron, neburon, bromacil, pyrazon, triflura-lin, diphenamid		1.5 × 11 m	Without crop	4 years	Horowitz et al. (1974a)
Chloropicrin	Loam of the tablelands	1.75 × 2.50 m	Winter wheat	1 year	Simon-Sylvestre (1974)
Simazine	Loam of the tablelands	3.2 × 10 m	Corn	1 year	Simon-Sylvestre and Chabannes (1974)
MCPA, 2,4,5-T, linuron, simazine	Loamy clay	2 × 25 m	Without crop	2 years	Torstensson (1974)
Trifluralin	Grassland and marshland, peat soil	100 m²	Cabbage	1 year	Tyunyaeva et al. (1974)
Benomyl, captan, thiram, dicloran, quintozene	Acid soil		Without crop	7 months	Wainwright and Pugh (1974)

TABLE III

Greenhouse Experiments

| Pesticide | Soil type | Soil weight | Experiment characteristics | | | Crop | Duration | Authors |
			Temperature (C)	Moisture (%)	Light			
Simazine	Two acid soils, two neuter soils	7 kg	20°			Without crop	3 weeks	Guillemat et al. (1960)
Dieldrin, DDT, aldrin, heptachlor, 2,4-D, simazine	Clayey loam, sandy loam	Lysimeter	Exterior conditions			Sudan grass, tomato, corn	2 years	Shaw and Robinson (1960)
Simazine, atrazine	Middle loam	Wagner pots (5 kg)				Corn, pea, oat		Kozlova et al. (1964)
Simazine	Tchernozem, clayey sand	4 kg	17°–20°	60		Without crop	6 weeks	Kulinska (1967)
Simazine	Sandy soil, clayey sand			30 or 50		Leguminous flowering plants		Hauce-Pacewiczowa (1970)
Simazine	Sand, garden soil	0.9 kg	20°	80	+	Corn	50 days	Kaiser and Reber (1970)
2,4-D, 2,4,5-T	Clayey soil, sandy garden soil	Mistcherlich pots (6 kg)	25°–30°			Without crop	4 months	Lozano-Calle (1970)

Compound	Soil	Amount	Temperature	Moisture[a]	Crop	Duration	Reference
N-Benzoyl-N-(dichloro-3,4-phenyl)-N,N'-dimethylurea	Loamy sand, calcareous soil	20 kg	15°–30°	60	Without crop	60 days	Catroux and Fournier (1971)
Atrazine, simazine, prometryn	Podzol	Wagner pots (5 kg)			± Corn, potato		Malichenko (1971)
Simazine	Heath soil	20 kg		60	Oat	2 years	Steenbjerg et al. (1972)
Simazine	Calcareous and humic soil	Mistcherlich pots 20 kg	Exterior conditions		Apple tree sowing	4 months	Wanic and Kawecki (1971–1972)
	Synthetic soil				Without crop	10 years	Horowitz and Blumenfeld (1973)
Atrazine + captan or thiram	Clayey loam		25°	+	Wheat, pea, corn	108 days	Houseworth and Tweedy (1973)
Linuron, monolinuron, dinoseb	Clayey sand		Exterior temperature	25, 35, 45, 55, and 65	White mustard	4 weeks	Kozaczenko and Sobieraj (1973)
Metoxuron	Loam		20°		Wheat	3 weeks	Simon et al. (1973)
Simazine, atrazine, cyanazine, 2,4-D, TCA	Clayey soil				Wheat	40 days	Desmukh and Shrikhande (1974)
2,4-D	Clayey loam			100	Without crop	32 days	Sharma and Saxena (1974)

[a] Moisture expressed as a percentage in relation to the soil's retention capacity.

17

Thus, Kulinska (1967) indicates temperatures between 17° and 20°C for pot experiments, Lozano-Calle (1970) 25° and 30°C, and Catroux and Fournier (1971) 15° and 30°C. Allott (1969) specifies 15°C for his trials, Kaiser and Reber (1970) 20°C in an air-conditioned room, and Houseworth and Tweedy (1973) 25°C. Nevertheless, greenhouse experiments are not carried out under constant conditions of temperature.

The *moisture* of the soils also varies considerably—a problem that will be mentioned in our discussion of *in vitro* experiments.

Some scientists expose their plants to *light:* 14 hours for wheat, peas, and corn (Houseworth and Tweedy, 1973); white light and 10,000 lux for corn (Kaiser and Reber, 1970). Some photodecomposition of the tested pesticides may take place on the surface, and this may modify the behavior of the pesticide toward the soil microflora (Horowitz and Herzlinger, 1974; Horowitz *et al.*, 1974b).

In the laboratory. Trials *in vitro* include all experiments carried out in the laboratory under controlled conditions—on a soil in a container (beaker, Erlenmeyer flask, column, Petri dish), or on a synthetic soil (Kaiser and Reber, 1970; Horowitz and Blumenfeld, 1973), or on a mineral medium in tubes or Petri dishes (Table IV). Here, also, a great diversity in the experimental procedures can be observed.

One of the first variables is the *condition of the soil,* as in the greenhouse experiments. Several microbiologists use a fresh soil sample for their experiments so that drying will not affect either the microorganisms or the chemical properties (Chandra and Bollen, 1961; Chandra, 1966; Agnihotri, 1971; Focht and Josseph, 1974). Bartha *et al.* (1967) keep their soil samples in a wet atmosphere, at 20°C. Others use a soil dried for one night only, which limits the mineralization of nitrogen and carbon (Wainwright and Pugh, 1973). In contrast, Torstensson (1974) experiments with a dried soil; Dubey and Rodriguez (1970) and Smith and Weeraratna (1975) also specify that the soil be air-dried.

The *soil container* may be completely shut (Grover, 1972; Vlassak and Livens, 1975); covered with a plastic film, which lets gas pass but prevents evaporation (Dubey and Rodriguez, 1970; Wainwright and Pugh, 1973, Tu, 1973); closed with a one-hole cork, which allows gaseous exchange (Chandra, 1966; Agnihotri, 1974); or open to the air. The moisture of the soils is checked frequently—every week (Bardiya and Gaur, 1970), every three days (Guillemat *et al.*, 1960), every other day (Freney, 1965), sometimes every day (Chandra and Bollen, 1961; Voets and Vandamme, 1970; Bliev, 1972), and in some cases once or twice a day (Kozaczenko and Sobieraj, 1973).

The *moisture* of the soil and the *temperature* may vary considerably. The moisture content may range between 10 and 100% of the retention capacity, and the temperature between 5° and 30°C. Each research worker seems to establish conditions of moisture and temperature at the beginning of an experiment without indicating the reasons for his choice. However, depending on the moisture of the soil, the activity of some microorganisms that are very sensitive to the oxygen concentration will be incresed or moderated for the same quantity of pesticide. Kozaczenko and Sobieraj (1973) report on the variation in herbicide toxicity in relation to the moisture of the soil, with a maximum obtained for 45–55% of the soil retention capacity.

   c. *The Soil Factor.* Very few microbiologists have shown any interest in the kind of the soil used in their experiments. In some reports a description of the soil is totally missing (see Tables I through IV). Others offer only a slight indication of the state of the soil—whether it is dry (Torstensson, 1974; Houseworth and Tweedy, 1973) or wet (Chandra and Bollen, 1961; Chandra, 1966; Focht and Josseph, 1974)—or its pH—whether it is neutral, acid, or alkaline (Guillemat *et al.*, 1960; Smith and Weeraratna, 1975). Some soils are described in general terms, such as arable soil (with reference to the pH, the CO_3 content, and the humus content) (Voets and Vandamme, 1970), garden soil (Kaiser and Reber, 1970), peat soil (Horowitz and Blumenfeld, 1973), meadowgrass (Namdeo and Dube, 1973), Ukrainia steppe (Fisyunov *et al.*, 1973), and meadow soils and moors (Tyunyaeva *et al.*, 1974).

   Certain factors, however, especially those relating to pesticide mobility, persistence, and toxicity, which can influence the microflora, may be partially explained by a knowledge of the soil type (Mashtakov *et al.*, 1962; Wiese and Basson, 1966; Pop *et al.*, 1968; Chopra *et al.*, 1970b; Helling, 1971; Van Faassen, 1974; Trzecki and Kowalski, 1974). The pH of the soil may also influence the effects of some pesticides. According to Smith and Weeraratna (1975), simazine stimulates nitrification in an acid soil, and delays it in an alkaline soil.

2. Factors Associated with the Pesticide

   Another source of variation in the effect of a pesticide may be found in the *additives* introduced into the formula. Thus, Saive and his colleagues (1974) found opposite actions in two commercial products, each containing 65% of zineb, but of different origin; whereas one of these pesticides increases the overall potential activity of the soil microorganisms, estimated by CO_2 evolution, the other inhibits it. Stanlake and Clark (1975) have found that a commercial preparation of malathion containing aromatic petroleum has a higher effect on pure cultures of bacteria than does malathion alone.

   Effects will also vary according to the *quantity* of pesticide used. A weak concentration of pesticide given in homeopathic quantities will stimulate some organisms, whereas a larger amount will cause inhibition. Research workers often use the normal quantities recommended. In order to amplify the phenomena, they have recourse to higher concentrations—twofold concen-

TABLE IV
Experiments *in Vitro*, in the Laboratory

Pesticide	Soil type	Experiment characteristics			Duration	Authors
		Temperature (C)	Moisture[a] (%)			
Nabam, dazomet	Wet soil	28° ± 1°	60		60 days	Chandra and Bollen (1961)
Simazine, atrazine	Fine air-dried sand	28°	15		28 days	Eno (1962)
	Mineral medium	28°			5–6 days	
2,4-DB, diallate, amitrole	Loamy soil	28°	33		56 weeks	Chandra (1964)
Atrazine, simazine	Pure culture	30°			3 days	Kolzlova *et al.* (1964)
Dieldrin, heptachlor	Wet soil	5°, 12°, 19°, and 26°	60		16 weeks	Chandra (1966)
Simazine	Mineral medium	20°			26 days	Kaiser and Reber (1970)
2,4-D, simazine	Mineral medium	28°			12 weeks	Szegi (1970)
TCMTB	Arable soil	28°	20		6 months	Voets and Vandamme (1970)
N-Benzoyl-N-(dichloro-3,4-phenyl)-N,N'-dimethylurea	Mineral medium				60 days	Catroux and Fournier (1971)

Pesticide	Soil type	Temperature	Moisture[a]	Duration	Reference
Simazine, 2,4-D, TCA	Podzol	25° ± 1°	60	60 days	Bliev (1972)
Picloram	Clayey soil	27°	41	172 days	Grover (1972)
Picloram, TCA	Light chestnut soil	18°–25°	60	7 days	Zharasov et al. (1972)
Benomyl	Arable soil (calcareous loam of the tablelands)			6 months	Raynal and Ferrari (1973)
Dichloropropene and other nematicides	Sandy loam	5° and 28°	60	42 days	Tu (1973)
Captan, thiram	Semiwet soil	28° ± 1°	66	28 days	Wainwright and Pugh (1973)
Thiram	Garden soil, loamy sand	18° ± 1°	50	6 weeks	Agnihotri (1974)
Biodegraded insecticides	Wet soil	22° ± 3°	100	50 days	Focht and Josseph (1974)
Dalapon, pebulate, cycluron + chlorbufam	Tchernozem, grassland	28°	60	63 days	Kiperman-Reznik and Grimalovski (1974)
Metribuzin	Loamy sand	28°	60	42 days	Lay and Ilnicki (1974)
Benomyl	Sandy and humus-bearing soil	15° and 20°	50	2 months	Van Faassen (1974)
Simazine, ioxynil	Acid, alkaline, air-dried	28°	33	6 weeks	Smith and Weeraratna (1975)

[a] Moisture expressed as a percentage in relation to the soil's retention capacity.

21

trations (Huge, 1970a; Catroux and Fournier, 1971), threefold and fourfold concentrations (Grossbard, 1971), fivefold concentrations (Malichenko, 1971), and even tenfold concentrations (Chandra, 1964).

In some cases, even higher rates have been applied. Thus, Mashtakov and his colleagues (1962) in their laboratory experiments have used up to the equivalent of 75 kg of atrazine per hectare, and Simon *et al.* (1973) have multiplied the normal rate of metoxuron by 50. With simazine, for example, the quantities used in greenhouse experiments may range from 0.030 kg (Hauke-Pacewiczowa, 1970) to 10 kg (Malichenko, 1971), be it so from 1 to 330 with all the intermediates—2 kg (Kozlowa *et al.,* 1964), 4 kg (Kulinska, 1967), and 6 kg (Guillemat *et al.,* 1960). It is therefore difficult to draw any conclusions when confronted with such variability in rates. With reference to this, Grossbard (1971) considers it impossible to compare field experiments and laboratory assays. In field experiments, inhibitory rates must be higher than those provoking the same symptoms in the laboratory, since the contacts between the microorganisms and the pesticide particles are less frequent.

The *method of application* also presents many variations, often associated with the mode of action of the product and with its solubility. Certain products are used at the soil surface, others are sprayed on the foliar surface, and yet other pesticides are incorporated into the soil. In addition to the difference in action due to the very nature of the product used, other effects were observed that varied according to the thickness of the treated soil. In surface treatments, the pesticides may become inert more quickly as a result of photodecomposition (Sokolov and Knye, 1973; Horowitz *et al.,* 1974b); on the other hand, only the top layers run the risk of being disturbed, and recolonization may occur more quickly. However, in the incorporation of toxic products, the rapidity of recolonization of the soil surface from the underlying layers and the rapidity of the return to the standard micropopulation will depend on the thickness of the treated soil.

For the same class of pesticides, differences in the method of application are found among the research workers, as may be noted from the following examples relating to herbicides: The product may be deposited on the surface soil by watering (Catroux and Fournier, 1971) or by pulverization (Shklyar *et al.,* 1961; Allott, 1969; Kolesnikov *et al.,* 1973; Simon–Sylvestre and Chabannes, 1974; Deshmukh and Shrikhande, 1974). The pulverization may be followed, three days later, by the passing of a rotary cultivator through the upper 15-cm layer (Camper *et al.,* 1973). Depending on the solubility of the herbicide, it is a question of the pulverization of a solution (Krumzdorov, 1974) or of a suspension (Fusi and Franci, 1972). The herbicide may be deposited at a depth of 3–4 cm (Yurkevich and Tolkachev, 1972) or of 0–15 cm (Malichenko, 1971); it may also be plowed in by a harrow (Ulasevich and Drach, 1971), to a depth of 15 cm (Ragab, 1974). The product may be closely mixed with the soil (this mode of application is particularly good for greenhouse and laboratory experiments) in the form of a powder (Eno, 1962; Wanic and Kawecki, 1971; Grover, 1972; Smith and Weeraratna, 1975), a suspension (Kulinska, 1967), or an emulsion (Voets and Vandamme, 1970), or in aqueous solution (Houseworth and Tweedy, 1973). To make it easier to incorporate the product into the soil, it may be dissolved in a solvent, which evaporates afterward. Solvents include acetone (Fusi and Franci, 1971), methanol (Walker, 1973), ethyl alcohol (Lay and Ilnicki, 1974) and petroleum ether and acetone (Tu, 1972).

There is considerable variation among workers in regard to the choice of *date of application* of a given product. Often no precise information is given on this subject in the literature. For simazine, for example, the field treatment may be given three weeks before the sowing of corn (Simon-Sylvestre and Chabannes, 1974), immediately after the sowing of corn (Mashtakov *et al.*, 1962), the day after (Yurkevich and Yolkachev, 1972), or three days after (Deshmukh and Shrikhande, 1974).

3. Factors Associated with the Experimenter

Finally, the experimenter is also responsible for some variation factors: the choice of the date of the analysis, of the microflora method of sampling the soil, and the choice of the methods of analysis.

There is no apparent uniformity in the choice of the *sampling dates* of soils for biological analyses, *in vivo* or *in vitro*. This is due not only to the great variety of products used and the different aims in view, but also to the preferences of the research workers. Some workers take very close soil samplings immediately after the application of the pesticide, but do not continue them for a long time. This formula seems appropriate when the soil microflora is no longer perturbed by the pesticide, if the latter has rapidly disappeared. Other experimenters, on the other hand, carry on their investigations longer in order to observe late reactions as well as residual products; extended investigations are also necessary when a succession of treatments has been given. Table V gives a survey of this diversity.

For experiments *in situ,* the *thickness of the analyzed soil* is not always specified. It may vary, depending on the products used (their solubility, their mode of application, etc.) and the requirements of the research workers. Thus, after a surface application of pesticides, it is logical that top-layer samples of soil be taken, as was done by the following authors: Shklyar *et al.* (1961), 0–1, 1–2 cm; Klyuchnikov *et al.* (1964), 0–5, 5–15, 15–25 cm; Bakalivanov and Nikolova (1969), 0–5, 5–15 cm; Huge (1970a) and Voets *et al.* (1974), 0–5 cm; Zharasov *et al.* (1972), 0–2, 2–10, 10–20 cm; and Simon-Sylvestre and Chaban-nes (1974) 0–5, 5–20 cm.

Some research workers first remove the top superficial centimeters (Kulinska, 1967; Huge, 1970b), thus eliminating the chief effects of the treatments; others do not separate the top layer from the deeper layers and then note attenuated

TABLE V
Diversity in the Dates of Soil Sampling for Biological Determinations

Pesticide	Biological determinations made after:	Authors
Simazine	0, 3 weeks	Guillemat *et al.* (1960)
Nabam, dazomet	0, 15, 30, 45, 60 days	Chandra and Bollen (1961)
Simazine, atrazine	0, everyday to 8 days, 14, 15, 21, 22, 27 days	Eno (1962)
Simazine, atrazine, 2,4-D	0, 15, 45, 75 days	Mashtakov *et al.* (1962)
2,4-DB, diallate, amitrole	0, 8, 28, 52, 56 weeks	Chandra (1964)
Dieldrin, heptachlor	0, 2, 4, 8, 16 weeks	Chandra (1966)
Maneb, anilazine	0, 1, 2, 4, 8, 16 weeks	Dubey and Rodriguez (1970)
Methabenzthiazuron	0, 1 day, 2, 3 weeks	Huge (1970a)
Prometryn, carbamate, 2,4-MCPB, dinoseb	0, 30, 40, 50, 60, 70 days	Kaszubiak (1970)
2,4-D, 2,4,5-T	0, 2, 4 months	Lozano-Calle (1970)
TCMTB	0, 1, 2, 3, 4, 5, 6 months	Voets and Vandamme (1970)
N-Benzoyl-N-(dichloro-3,4-phenyl)-N-N'-dimethylurea	0, 5, 15, 60 days	Catroux and Fournier (1971)
Atrazine, simazine, prometryn	0,1 hour	Malichenko (1971)
Simazine, prometryn, 2,4-D	0, 15, 30, 60, 100 days	Kozyrev and Laptev (1972)
2,4-D simazine	0, everyday to 15 days, 40 days	Yurkevich and Tolkachev (1972)
TCA, EPTC, cycluron + chlorbufam,	0, 10, 30, 60 days	Zharasov *et al.* (1972)
Picloram, TCA	0, 7 days	
Paraquat	0, 11 months	Camper *et al.* (1973)
Atrazine + captan or thiram	0, 2, 16, 44, 76, 108 days	Houseworth and Tweedy (1973)
Sodium chlorate	0, 5, 27, 150 days	Karki *et al.* (1973)
Dinoseb	0, 2, 3, 5, 7, 14, 21 days	Szember *et al.* (1973)
Captan, thiram	0, 7, 14, 21, 28 days	Wainwright and Pugh (1973)
Simazine, atrazine, cyanazine 2,4-D, TCA	0, 10, 20, 40 days	Deshmukh and Shrikhande (1974)
Biodegraded insecticides	0, 2, 4, 6, 13, 15, 17, 20, 22, 29, 36, 38, 41, 43, 50 days	Focht and Josseph (1974)
Cycluron + chlorbufam dalapon, pebulate	0, 5, 8, 15, 22, 36, 63 days	Kiperman-Reznik and Grimalovski (1974)
2,4-D	0, 2, 8, 14, 20, 26, 32 days	Sharma and Saxena (1974)
Atrazine (after eight applications)	0, 1, 4, 9, 21 months	Simon-Sylvestre (1974)
Simazine	0, 20, 50, 113, 190 days	Simon-Sylvestre and Chabannes (1974)
Trifluralin	0, 2, 6, 14 weeks	Tyunyaeva *et al.* (1974)
Benomyl, captan, thiram dichloran, quintozene	0, 28, 56, 84 days	Wainwright and Pugh (1974)
Atrazine	0, 2 weeks, 11 months	Voets *et al.* (1974)

actions. Kozyrev and Laptev (1972), Tyunyaeva *et al.* (1974), 0–10 cm; Shklyar *et al.* (1961) (root zone), 8–10 cm; Namdeo (1973) and Ragab (1974), 0–15 cm; and Karki *et al.* (1973), 0–20 cm. Kulinska (1967) took the soil from the pots and mixed it before taking samples.

When pesticide treatments are given in depth or in a series of surface applications, it is important to carry out biological analyses over a greater soil depth (sampling with a borer)—for example, 0–40 cm (Belohov, 1974), or 0–25 cm (Simon-Sylvestre, 1974).

4. Conclusions

After presenting such a diversification of experimental protocol, it is difficult to express an opinion on their value or to recommend one over another. The best procedures probably take the practical conditions into consideration, but difficulties then arise in determining and measuring the secondary effects; thus, when working under environmental conditions, investigators often magnify the different biological processes that are most favorable to both the microflora and the evaluation of these effects. Less favorable conditions would perhaps make the research easier by increasing the differences between the pesticide-treated soil and the control soil. In any event, we must not limit our study to a single biological process in order to determine the effect of a pesticide on the soil microflora. A large body of facts is essential for proper definition of the scope of the problem.

C. CONCLUSIONS

In view of the diversity of the experimental protocols and of the methods of biological analysis used to study the secondary effects of pesticides on the soil microflora, it is difficult for the research worker to make a choice.

It seems increasingly necessary, first on a national basis and then on an international level, to standardize some research protocols that would make it possible, under specific conditions, to put pesticides to the test, to compare them, and then to establish a scale of values that would be adaptable whenever the synthesis of a new product is carried out by the laboratories of commercial firms. The need for such a program has already been expressed by several research workers, who must choose from the microbiological research protocols those that will provide the most information about the effects of pesticides on the soil microflora. In this regard, Grossbard (1970a) indicated that, for the herbicides, four studies are the minimum: CO_2 evolution, mineralization and symbiotic fixation of nitrogen, and counts of microorganisms. Grossbard has since added the study of cellulolysis, a very important process in humus formation, and a process that is solely biological.

Merenyuk *et al.* (1973) suggested only three studies: nitrification, cellulose decomposition, and count of the total microflora. Novak (1973) prefers to complete nitrification and cellulolysis by ammonification and respiratory tests. Macura (1974) uses only nitrification and cellulose breakdown, which he considers to be the most sensitive and most important processes. These two studies, in fact, are included by all microbiologists.

III. Effects of Pesticides on the Microorganisms and on the Total Activity of the Soil

A. COUNTS OF THE MICROORGANISMS

Considerable research has been done on total population counts of bacteria, fungi, and actinomycetes after the application of pesticide treatments. However, the results have been inconclusive and have not provided adequate information on the ecology of these various microorganisms.

1. Bacteria

Certain *herbicides* stimulate the growth of bacteria because they are easily degraded by the bacteria. Examples are paraquat (Tu and Bollen, 1968), endothal, which is degraded in a few weeks (Novogrudskaya and Isaeva, 1965), and calcium cyanamide, which, when applied at doses lower than normal field rates, is active for nine months (Audus, 1970). Some substituted ureas, such as linuron, and the combination cycluron + chlorbufam, are also stimulating in normal doses (Kozaczenko, 1974); 2,4-D has the same effect in concentrations up to 4–5 ppm; at higher concentrations the effect is reversed. Sharma and Saxena (1974) believe that this stimulation of bacteria and also of fungi is only an indirect action of the product, which, by eliminating the actinomycetes, decreases the competition between the other microorganisms. Some herbicides belonging to this same phenoxyacetic group are harmless toward the bacterial population when applied at normal field rates (according to Lozano Calle, 1970), with 2,4-D and 2,4,5-T in clayey soils and sandy soils. According to some authors, these products become stimulating *in vitro* at doses 100- and 1000-fold higher than normal, whereas they depress bacterial numbers when applied *in situ* at the same concentrations.

Other herbicides have no effect on bacteria when applied at normal rates; these include sodium chlorate (Karki and Kaiser, 1974); picloram (Grover, 1972); propachlor and alachlor (Enkina and Vasilev, 1974); phenobenzuron (Catroux and Fournier, 1971); trifluralin (Tyunyaeva *et al.,* 1974); cycluron + chlorbufam (mentioned previously as having a positive effect), endothal, dalapon, trichloracetate, chlorazine cycloate, and lenacil (Chulakov and Zharasov, 1975); and DNOC, propham, pyrazon, TCA, DCPA, PCP,* and some substituted ureas (Audus, 1970).

Finally, there is a herbicide category that causes depressive effects; dinoseb, which is more toxic than bentazone, reduces the bacterial population (Torstensson, 1975), but only at high rates. In some cases, the initial depressive effects are followed by an increase in the bacterial number beyond the normal level. This delayed stimulation is caused by the adaptation time of the bacteria and by the increase in the environment of nutrients that come from weeds killed by the treatment. According to some workers, it can also be explained by the utilization of the herbicides as substrates, but only for herbicide quantities above the normal rates of field applications. This group includes dinoseb, chlorazine, and cycluron + chlorbufam (Nepomiluev et al., 1966). This depressive action does not last long, with a few exceptions; dinoseb decreases the bacterial population for as long as three months, and dalapon and EPTC are effective on alfalfa for two months, without, however, affecting the nitrifying bacteria.

The triazines often have no effect on the soil bacteria, as was shown in the following studies: (a) Atrazine: in long-term trials on apple trees (Voets et al., 1974) and in pots with wheat, bean, and corn (Houseworth and Tweedy, 1973); the stimulating effect observed in this case is indirect and is not due to the atrazine but to the fact that the degradation of the weeds killed by this product leads to an increase in the bacterial and fungal populations. (b) Simazine: in pots and in situ (Kulinska, 1967); in pots and in the laboratory (Freney, 1965; Kaiser and Reber, 1970). (c) Simazine and atrazine: in the laboratory (Eno, 1962; Amantaev et al., 1963), providing (with reference to the work of Eno) that the normal rates are used.

These two triazines have no effect in field experiments (Zubets, 1973; Kruglov et al., 1975a). Kruglov and co-workers, however, add that this conclusion is valid only for a single application. The superposition of two- or three-year treatments leads to a depressive action.

In some cases, triazines may also result (a) in stimulation—with methoprotryne in wheat rhizosphere (Micev, 1970); or with atrazine for two months and in proportion to the rate used (Percich, 1975); or (b) in an evanescent depressive effect, followed or not by stimulation—in sandy soil with atrazine and simazine (Klyuchnikov et al., 1964); in apple orchards (Kuryndina, 1965); in forest nurseries (Milkowska and Grozelak, 1966); in strawberry plants with simazine, applied between the rows (Bakalivanov and Nikolova, 1969); in corn with simazine (Simon-Sylvestre, 1974); in tomatoes for 100 days after an application

*PCP—product used for potato tops.

of metribuzin (Velev and Rankov, 1975); or in a perennial crop after three years of treatment with atrazine (Kruglov *et al.*, 1975b).

This negative action may also persist for one year, after eight consecutive years of atrazine application, at normal doses, as counts carried out in an apple orchard have shown (Simon-Sylvestre, 1974).

Some workers have mentioned a higher sensitivity of the spore-forming bacteria to the toxic actions of the triazines.

The *insecticides* have, in general, no effect on bacterial numbers when applied at normal rates, according to counts carried out fifteen days after treatment (work of Pathak *et al.*, 1960-1961, on DDT, chlordane, and aldrin) and one month after the application of chloro-insecticides (Eno and Everett, 1958).

Focht and Josseph (1974) have not observed any modification in the microflora in general, even for tenfold rates of acephate and methamidophos. At very high concentrations, however, Audus (1970) reports that the insecticides of the chlordane group have a selective action in inhibiting completely the grampositive bacteria. Some other depressive effects are reported in the literature, but they persist for only two weeks with the organophosphorus insecticides (Tu, 1970) and for sixteen weeks with lindane (Tu, 1975).

The number of bacteria increases in general after a *fungicide* treatment, probably as a result of the elimination of fungi and a decrease in competition. Several papers mention this temporary increase as occurring on the seventh day after a treatment with thiram (Agnihotri, 1974); during the second week after a treatment with benomyl (Yakolev and Stenina, 1974); and on the twenty-eighth day after a treatment with benomyl, captan, thiram, dicloran, formaldehyde, or quintozene (Wainwright and Pugh, 1974).

Van Faassen (1974) also finds an increase in the number of bacteria after a treatment with benomyl, and Wainwright and Pugh (1975), in assays *in vitro* with captan, reveal the influence of the concentration of the product on the amplitude of the bacterial growth and on the date of maximum growth. Hofer *et al.* (1971), Ponchet and Tramier (1971), and Siegel (1975), on the other hand, do not observe any variation in the bacterial population after a treatment with benomyl. Chandra and Bollen (1961) report an opposite effect—a decrease in the number of bacteria for 30 days, and a recovery after 60 days.

The soil *fumigants* (Audus, 1970) cause a decrease in the bacterial population for varying periods of time: for several days with ethylene dibromide, for a few weeks with DD, and for as long as six months with chloropicrin and formaldehyde. This decrease is followed immediately by a great increase in the number of bacteria, and later by a return to a normal population. The same result—namely, a temporary depressive effect—was obtained in the assays of Corden and Young (1965) with metam-sodium and those of Tu (1972) with four nematicides: DD, carbofuran, fensulfothion, and methylisothiocyanate.

2. Fungi

It is equally difficult to generalize about the results obtained in research on soil fungi. This section will deal only generally with the action of pesticides on these microorganisms. The special case of the pathogenic fungi is the subject of a specific study discussed in another section (Section V).

The *herbicides*, in normal doses, often do not modify the soil fungal population. This is found for trifluralin (Tyunaeva *et al.*, 1974), NaClO$_3$ (Karki *et al.*, 1973), and atrazine and simazine (Eno, 1962; Amantaev *et al.*, 1963; Voets *et al.*, 1974).

Some depressive effects have been reported, however, in laboratory experiments as well as in field assays with propanil (Oleinikov *et al.*, 1975), propachlor (Husarova, 1972),2,4-dichlorophenoxyacetic acid (Mashtakov *et al.*, 1962; Abueva and Bagaev, 1975), pyrazon (Zharasov *et al.*, 1972), cycluron + chlorbufam (Zharasov *et al.*, 1972; Kozaczenko, 1974), and even simazine and atrazine (Mashtakov *et al.*, 1962; Bakalivanov and Nikolova, 1969; Malichenko, 1971; Simon-Sylvestre, 1974) and prometryn (Malichenko, 1971). These depressive effects decrease in time (Malichenko, 1971; Simon-Sylvestre, 1974). Bakalivanov and Nikolova (1969) show a quick recovery to a normal population after ten days. Some negative actions are also observed after the use of higher doses of certain herbicides such as paraquat (Camper *et al.*, 1973) and substituted ureas (Grossbard and Marsh, 1974).

In contrast to these findings, a stimulation of the fungal population is reported in the studies of Kaszubiak (1970) with prometryn, dinoseb, and 2,4-DB, of Micev (1970) with methoprotryne, of Zubets (1973) with simazine and atrazine, during the year following the treatment, and of Sharma and Saxena (1974) with 2,4-D at a definite concentration.

The *insecticides*, in general, affect the soil fungi in the same way as they do the soil bacteria. No action is seen, in soils of different types, with chloro-insecticides (Pathak *et al.*, 1960–1961) or with lindane alone, even with high doses (Tu, 1975). However, Eno and Everett (1958) report a stimulation with dieldrin one month after treatment, and Tu (1970) reports a brief decrease (after one to two weeks) with organophosphorus insecticides.

On the other hand, Cowley and Lichtenstein (1970) describe an inhibition of the growth, on Czapek nutrient media, of several fungal species, isolated from prairie soils, in the presence of aldrin, lindane, parathion, or carbaryl. This inhibition decreases with the addition of yeast extract, asparagine, or some ammonium salts to the culture media; it is completely suppressed with carbaryl and is reduced to a large extent with aldrin.

As would be expected, the soil fungi are affected by a *fungicide* treatment, as Agnihotri (1974) demonstrated with thiram, and Foster (1975) and Siegel (1975) observed with benomyl. This decrease may be selective; thus, Pugh *et al.* (1973) found the cellulolytic fungi to be less tolerant than other fungi to the organomercuric products. Other workers believe, in particular for benomyl, that this is a fungistatic action and not a fungicidal one (Hofer *et al.*, 1971; Ponchet and Tramier, 1971; Raynal and Ferrari, 1973).

Actually, this decrease in the number of fungi in the soil, which occurs even with small doses of the fungicide, is only temporary. It is maximum after 2 days with captan, in laboratory assays, and after 3–15 days with captan, dicloran, drazoxolon, and triarimol, in the field (Wainwright and Pugh, 1975). After 2 weeks with 300 kg of benomyl per hectare (Yakolev and Stenina, 1974), 74% of the fungi have disappeared, and only three species survive. The recovery to a normal population takes place after 28 days (Wainwright and Pugh, 1974), with a treatment using benomyl, captan, thiram, dicloran, and quintozene, and after 60 days (Chandra and Bollen, 1961) with nabam and dazomet. The inhibitory effect of captan in all the fungi lasts longer than that of thiram (Houseworth and Tweedy, 1973).

Paradoxically, fungicides may also have a positive effect on fungi. These organisms are stimulated by high doses of benomyl (Avezdzhanova *et al.*, 1976). Wainwright and Pugh (1975) observed that four fungicides (captan, dicloran, triarimol, and drazoxolon), at normal field rates, lead to a reduction of fungi 3 and 15 days after the treatment; then, after 157 days, the fungal population increases, sometimes beyond the normal level.

The *nematicides* decrease the number of soil fungi for 2 weeks in a sandy silt; this occurs with ethoprop, dichloropropene, and methylisothiocyanate (Tu, 1973). This reduction is temporary, however, with metam-sodium (Corden and Young, 1965), with carbofuran, DD, fensulfothion, and methylisothiocyanate (Tu, 1972), and with metam-sodium and methylisothiocyanate (Welvaert, 1974); it is incomplete with methylisothiocyanate and dazomet (Welvaert, 1974).

3. Actinomycetes

The actinomycetes are, in general, very tolerant to most groups of *herbicides* applied at normal rates. They are not affected by the phenoxy acids, DNOC, $NaClO_3$, calcium cyanamide, or substituted ureas (Audus, 1970), phenylureas (Doxtader, 1968), linuron ± lenacil (Balezina and Tretyakova, 1974), trifluralin (Tyunyaeva *et al.*, 1974), paraquat (Camper *et al.*, 1973), pyrazon (Mezharaupe, 1967), simazine (Kulinska, 1967), or simazine and atrazine, alone or mixed (Zubets, 1973), during either the first year or the second year.

Depressive effects are mentioned, however, with dalapon and EPTC under alfalfa (Rakhimov and Rybina, 1963), with methoprotryne under wheat (Micev, 1970), and with simazine used between the rows of strawberry plants (Bakalivanov and Nikolova, 1969), but only temporarily. Owing to the fact that simazine may be used as a source of both carbon and nitrogen, the initial depressive effect is followed by a stimulation of the growth of actinomycetes with high concentrations of 2,4-D (Sharma and Saxena, 1974) or with a series of treatments for five years in an apple orchard with the mixture simazine + amitrole (Teuteberg, 1968) or for seven years, under corn, with propachlor and atrazine (Husarova, 1972).

Stimulating effects are also observed with prometryn, dinoseb, and 2,4-DB (Kaszubiak, 1970) and with chlorpropham (CIPC) under cotton (Taha *et al.,* 1972).

Very few papers deal with the effects of the *insecticides* on the actinomycetes. According to Audus (1970), these germs are resistant to DDT, BHC, toxaphene, aldrin, dieldrin, and parathion applied at normal rates, and Tu (1975) confirms this statement for lindane. Only in pure cultures do the actinomycetes show some sensitivity.

The *fungicides* and *soil fumigants* are in general toxic toward the actinomycetes, even at normal rates, as indicated by the work of Corden and Young (1965) with metam-sodium, nabam, and dazomet, and that of Agnihotri (1971) with captan. A temporary effect was observed by Agnihotri (1974) with thiram and by Siegel (1975) with benomyl.

Only Wainwright and Pugh (1975), in laboratory experiments, report a stimulating effect with a low dose of captan, and Van Faassen (1974) a lack of action with benomyl.

4. Algae

A last group of organisms, somewhat apart in the soil micropopulation, deserves to be mentioned. These are algae, of little importance in cultivated soils and studied only by some ecologists. Their research has usually been limited to the effects of herbicides the inhibitory power of which is in general proportional to their concentration (Wright, 1972). In fact, these data have made it possible to perfect a technique, built on the sensitivity of the green algae, that detects the presence of certain herbicides in the soil according to the quantity of chlorophyllian pigment extracted. A current publication (Lefebvre-Drouet and Calvet, 1978) will contribute to the precision of this technique, already used by several research workers.

All the triazines exert a negative action on algae, more significant after one application on a light soil than after five treatments on a heavy soil (Pantera, 1970). Kruglov and Kvyatrovskaya (1975) report that the sensitivity of *Chlorella* to phenylureas depends on the chemical structure of the herbicide and on the soil content in humus. At very small concentrations (0.001–0.01 ppm), however, substituted ureas and atrazine have a stimulating action on algae. At higher concentrations (0.1–1 ppm) the growth of the algae is inhibited (Pillay and

Tchan, 1972). Lenacil, alone or mixed with linuron, would be toxic, whereas monolinuron, associated with linuron or alone, would not (Balezina and Tretyakova, 1974). According to Kirkwood and Fletcher (1970), MCPB inhibits growth, respiration, and phosphorus absorption more strongly than does MCPA, in the case of three unicellular algae. Phenazone, at high rates, also reduces the growth of algae (Pociene et al., 1974), whereas with 4 kg/ha there is no effect. *Chlorella* are not affected by MSMA in liquid medium, but photosynthesis is stopped by a 2-hour exposure to 5×10^{-6} M prometryn or diuron and to 5×10^{-7} M fluometuron (Davis et al., 1976).

The sensitivity of algae varies, depending on the herbicide, but generally *Chlamydomonas* and *Chlorococcum* are more sensitive than *Chlorella* and *Nostoc*, according to the findings of Cullimore and MacCann (1977) on the effects of four herbicides (2,4-D, trifluralin, MCPA, and TCA) on algae isolated from a grassland loam soil. A herbicide treatment, in the top layer of the soil, is followed by a reduction in the cell numbers of sensitive algae and an increase in the population of *Chlorella*. The authors also note an overall reduction in cell numbers for the algae, growing preferentially on a nitrogen-free medium. With metribuzin, Arvik et al. (1973) find the same classification—*Chlamydomonas* the most sensitive, and *Chlorella* the most resistant.

It is difficult to draw conclusions from such varied results on the counts of the microorganisms of the soil. The statement of Kaszubiak (1970), "Bacteria were stimulated or depressed by all herbicides, depending upon the preparation, its dose, and sampling time," although it concerns only bacteria and herbicides, very well sums up the complexity of the facts, which characterizes all the studies carried out on soil microorganisms and pesticides.

However, other causes of variation must be added:

Nature of the soil and its composition: According to Karki et al. (1973), $CaClO_3$ has a more marked effect on acid soils. On the other hand, its harmful actions are less important in soils that are rich in organic matter, for greater amounts of the chemical products are adsorbed (Fusi and Franci, 1971; Beckmann and Pestemer, 1975; Rankov and Velev, 1975; Vlassak and Livens, 1975). The addition of silica and clay to a light soil decreases the inhibitory effect observed in soil microorganisms (except in fungi) after a treatment with atrazine or prometryn (Panterowa et al., 1975).

Climatic conditions: In a wet year, the sugar-beet herbicides, such as cycluron + chlorbufam, have a stronger effect on the soil fungi (Zharasov et al., 1972).

Duration of the treatment: Sobieszczanski et al. (1975) mention an adaptation of bacteria to certain herbicides—atrazine, simazine, prometryn—used continuously.

B. TOTAL ACTIVITY OF THE SOIL

1. Respiratory Activity

As with the counts of the microorganisms, the results for total respiratory activity vary considerably, depending on the pesticides studied and the experimental conditions.

Some *herbicides* do not affect this activity when they are applied at normal field rates; these include simazine and atrazine (Eno, 1962), simazine (Kulinska, 1967), and triallate and MCPA (Grossbard, 1971). Even after a series of seven to eight treatments, picloram (Grover, 1972), metoxuron (Grossbard and Marsh, 1974), $NaClO_3$ (Karki and Kaiser, 1974), and 2,4-D and 2,4,5-T (Ruffin, 1974) have no effect; this is found also with isoproturon and triazophos at hundredfold concentrations (Neven *et al.*, 1975).

However, some of these herbicides partially inhibit the respiration of the microflora of the soil if they are used at high rates—for example, simazine (Kulinska, 1967), metoxuron and linuron (Grossbard and Marsh, 1974), and dinoseb acetate (Neven *et al.*, 1975)—or if the treatments are repeated—for example, simazine and linuron (Grossbard, 1971).

Some herbicides inhibit the respiratory activity of the soil. One of these is monuron (Tolkachev, 1974). This inhibition disappears in time, however, and stimulation takes place. Other herbicides, inhibitory at normal doses, become favorable at high concentrations, such as dalapon and 2,4-D (Bliev, 1973) in podzols and lawn soils, and pyrazon (Lauss and Danneberg, 1975).

Other herbicides stimulate the gaseous exchanges in the soil. Examples of these are simazine and ioxynil (Smith and Weeraratna, 1975). This stimulation is sometimes followed by inhibition—for example, after treatment with chloroxuron, diuron, fluometuron, metobromuron, or monuron (Grossbard and Marsh, 1974); dalapon, 2,4-D, TCA, or simazine (Kozlova *et al.*, 1974) (assays *in vitro;* or simazine or atrazine (Kruglov *et al.*, 1975a). The negative action appears only after the third or fourth treatment.

Whereas DDT and HCH have no effect on the respiratory activity of the soil when applied in normal doses (Drouineau, *et al.*, 1947), nor do DDT, dieldrin, and malathion at tenfold concentrations, (Ruffin, 1974), other *insecticides* lead to an increase in the oxygen uptake for nine weeks, as shown in Tu's assays (1975) with lindane, or an increase in CO_2 evolution, as found in the experiments of Eno and Everett (1958) with toxaphene and dieldrin and those of Tu (1970) with organophosphorus insecticides. In contrast, an inhibition of the respiration is found with mephosfolan (Purushothaman *et al.*, 1974).

Varied actions of the *fungicides* are also mentioned, from the absence of effect, observed in the work of Hofer *et al.* (1971) and of Van Faassen (1974) with benomyl, to the inhibition of respiration, also found with benomyl (Weeks and Hedrick, 1975) and with captan (Agnihotri, 1971). In this last example, the inhibition is followed by a stimulation, probably resulting from the use of the decomposition remains of the captan by the microorganisms.

For the *nematicides*—carbofuran, fensulfothion, and ethoprop—the oxygen uptake increases with the concentration of the product used in the soil (Tu, 1972, 1973).

2. Enzymatic Activity

Experiments on enzymatic activity are still very limited, but already the results show considerable diversity. Most studies concern dehydrogenase activity, associated with the total respiratory activity, according to some workers. For Ulasevich and Drach (1971), one normal application of atrazine to a cultivated field, and two applications to a bare soil, are enough to increase the dehydrogenase activity, whereas the use of the same product for fifteen years, in apple orchards, leads to a decrease in all enzymatic activity (Voets *et al.*, 1974). In the assays of Hofer *et al.* (1971) and Van Faassen (1974), benomyl remains without effect on the activity of the dehydrogenases. On the other hand, sodium chlorate (Karki and Kaiser, 1974) and mephosfolan (Purushothaman *et al.*, 1974) inhibit this enzyme group.

In regard to enzymatic activity in general, a whole range of results is possible, from no effect with simazine at normal doses (Kulinska, 1967) or with benomyl (Hofer *et al.*, 1971), to inhibition with ethofumesate and pyrazon (Verstraete and Voets, 1974), to stimulation with aldicarb (Verstraete and Voets, 1974).

Urease is sometimes the only enzyme affected (Zubets, 1973) by simazine or atrazine, alone or mixed, and sometimes it is the only one remaining untouched after treatment with paraquat (Giardina *et al.*, 1970). Lethbridge and Burns (1976) also find an inhibition of the soil urease by organophosphorus insecticides; malathion is the least potent inhibitor of the three insecticides studied (fenitrothion and phorate are the other two). Many of the microorganisms, inhibited at the beginning of the experiment (the first 48 hours), develop a tolerance after 21 days, if the application rate is not too high.

Catalase activity increases with dalapon and 2,4-DA in the laboratory as well as in soils of varied nature, while the activity of amylase remains constant (Bliev, 1973).

This range of pesticide action on soil enzymes is clearly indicated in the research of Karanth *et al.* (1975) with fenaminosulf; the inhibition caused by this fungicide depends on its concentration and on the incubation period. Thus, invertase, urease, and tryptophanase are stimulated by 20 ppm of fenaminosulf, but are inhibited by 100 and 200 ppm.

C. CONCLUSIONS

The great diversity of experimental protocols, both *in situ* and in the laboratory, leads to highly varied conclusions regarding the effects of pesticides on the soil microflora—ranging from one extreme to the other, from inhibition to stimulation. It is therefore difficult to classify these results or to express them in generalities, or even to try to elucidate the part played by each factor responsible for the effects observed. In the laboratory studies, carried out under controlled conditions, the observation and evaluation of the microorganisms are more precise and the facts are more reproducible, but the concentrations of the products used are often higher than those applied in agricultural practice, and the methods of application are so different that it is difficult to compare the results with those obtained in field experiments.

This diversity in the effects of pesticides on the soil microflora is increased even more by the great differences in the chemical formulas of the products—sometimes quite simple and mineral, sometimes complex and organic.

IV. Effects of Pesticides on the Biological Cycles of the Soil

The preceding section presented data on the major groups of soil microorganisms and total soil activity. This section deals with more specific questions. It is concerned with the consequences of pesticide treatments on the biological cycles of the soil, the importance of which in numerous processes of synthesis and mineralization has already been shown.

A. THE CARBON CYCLE

Microbiological studies on the microorganisms active in the carbon cycle, especially the cellulolytic organisms, relatively are limited. Nevertheless, the facts revealed are fairly complex.

Among the *herbicides* that have no effect on the cellulolytic organisms when applied at normal doses, Audus (1970) mentions the synthesis phytohormones, monuron, the substituted phenols, cycluron + chlorbufam, DCU, and TCA. In an earlier report, Shklyar (1961) mentioned 2,4-D, simazine, monuron, dinoseb, and chlorpropham (CIPC), used in pre-emergent applications. Later research by Micev (1970) and by Wolf *et al.* (1975) added methoprotryne and bromacil to this list.

A reduction in the decomposition of cellulose appears, however, in experiments *in situ* with simazine and atrazine (Klyuchnikov, *et al.*, 1964), carbaryl (Atlavinyte *et al.*, 1974), linuron (Grossbard and Marsh, 1974), certain triazines, used for seven years, linuron, monolinuron, and cycluron + chlor-bufam (Kozaczenko and Sobieraj, 1973) (pot experiments), and atrazine in a long-term experiment (fifteen years) in apple orchards (Voets *et al.*, 1974). A similar reduction is found in experiments in pots with linuron, monolinuron, and cycluron + chlorbufam (Kozaczenko, 1974). Mainly the *Cytophaga* are affected.

Klyuchnikov *et al.* (1964) found the inhibitory action to be more severe with atrazine than with simazine, which is less water-soluble and penetrates less deeply into the soil. These workers, however, note an action of these two her-bicides down to 25–35 cm in light soils. This decrease in activity never results from a reduction in the number of cellulolytic organisms. A modification of the species may occur after eight sprayings of atrazine (Simon-Sylvestre, 1974). The initial depressive effect disappears in certain cases, followed by a cellulose decomposition greater than that found in untreated soils.

An immediate stimulation of cellulolytic activity is also mentioned with 2,4-D (Abueva and Bagaev, 1975), dalapon, or TCA at high rates and mixed with 2,4-D (Kozlova *et al.*, 1974), atrazine (Percich, 1975), linuron (Miklaszenski, 1975), and propachlor (Rankov and Velev, 1975) at low concentrations. The opposite effect is observed after high doses.

In pure cultures, Audus (1970) found that certain herbicides have no effect even at very high concentrations; these include atrazine, simazine, dalapon, diuron, and chlorthal. Others become harmful only at concentrations of 0.1% and completely inhibit the cellulose decomposition; dicamba, tricamba, and 2,3,6-TBA are examples. Sometimes toxicity appears at lower rates—for example, with propanil, DMPA, 2,4-D, and paraquat (Szegi, 1970). For 2,4-D, a slight inhibition is followed by a stimulating effect. Grossbard (1974) also mentions glyphosate, metoxuron, paraquat, and amitrole as products that cause cellulose breakdown, in pure cultures as well as under field conditions.

Only at very high concentrations do the *insecticides* affect the cellulolytic organisms in the soil (Audus, 1970). To detect an effect on their number, the normal dose must be multiplied by five with DDT, and by seventy with parathi-on; one must use rates twenty times as high with chlordane, forty times as high with demeton, and two hundred times as high with dieldrin, to affect cellulose decomposition. Audus also mentions a positive action on the cellulolytic popula-tion with heavy applications of BHC and parathion.

Audus (1970) demonstrates a negative action, even at normal rates, of the *fungicides* and the *soil fumigants* DD and methyl bromide on cellulose decompo-sition, and a tolerance to metam-sodium in the cellulolytic fungi. On the other hand, the studies of Simon-Sylvestre (1974), carried out *in situ* on two different soils, show that DD does not affect the number of aerobic cellulolytic organisms, whereas metam-sodium decreases it considerably; by 27 and 33 days after the treatment, only 41% and 68%, respectively, of the cellulolytic population remain in the soil. Chloropicrin may also give varied results, even in the same soil, depending on the conditions of soil preparation (Simon-Sylvestre, 1974).

B. THE NITROGEN CYCLE

The important role played by nitrogen in plant nutrition has given rise to numerous studies on the effects of pesticides on the different links of the biological cycle of this element.

3. Ammonifying Bacteria

The ammonifying bacteria are relatively tolerant to pesticides, probably owing to the great heterogeneity of the bacterial species found within this group. As with the other microorganisms, however, the various studies show a diversity in the effect of the pesticides on them. Actually, the same product may have different effects, depending on the experimental conditions; this can be seen in several groups of results.

No effect on the ammonifying bacteria has appeared with the following *herbicides* used at the recommended doses: simazine under corn (Freney, 1965; Simon-Sylvestre and Chabannes, 1975), sodium chlorate and pyrazon (Apltauer and Skopalikova, 1966), paraquat (Tu and Bollen, 1968), DNOC, often toxic to other groups (Audus, 1970), trifluralin (Tyunyaeva *et al.*, 1974), 2,4-D, simazine, atrazine, and cyanazine (Deshmukh and Shrikhande, 1974), and substituted ureas (Grossbard and Marsh, 1974). Similar results were obtained with repeated treatments (Horowitz *et al.*, 1974a) with about ten herbicides, including substituted ureas and simazine, in soils that were not cultivated but were watered by spraying.

When certain products are applied at high rates, however, a depressive effect may occur. Balicka and Sobieszczanski (1969) have observed this with triazines, substituted ureas, pebulate, and EPTC. Harmful effects on these bacteria are mentioned also by Audus (1970), in studies on TCA, propham, and fenuron, and PCP used in water-logged paddy soils.

Several other herbicides should be included in this survey:

(*a*) 2,4-D and 2,4-MCPA (Bertrand and De Wolf, 1972), but only temporarily.

(*b*) Sodium chlorate (Karki and Kaiser, 1974).

(*c*) Metribuzin on tomatoes (Velev and Rankov, 1975). The effect, visible 20 days after the application, disappears 100 days later.

(*d*) Oxamyl and phenmedipham (Vlassak and Livens, 1975). With phenmedipham, the harmful effect does not appear in soils rich in organic matter. A depressive action may also occur in the case of repeated herbicide treatments.

(*e*) Simazine and linuron (Grossbard, 1971).

(*f*) Atrazine and propachlor (Husarova, 1972), with a monocultivation of corn.

Finally, a third herbicide category favors both the growth and the activity of the ammonifying bacteria. This group includes the following:

(*a*) Simazine and atrazine used *in situ* (Kozlova *et al.*, 1974).

(*b*) Dinoseb (Szember *et al.*, 1973), but the effect is low.

(*c*) Prometryn and fluometuron, used on cotton (Darveshov, 1973).

(*d*) MCPA and 2,4,5-T (Torstensson, 1974). The stimulation by these two products is light; it exists neither with linuron nor with simazine, used under the same conditions.

(*e*) 2,4-D (Abueva and Bagaev, 1975).

Long-term applications of atrazine (Voets *et al.*, 1974) and simazine (Kolcheva *et al.*, 1974) also lead to an increase in the ammonifying bacteria, but this effect is only temporary with atrazine.

The effect of *insecticides* on the ammonifying bacteria has not often been studied. Audus (1970) reports demeton to be the most toxic insecticide, as it inhibits ammonification even at levels below field rates. We must also mention in this category two new insecticides, monocrotophos and methidathion, which are depressive if they are applied at high doses (Idris, 1973), as well as HCH in the presence of ammonium nitrate (Kir *et al.*, 1974).

Aldrin (Ross, 1952) and acephate and methamidophos (Focht and Josseph, 1974) have no effect on the ammonifying bacteria, nor has DDT, according to Jones (1952), although Ross (1952) mentions stimulation by DDT and also by chlordane in one trial only. The application of organophosphorus insecticide (Tu, 1970) also stimulates ammonification.

The literature on *fungicides* and *nematicides* gives examples of the stimulation of ammonifying bacteria with the following:

(*a*) Chloropicrin and ethylene dibromide (Audus, 1970).

(*b*) Carbofuran, DD, fensulfothion, and methylisothiocyanate (Tu, 1972).

(*c*) Captan, thiram, and an organomercuric at low doses, in experiments *in vitro* (Wainwright and Pugh, 1973). At higher concentrations of thiram and the organomercuric, the stimulation changes into inhibition.

(*d*) Benomyl, captan, thiram, dicloran, and quintozene, *in situ,* at twice the field rate (Wainwright and Pugh, 1974).

(*e*) Thiram (Agnihotri, 1974). The ammonification increases for six weeks, its intensity depending on the amount of thiram used.
(*f*) Benzenehexachloride (Akotkar and Deshmukh, 1974).

Or we may find cases of inhibition—for example, with maneb and anilazine, used at high rates *in vitro* (Dubey and Rodriguez, 1970), or with metam-sodium, which suppresses peptone ammonification for a certain time (Roa, 1959).

2. Nitrifying Bacteria

The nitrifying bacteria responsible for ammonia oxidation in nitrates are both the most studied organisms, owing to their agronomic importance, and the most sensitive microorganisms to stress of all kinds. They are indeed scarce and are provided with very complex and probably very frail enzymes.

Numerous studies have been carried out on these nitrifying bacteria, but, owing to the diversity of the results obtained, their classification is not a simple matter.

According to some studies, the use of certain *herbicides* at normal doses has no consequence either on the nitrifying bacteria or on nitrification. To this group belong the *triazines*, including simazine and atrazine, in a study *in vitro* (Eno, 1962); simazine, *in situ* as well as *in vitro* (Freney, 1965); simazine and atrazine in a tchernozem soil (Kudzin *et al.*, 1973); simazine and prometryn, even after repeated applications (Horowitz *et al.*, 1974a), and simazine (Torstensson, 1974): This group also includes the *substituted ureas*, such as linuron (Kudzin *et al.*, 1973); linuron, monolinuron, and cycluron + chlorbufam in a pot assay with white mustard (Kozaczenko and Sobieraj, 1973); neburon and fluometuron, even when used several times in the same field (Horowitz *et al.*, 1974a); picloram (Grover, 1972); and dalapon and paraquat (Namdeo and Dube, 1973).

For these same products several workers obtain depressive results with other experimental protocols. Thus, nitrification is inhibited by the *triazines*, including simazine and atrazine in liquid medium, although this depressive effect does not appear in the soil (Balicka and Sobieszczanski, 1969); simazine and atrazine during the first year (Zubets, 1973); atrazine in a long-term assay (Voets *et al.*, 1974); prometryn in liquid medium (Balicka and Sobieszczanski, 1969; Voinova *et al.*, 1975); and metribuzin (Velev and Rankov, 1975), after which recovery to a normal population takes place after 100 days. Inhibition of nitrification is also found with the *substituted ureas*, including linuron, monolinuron, and chloroxuron, in liquid medium (Balicka and Sobieszczanski, 1969); monolinuron (Szember *et al.*, 1973) (the activity of the nitrifying bacteria is still inhibited twelve months after the application); linuron, only immediately after the treatment (Torstensson, 1974); and diuron, fluometuron, metobromuron, monuron, metoxuron, and linuron, only when they have been used in high concentrations (Grossbard and Marsh, 1974).

We must also mention—and this list is not exhaustive—a harmful effect with the synthesis phytohormones, including (a) 2,4-DB (Chandra, 1964). The toxic effect of the herbicide decreases only eight weeks after the treatment. (b) MCPA and 2,4,5-T (Torstensson, 1974). The sensitivity of the nitrifying bacteria appears only at high rates. (c) 2,4,5-TP (Cervelli et al., 1974). Stimulation takes the place of inhibition at higher doses for this last product.

Harmful effects are also caused by pyrazon (Lauss and Danneberg, 1975) at high concentrations and the carbamates diallate (Chandra, 1964) and diallate and chiefly phenmedipham, used at normal rates under sugar beet (Livens et al., 1973). This toxicity has an adverse effect on plant nutrition and leads to a decrease in the quality of the sugar-beet juice. Other herbicides with toxic effects include trifluralin (Tyunyaeva et al., 1974), which is toxic for two weeks after application; sodium chlorate, which reduces the nitrifying bacteria for nearly a year and strongly inhibits the oxidation of nitrites to nitrates (Audus, 1970); and calcium cyanamide, which at normal doses almost wholly eliminates nitrifiers (Audus, 1970).

In contrast to the depressive effects on the growth and activity of the nitrifying bacteria, observed after a herbicide treatment, many stimulating actions have also been noted. Such actions were observed with the triazines applied at normal doses in studies by Amantaev et al. (1963) with simazine and triazine, both in the laboratory and in a light chestnut and an irrigated soil; by Kozlova et al. (1964) in situ with simazine; by Kulinska (1967) with simazine in situ or in pot experiments—at high rates the nitrification is reduced; by Smith and Weeraratna (1974) with simazine, in the presence, or not, of N serve;* by Darveshov (1973) with prometryn used in cotton fields; by Reichlova (1975) with terbutryn in vitro—after 24 weeks of incubation, the nitrification is slightly inhibited; by Smith and Weeraratna (1975) with simazine, which, like ioxynil, increases nitrification in an acid soil and delays it in an alkaline one; by Kruglov et al. (1975b) with atrazine used for three years; and by Kolcheva et al. (1974) with atrazine used for twelve years. Stimulating effects were also reported with the amides propachlor and alachlor (Enkina and Vasilev, 1974) in the sunflower rhizosphere; with the sodium salt of 2,4-D (Abueva and Bagaev, 1975); with pyrazon and pebulate in irrigated soils (Urusbaev, 1975); and with chlorpropham (CIPC) and its metabolites in the cotton rhizosphere (Taha et al., 1972).

Some studies carried out in pure cultures, the results of which are reported by Audus (1970), lead to a variety of conclusions about herbicides and nitrification. 2,4-D and TCA at normal doses have no effect, nor does picloram. On the other

*N serve = an inhibitor of autotrophic nitrification

hand, chlorpropham (CIPC) and EPTC, at field rates, inhibit nitrite oxidation by *Nitrobacter* and stop it completely at 150 ppm. *Nitrobacter* is more sensitive to monuron than is *Nitrosomonas* in pure culture.

With the majority of *insecticides* used at field rates, no action has been recorded on the growth or the activity of the nitrifying bacteria. This category includes DDT, chlordane, and aldrin in sandy soils (Ross, 1952); DDT, chlordane, and aldrin in incubation assays (Pathak *et al.*, 1960–1961); DDT in a study *in vitro* (Jones, 1952); DDT, chlordane, aldrin, dieldrin, heptachlor, lindane, and toxaphene, with five annual applications (Martin *et al.*, 1959); DDT, aldrin, dieldrin, and heptachlor, in lysimetric studies (Shaw and Robinson, 1960); and acephate and methamidophos, even with a tenfold rate (Focht and Josseph, 1974).

Certain insecticides cause depression when used at high concentrations. These include lindane, dieldrin, and aldrin (Bardiya and Gaur, 1970). The nitrification is inhibited by doses of 25 ppm. The toxicity varies according to the products; it lasts for a week with lindane, for two weeks with dieldrin, and for three weeks with aldrin. High concentrations of monocrotophos and methidathion also cause depression (Idris, 1973).

Even at the recommended doses, Chandra (1966) finds a negative effect with dieldrin and heptachlor on nitrification. This inhibitory action, varying with different soils, is stronger for dieldrin, and decreases in the course of time and in relation to all the factors that stimulate nitrification. Tu (1970) also records a slight reduction in nitrification with four organophosphorus insecticides. Garretson and San Clemente (1968) have studied in pure culture the effects of several insecticides on the activity of *Nitrosomonas europaeus* and *Nitrobacter agilis*. The toxicity, determined either by a delay in nitrification or, in severe cases, by complete inhibition, depends on the insecticide and its concentration; aldrin and parathion are most harmful toward *Nitrobacter,* and lindane and malathion are most harmful toward *Nitrosomonas*.

Audus (1970) also relates some increases in nitrification with high rates of lindane, heptachlor, parathion, and disulfoton.

Nitrifying bacteria are affected considerably by the *fungicides* and the *soil fumigants*. All the published reports mention these depressive effects which are very noticeable, even at normal doses. Among the harmful products, the following may be listed.

(*a*) Ferbam (Jaques *et al.*, 1959). The action, studied by percolation, is visible in the two phases of nitrification.

(*b*) Nabam and dazomet (Chandra and Bollen, 1961). Nitrification is totally suppressed for 30 days; then the effect of the products decreases.

(*c*) Anilazine and maneb (Dubey and Rodriguez, 1970). The inhibition, noted only on the bacteria that oxidize ammonia, lasts a shorter time when the environmental conditions are favorable to nitrification. Maneb is more harmful, and its effect lasts even after the disappearance of the product.

(*d*) Captan (Agnihotri, 1971). There is a reduction in nitrification for two to three weeks, depending on the fungicide concentration.

(*e*) Fensulfothion, carbofuran, DD, and methylisothiocyanate (Tu, 1972).

(*f*) Captan, thiram, and an organomercuric (Wainwright and Pugh, 1973). These three products, which have no effects or are sometimes slightly stimulating at low concentrations, cause depression at high rates.

(*g*) Organomercuric, at high doses (Pugh *et al.*, 1973).

(*h*) Thiram (Agnihotri, 1974). The inhibition, variable with the concentration of the fungicide, may last for five weeks.

(*i*) Benomyl (Van Faassen, 1974). In pure cultures, the first phase of the nitrification is delayed, and the second is inhibited at a high concentration of the fungicide. On the other hand, *in vivo* the total mineral nitrogen seems to increase.

(*j*) Benomyl, captan, thiram, dicloran, formaldehyde, and quintozene (Wainwright and Pugh, 1974). A definite effect may be observed on the nitrates, even twelve weeks after the application of one of these fungicides. The lowest rate of nitratification is found after four weeks with formaldehyde; quintozene is the least harmful.

(*k*) Metam-sodium, DD, and dazomet, in a model assay (Markert and Kundler, 1975).

This depressive effect of the fungicides and the soil fumigants does not last. Nitrification becomes normal after varying periods of time, depending on the products—from 17 days with zineb to 120 days with chloropicrin (Audus, 1970).

To sum up this section on nitrifying bacteria, we refer to the laboratory studies of Bartha *et al.* (1967) on different pesticides and their interpretation of their data on nitrification. They find a relationship between the chemical configuration and the endurance of these products and their effects in the soil. With the inhibitory products, several levels of reaction may be noted.

(*a*) The toxic effect decreases with time, indicating that either these products undergo a transformation (microbiological degradation, for example), or a resistant nitrifying population develops in the soil. Atrazine, EPTC, malathion, and parathion are examples.

(*b*) The inhibition of nitrification remains constant throughout the test period. Such products are chemically and biologically stable. Chloranocryl, diphenamid, fenuron, TPC, and monuron are examples.

(c) The toxicity increases with time, suggesting the formation in the soil of a product more harmful than the initial pesticide. Chlorpropham (CIPC), diuron, and linuron are examples.

3. Nitrogen-Fixing Bacteria

Although the aerobic nitrogen-fixing bacteria are not as interesting as the ammonifying and nitrifying bacteria, they have been the subject of considerable research, with a wide range of results.

   a. *Nonsymbiotic Bacteria.* The growth of aerobic, nonsymbiotic nitrogen-fixing bacteria is not affected by most *herbicides* when they are used at the recommended doses. Audus (1970) mentions phenolic compounds, synthesis phytohormones, substituted ureas, maleic hydrazide, TCA, dalapon, and ioxynil. Amantaev *et al.* (1963) report similar findings with simazine and atrazine, and Bertrand and DeWolf (1973) with the mixture 2,4-D + MCPA, the behavior of which varies with the species of nitrogen-fixing bacteria. Not very active toward *Azotobacter,* this mixture stimulates *Clostridium* at low doses, and inhibits them at high concentrations. Other herbicides with no effect include isoproturon and triazophos (Neven *et al.,* 1975) and cycluron + chlorbufam, endothal, and dalapon in irrigated soils (Urusbaev, 1975).

   Some herbicides, however, have a depressive effect on the *Azotobacter* number under certain conditions and at rates not much above normal field applications (Audus, 1970). These are PCP, DNOC, siduron, simazine, and atrazine, under corn, for *Azotobacter,* and propham (IPC) for *Clostridium.* A negative effect is also reported with simazine used between the rows of strawberry plants (Bakalivanov and Nikolova, 1969); with linuron, monolinuron, and cycluron + chlorbufam (Kozaczenko and Sobieraj, 1973); with paraquat, the effect of which depends on the nature of the soil (Szegi *et al.,* 1974); with dinoseb acetate at 10 ppm (Neven *et al.,* 1975); with pyrazon and pebulate, which are inhibiting for two months (Urusbaev, 1975); and with linuron and chlorpropham (CIPC) (Wegrzyn, 1975).

   Other workers have demonstrated positive effects of the herbicides on the number of nonsymbiotic nitrogen-fixing bacteria. Audus (1970), in reporting on the action of these herbicides on *Azotobacter,* describes the positive effects of simazine and atrazine in corn crops on loamy soils; of atrazine, trietazine, and prometryn in the top layer of the soil under peas at emergence (the effect is reversed when the peas come into flower); and of MCPA in the oat-plant rhizosphere. Simazine in a leached tchernozem soil has a similar effect on *Clostridium* (Audus, 1970). A stimulating effect on the number of nitrogen-fixing bacteria is also found with atrazine on both *Azotobacter* and *Clostridium* after the first application—the effect is reversed after the second application (Ulasevich and

Drach, 1971); with 2,4-D up to a certain concentration, above which the number of *Azotobacter* decreases (Sharma and Saxena, 1974); and with prometryn (Wegrzyn, 1975). There is also a stimulating effect on the diameter of *Azotobacter* colonies with atrazine in trials *in vitro* (Aliev *et al.*, 1973), and on nitrogen fixation with dalapon and 2,4-DA for the first forty days (Bliev, 1973).

Limited research has been carried out on the action of the *insecticides* on nonsymbiotic nitrogen-fixing bacteria. In general, no effect is found at normal doses, as indicated by the experiments of Drouineau *et al.* (1947) with DDT and HCH; of Jones (1956) *in vitro* with DDT, chlordane, dieldrin, aldrin, and endrin; of Pathak *et al.* (1960–1961) *in situ* with DDT, chlordane, and aldrin; and of Mendoza (1973) with DDT and menazon.

Depending on the concentration of the insecticide, some cases of depression have been reported in experiments carried out *in vitro* with malathion, dimethoate, and carbaryl (Mendoza, 1973) and phoxim (Eisenhardt, 1975), and *in situ* with metasystox, DDT, and methoxychlor (Brenner *et al.*, 1974). The recovery to a normal population occurs after three weeks.

In the category of *fungicides*, only the reports of Jones (1956) and of Pathak *et al.* (1960–61) mention no effect with HCB on nonsymbiotic nitrogen-fixing bacteria. On the other hand, chloropicrin and the nematicides DD and metamsodium clearly decrease the number of these bacteria in the soil (Simon-Sylvestre, 1974).

b. Symbiotic Bacteria (Rhizobium).   The effect of pesticides on *Rhizobium* may be divided into two types of action, which may be independent of each other—one on the bacteria itself and on its growth, and the other on the host plant, its infestation, the phenomenon of root nodule formation, and nitrogen fixation.

Almost all the literature on *herbicides* and *Rhizobium* records a harmful action. There is possibly an inhibitory effect on the growth of *Rhizobium,* of varying intensity depending on the strain, as shown in the experiment of Pantera (1974) with lupine and linuron. Dalapon, decreases the growth of *Rhizobium in vitro,* but does not affect nodulation of the alfalfa (Lakshmi–Kumari *et al.,* 1974).

Dinoseb, which is more toxic than bentazon, delays the growth of *Rhizobium* only when applied at doses above normal (Torstensson, 1975).

According to Grossbard (1975), nodulation and nitrogen fixation are slightly reduced with atrazine, dinoseb, asulam, and linuron. The effect is greater with 2,4-DB, alone or strengthened with dalapon (Garcia and Jordan, 1969). This decrease in nodulation and nitrogen fixation is the result of damage caused by 2,4-DB on the plants and of the abnormal growth of the roots. A reduction is also noted with simazine (Hauke-Pacewiczowa, 1970). Paromenskaya (1975) reports that simazine in toxic doses prevents the reduction of the amide nitrogen in the plant and disturbs the ammonium assimilation.

In pure culture, toxicities appear at higher doses than those corresponding with normal rates, (Audus, 1970). DNOC, dinoseb, pyrazon, diuron alone or mixed with propham (IPC), and linuron are the most harmful herbicides, and dalapon, simazine, and prometryn are the least toxic.

Exceptions to these findings have been noted. Some treatments with no effects were recorded by Audus (1970) on *Rhizobium* with *insecticides* at normal doses. The results of Suriawiria (1974) on a study with soya indicate that the root nodules have not decreased after a treatment with endrin, even at high concentrations, and Mendoza (1973), studying the effects of DDT and menazon on *Rhizobium trifolii,* found the growth is not affected at any dose.

The insecticides in general have a depressive action on *Rhizobium* itself, the sensitivity of which varies with the strain (Brakel, 1963). *Rhizobium* isolated from *Trifolium* is very sensitive, whereas the strain isolated from alfalfa is resistant. According to Brakel, lindane causes a more severe reaction than aldrin or parathion. The growth of *Rhizobium trifolii* is also inhibited by malathion, carbaryl, and dimethoate (Mendoza, 1973).

The insecticides also have a negative effect on nodulation. Goss and Shipton (1965) report that, in their trials of leguminous inoculation, dimethoate, used even a month before the inoculation, causes damage and prevents nodulation. The results are disastrous also with aldrin, dieldrin, chlordane, DDT, BHC, lindane, and parathion. A depressive effect on nitrogen fixation is also observed. When the phoxim concentration in the synthetic medium varies from 10 to 1000 ppm, the percentage of fixed nitrogen decreases from 56 to 7%, as compared with the standard sample (Eisenhardt, 1975).

With the *fungicides* and the *soil fumigants,* the effects on *Rhizobium* depend on the strain and on the product (Audus, 1970). Ethylene dibromide applied at the normal rate considerably decreases nodulation. The copper salt of 8-hydroxyquinoline is toxic only to some strains.

In a nematology study on soya, Reddy *et al.* (1975) record a lack of action of benomyl on nodulation, contrary to findings for aldicarb, oxamyl, and carbofuran. Moreover, ethylene dibromide and DD have been reported to improve nodulation in soya bean, but the mode of action is not known.

4. Denitrifying Bacteria

Microbiologists have not shown much interest in the denitrifying microflora, probably because the conditions that are most often met in the soil do not favor these anaerobic organisms.

In reporting on the *herbicides,* Guillemat *et al.* (1960) find no effect of simazine on the denitrifying bacteria, even at rates as high as two hundred times the normal field rates. Kozlova *et al.* (1964), on the other hand, record a positive effect in field experiments with corn and lupine, Torstensson (1974) a stimulation of the denitrifying bacteria with MCPA and 2,4,5-T, and Kuryndina (1965) a depressive effect after three consecutive years of application of simazine in an orchard at rates slightly above normal. Dinoseb (Szember *et al.*, 1973) and sodium chlorate (Karki and Kaiser, 1974) also inhibit the denitrifying microflora. 2,4-D has a depressive effect in pot experiments with flax (Abueva and Bagaev, 1975). Atrazine was inhibiting in a long-term trial with apple trees (Voets *et al.*, 1974), but recovery to a normal population was seen twelve months after the last treatment.

Other workers record a temporary depressing action of some herbicides on the denitrifying bacteria—Lobanov and Poddubnaya (1967) after a normal application of DCU and TCA under sugar beet, and Urusbaev (1975) for a month, after a treatment with pyrazon or pebulate.

The *insecticides* are in general non-toxic to the denitrifying bacteria, even at rates up to one hundred times the normal rates (parathion) or in water-logged soils (BHC) (Audus, 1970).

The denitrifying bacteria are more affected by the *fungicides* and the *nematicides* (Audus, 1970). Nabam and maneb are more toxic than ferbam, thiram, and ziram; the inhibition in the latter group is proportional to the number of dithiocarbamate radicals. In contrast, chloropicrin in paddy soils, ethylene dibromide, and DD (the latter at normal rates) increase the denitrifying population.

C. THE SULFUR CYCLE

Very few studies have been carried out on the behavior of the sulfur-oxidizing bacteria in response to *herbicides*. However, sulfur oxidation is not affected by 2,4-D, MCPA, maleic hydrazide, or ammonium sulfamate, even at concentrations well above field rates (Audus, 1970). Paraquat, on the other hand, has a slightly harmful action (Tu and Bollen, 1968).

The *insecticides,* in general, have no effect on the sulfur-oxidizing bacteria. This has been shown with DDT (Jones, 1952) and with acephate and methamidophos (Focht and Josseph, 1974) even at concentrations ten times the normal rate. However, an application of 100 $\mu g/g$ of diazinon leads to an increase in sulfur oxidation of about 15%, and thionazin and chlorpyrifos, at the same concentration, cause a decrease ranging from 12 to 17% (Tu, 1970).

Elementary sulfur oxidation is also slower with some *nematicides,* such as fensulfothion, methylisothiocyanate, DD, and carbofuran (Tu, 1972).

D. THE PHOSPHORUS CYCLE

Only a few workers mention the effects of pesticides on the bacteria of the phosphorus cycle. Among the *herbicides,* simazine and 2,4-D reduce the growth of these bacteria (Yurkevich and Tolkachev, 1972), but trifluralin has no effect on phosphorus mineralization (Tyunyaeva et al., 1974).

The organophosphorus insecticides do not affect the mineralization of the organic phosphorus of the soil (Tu, 1970), whereas after a treatment with DDT, chlordane, or aldrin the content of available phosphorus in the soil increases (Pathak et al., 1960–1961).

Phosphorus mineralization is also stimulated after a fungicide treatment, at least with the two products studied—HBC (Pathak et al., 1960–1961) and thiram (Agnihotri, 1974).

E. THE MANGANESE CYCLE

Smith and Weeraratna (1975), in one of their studies, record that the biological action of manganese is delayed by a treatment with 10 ppm ioxynil and increased by simazine applied at the same rate.

F. CONCLUSIONS

Attention must be drawn again to the multiplicity of facts presented in this section, representing the most outstanding results recorded in recent literature on the biological cycles. It is difficult to use all these data in their present form to predict the eventual consequences of the application of a pesticide on the biological cycles, and more generally on the soil microflora. Yet, the part played by soil microorganisms is important in agriculture. Therefore, it is essential that we standardize the experimental protocols and the methods of biological measurement if we are to obtain comparable data.

V. Action on Pathogenic Microorganisms

The activity of the soil microflora is both the cause and the consequence of the regulative phenomena that make it possible to maintain in the soil, for given ecological data, a certain biological balance, which generally prevents the excessive proliferation of certain organisms, and assures the protection of others. Two of these mechanisms particularly concern the life and fertility of soils: the phytopathogenic action of some microorganisms, and the antagonistic or synergistic relationships of these microorganisms with the other species.

The often noticeable specificity of these actions establishes, moreover, a great sensitivity to environmental factors and to the agrochemical substances that have a strong biological activity—pesticides, for instance. Thus, the side effects of these biocides and their favorable or unfavorable consequences on the pathogenic microflora must not be disregarded. Furthermore, the possibility of adaptation of these microorganisms and their increased resistivity to the pesticides is of great importance.

The purpose of this section is to examine these two points concerning the relationships between the pathogenic microflora and the pesticides—first in a general sense, with reference to their agronomic interest, and then at the level of the phenomena involved, with emphasis on the direct effects of pesticides on pathogens and on their antagonists in the soil.

Properly speaking, a few of the references given here do not deal with "soil" microorganisms. However, we have mentioned them when they contribute to an understanding of the described phenomena.

A. THE INFLUENCE OF PESTICIDES ON PLANT DISEASES

1. Field Observations and Experiments

For about twenty years, we have known that herbicides and insecticides, as well as fungicides, can influence nontarget organisms and plant diseases. However, few reports concern observations or experiments made under strictly agronomic conditions.

a. Agronomic Experiments. The antifungistic activity of some *herbicides* has often been shown. As early as 1949, Guiscafré–Arrillaga tested 2,4-D in the prevention of postharvest decay of oranges. This herbicide showed a strong inhibitory action against *Penicillium digitatum* and *Phomopsis citri*. Chappell and Miller (1956) observed that the use of dinoseb for weed control in a peanut field resulted in the improvement of the vegetative and sanitary states of the plants. In the same way, Campbell (1956) and Horner (1965) indicated that mint rust could be controlled by dinitramine or dinoseb, respectively. With dinoseb, Olsen (1966), who was working on the chemical control of potato wart disease (*Synchytrium endobioticum*) in naturally infested field plots, controlled the disease with rates of application similar to those used for weed control. In contrast to these findings, Nilsson (1973) has demonstrated the influence of oxitril (a mixture of dichlorprop, MCPA, ioxynil, and bromoxynil) on take-all (*Gaeumannomyces graminis*) and eyespot disease (*Cercosporella herpotrichoïdes*) of winter wheat. Both diseases were significantly higher in the treated plots.

Relatively few studies have been carried out on the effects of *insecticides* on plant diseases, probably because they are not generally employed as direct soil treatments. Richardson (1957) has studied the action of some insecticides against *Helminthosporium sativum* on barley seedlings; maleic hydrazide and heptachlor increased the severity of infection, but root rot was reduced by aldrin, endrin, and chlordane. Forsberg (1955) obtained a reduction in gladiolus scab (*Pseudomonas marginata*) by application of aldrin, lindane, or heptachlor to the soil. Edgington and McEven (1975) observed the effect of insecticides in combination with fungicides on onion smut (*Urocystis cepulae*). In particular, ethion ensured control even in the absence of fungicides.

Some reports concerning *nematocides* can also be mentioned here. Thus, DBCP (1,2-dibromo-3-chloropropane) decreased infestation of taxus roots by *Rhizoctonia* (Miller and Ahrens, 1964). Ethoprop treatments at flowering on groundnut reduced the damage caused by *Sclerotium rolfsii* (Rodriguez–Kabana *et al.*, 1976a). Likewise, field studies on peanuts revealed that applications of fensulfothion lowered the damage by *Sclerotium rolfsii* during the early part of the season (Rodriguez–Kabana *et al.*, 1976b).

Moreover, contrary to what might be expected, the use of *fungicides* does not always lead to a diminution of the global soil fungic population. Thus, according to Wainwright and Pugh (1974), field application of six fungicides at twice the normal rate increases both the bacterial and the fungal numbers. Furthermore, specific fungicides can affect the balance between the species in the soil. With regard to pathogens, the incidence of damping-off of cucumbers was consistently higher in plots treated with PCNB than in an untreated control (Haglund *et al.*, 1972). In the same way, the use of benomyl for control of *Cercospora* on peanuts gave a consistently high level of white mold damage due to *Sclerotium rolfsii*. (Backman *et al.*, 1975). Finally, it has been shown that some fungicides, such as oxycarboxin and carboxin give good control of bacterial blight of cotton (Nayak *et al.*, 1976).

   b. Experiments under Artificial Conditions. In order to facilitate their experiments, many researchers prefer to promote the disease artificially by inoculating the soil with the pathogen. Some recent results obtained under such conditions are reported in Table VI.

Sometimes the inoculum is added to a soil previously disinfected or sterilized. Generally, doses of inoculation far exceed normal field populations. This may be of great importance. For example, Chandler and Santelman (1968) have shown that in the fields, on cotton seedlings, injurious interactions occurred between nitralin and *Rhizoctonia solani* only if there was a high level of the fungus in the soil.

The time of inoculation of the soil is also of interest. Such is the case concerning the effectiveness of treatment with 2,4-D against the *Fusarium* wilt disease of tomato. When the inoculation preceded the treatment, susceptibility to disease was not appreciably affected. But in plants inoculated after treatment with pesticides, the disease was reduced (Davis and Dimond, 1953). In contrast, the increasing incidence of the *Rhizoctonia* damping-off due to diphenamid was reduced by a delayed application of the pesticide (Eshel and Katan, 1972).

TABLE VI
Effect of Some Herbicides on Plant Diseases in Soil Artificially Infested

Pesticide	Diseases	Effect[a]	Authors
Dinoseb	Root rot of pea (*Aphonomyces euteiches*)	D	Jacobsen and Hopen (1975)
Diphenamid	Damping off in pepper (*Rhizoctonia solani*)	I	Katan and Eshel (1974)
	Damping-off in tomato seedling (*Rhizoctonia solani* and *Pythium aphanidermatum*)	D,I	Batson and Cole (1975)
Trifluralin or dinitroanilene herbicides	Seedling disease in cotton (*Rhizoctonia solani*)	I	Neubauer and Avizohar-Hershenson (1973)
	Rhizoctonia, Verticillium, Fusarium disease on eggplant, tomato, pepper	D	Grinstein *et al.* (1976)
	Root rot of pea (*Aphanomyces euteiches*)	D	Harvey *et al.* (1975) Jacobsen and Hopen (1975)
	Club root on cabbage (*Plasmodiophora brassicae*)	D	Buczacki (1973)
Diuron	*Verticillium* wilt in cotton	D	Minton (1972)
EPTC	Root rot caused by *Fusarium solani* on navy bean seedlings	I	Wyse *et al.* (1976b)
Triazine herbicides	Mildew on wheat (*Erysiphe graminis*)	D,I	Heitefuss and Brandes (1970)
Atrazine	Root rot of pea (*Fusarium solani* f. sp. *pisi*)	I	Percich and Lockwood (1975)
Atrazine	Corn seedling blight (*Fusarium roseum* f. sp. *cerealis*)	I	Percich and Lockwood (1975)

[a] D—decrease of the disease. I—increase of the disease.

It may be of interest to determine if there is a relationship between the increase or decrease of the diseases in the treated crops and the type of pesticide or pathogen involved. Few authors have compared the activity of a series of products toward a disease under well-determined experimental conditions. Cesari and Rapparini (1969) determined the fungicidal activity of about twenty-five chemicals used for selective weed-killing against *Fusarium oxysporum* f. sp. *pisi* and *Rhizoctonia solani* in greenhouse tests with pea or bean used as the test plant. Only the herbicide dichlobenil proved to be effective against the two pathogens.

But the experimental conditions (especially the level of inoculation) did not allow the authors to observe small fungistatic effects or stimulating effects. Generally all the results are obtained under a great variety of experimental conditions. Moreover, these conditions may act in an opposite way on the various factors that control the disease, and it is not possible to infer general rules. Thus, urea and triazine herbicides enhanced mildew on wheat (*Erysiphe graminis*) (Heitefuss and Brandes, 1970) but reduced infection by *Cercosporella herpotrichoïdes* (the eyespot of wheat) (Heitefuss and Bodendörfer, 1968). High rates of trifluralin aggravate infection by *Rhizoctonia solani* on cotton seedlings (Chandler and Santelman, 1968). But according to Buczacki (1973) the same herbicide lowers the incidence of *Plasmodiophora brassicae* on cabbage. Foliage application of 2,4-D on tomato decreased the incidence of *Fusarium oxysporum* f. sp. *lycopersici* (Davis and Dimond, 1953) but soil application of 2,4-D before root inoculation on young plants promoted it (Richardson, 1959). In the same way, 2,4-D increased the incidence due to *Helminthosporium sativum* on wheat (Richardson, 1959) but decreased it on barley (Richardson, 1957). The application of the urea and triazine herbicides in pot tests after the inoculation of wheat with *Erysiphe graminis* resulted in a temporarily reduced disease level. But different results were obtained as soon as the plant had overcome the impact of the pesticides on its growth, and more mildew developed (Heitefuss and Brandes, 1970). Finally, with diphenamid Batson and Cole (1975) reported a decrease in damping-off of tomato seedlings by *Rhizoctonia solani* and *Pythium aphanidermatum* with an application of 3.32 kg/ha, and a stimulation of disease with 6.72 kg/ha.

2. Mechanisms Involved in the Relationships between Pesticides and Plant Diseases

Several reviews on the influence of herbicides on plant diseases have been published, particularly by Kavanagh (1969, 1974) and by Katan and Eshel (1973). Each of them considered the various interactions occurring between the plant, the pesticide, the pathogen, and the surrounding microorganisms. We shall mention only briefly the importance of the effects of pesticides on the sensitivity of plants to pathogens; we shall make a more detailed study of the effects of pesticides on the pathogenic microflora.

a. Modification of the Sensitivity of the Host. The sensitivity of plants to pathogens can be modified by some pesticides. On this subject, Van Der Zweep (1970) indicates three different aspects that must be considered: first, the effect of the products on the treated plants; second, the consequences of this effect on the organisms causing the disease; and finally, the consequences for the development of the disease and the expression of its symptoms.

For example, a few herbicides can cause an increase in the exudation of carbohydrates through the roots of a plant. Such an effect has been mentioned with the use of picloram on young corn seedlings (Lai and Semeniuk, 1970). In the same way, pyrazon used as a herbicide of pre-emergence on sugar beet causes an increase in the exudates of glucose and in the permeability of the cell membrane. This could explain the strong increase in damping-off of sugar beet that occurs in plants grown in an infested and herbicide-treated soil (Altman, 1973). Moreover, artificial defoliation of sugar maple trees leads to chemical changes in their roots and predisposes them to attacks of *Armillaria mellea* (Wargo, 1972).

Some other types of action are sometimes described. For example, diphenamid delays the emergence of tomato seedlings, which increases the negative effects on seeds and causes an increase in postemergence damping-off in soils infested with *Rhizoctonia solani* (Cole and Batson, 1975).

   b. Direct Effects on Pathogens.   In addition to the fungicides, many pesticides can cause a direct effect on pathogenic fungi or on bacterial parasites of plants. Studies on the development of microorganisms are often carried out *in vitro* on artificial liquid or agar media, or in soil suspensions or samples of sterilized soils. Table VII gives a partial list of these studies, particularly the most recent ones. The characteristics studied most often are mycelium growth and the germination of spores in a synthetic medium or the production of CO_2 in soil tests.

The main parameters involved are as follows.

Influence of the microorganism.   In regard to the sensitivity of pathogenic and saprophytic microflora to pesticides, Valaskowa (1968) noted that pathogenic microorganisms are more easily inhibited than the others, and she considered a practical application of this observation. In contrast, Richardson (1970) observed that a much larger proportion of saprophytic than of pathogenic strains was tolerant of or was even stimulated by atrazine. The activity of the products toward the pathogens also varies according to the species and sometimes according to the strains studied. Thus, Ebner (1965) tested six herbicides and noticed that diuron and linuron acted as fungicides *in vitro* against *Rhizoctonia solani, Colletotrichum lindemuthianum, Rhizopus nigricans,* and *Septoria apii,* but not against *Aspergillus niger* or *Alternaria solani.* However, depending on the physiological process studied, opposite results can sometimes be obtained with the same organism. Thus, 20 ppm of prometryne stimulates the production of CO_2 by *Fusarium oxysporum* f. sp. *vasinfectum* when inoculated in a sterile soil, but inhibits the germination of spores (Chopra *et al.,* 1970a).

Influence of pesticide properties.   Unfortunately, few studies enable the researcher to compare the effect of several pesticides on the same pathogen under similar conditions. Valaskowa (1968) compared fifteen fungi with respect to their sensitivity to sixteen commercial herbicides. She ascertained that contact herbicides (pentachlorophenol, ammonicum dinitrocresol, dinoseb, etc.) have a stronger inhibiting effect than systemic herbicides (pyramin, atrazine, or dalapon). Ercegovich *et al.* (1973) studied the effects of thirty-eight substituted triazines on mycelium growth and the formation of sclerotes by *Sclerotium rolfsii*. They came to the conclusion that none of the tested products was fungicidal to this organism at a concentration of 100 mg/l in agar medium. Only one chemical was highly fungistatic but eleven others showed a certain activity. No simple relationship between the molecular structure and the fungitoxicity was proved by the authors. Finally, only those chemicals that were very active as herbicides were also active against the fungus. However, some chemicals that were active as herbicides did not have a marked action against the fungus. These studies can be compared with those of Bozarth and Tweedy (1971), who conducted the same experiments under similar conditions. Fluometuron, ipazine, metobromuron, and, to a lesser degree, trifluralin, atratone, and atrazine appear to inhibit appreciably the radial growth of *Sclerotium rolfsii* in an agar medium-containing 100 mg/l of pesticide. Simazine, however, shows no apparent effect.

In the same way, Ujevic and Kovacikova (1975) tested *in vitro* the action of eight herbicides and noticed that, as opposed to pyrazon, dinoseb used in high concentrations is not effective against *Fusarium oxysporum, Verticillium alboatrum,* and *Botrytis tulipae,* even though the fungitoxicity of this product against *Fusarium oxysporum* f. *lycopersici* and *Fusarium oxysporum* f. *conglutinans* had been demonstrated earlier by Chappell and Miller (1956).

The effects vary greatly with the dose. Products that inhibit when used in large quantities may have a stimulating action on the pathogen when used in smaller quantities. This is the case of trifluralin with *Sclerotium rolfsii* (Rodriguez-Kabana *et al.,* 1969) and of fluometuron with *Sclerotium rolfsii* (Bozarth, 1969). Richardson (1970) compared the effect of atrazine at concentrations ranging from 4.4 to 140 ppm on twenty-nine fungi strains. He observed that growth of *Aspergillus fumigatus, Chaetomium funicola, Fusarium solani, Fusarium sporotrichioides,* and *Trichoderma viride* was stimulated by the lower doses and retarded by the higher doses. A single strain, *Epicoccum nigrum,* had its growth stimulated with every dosage.

The drawback of many comparisons among pesticides is that they do not take into account the solubility of the compounds in the medium used. Some of the products studied are slightly soluble, and one may expect that the concentrations chosen are sometimes excessive and without true significance with regard to their toxicity. In fact, one may suggest in numerous cases a probable relationship between the apparent activity and the solubility of the substances. For instance, among the four herbicides tested by Bozarth and Tweedy (1971), trifluralin, which is only slightly soluble in water (less than 1 mg/l), does not increase noticeably its toxicity against *Sclerotium* when its concentration in the medium changes from 10 to 100 mg/l. In contrast, the activity of the three other products—atrazine, fluometuron, and metobromuron—which are more soluble

TABLE VII

Recent Studies on the Effect of Herbicides on Fungi

Herbicide	Concentration (ppm)	Medium[a]	Strain	Effect[b]	Authors
Triazines					
Atrazine	50	SM	*Diplodia maydis*	I, radial growth	Houseworth and Tweedy (1972)
	5–140	SM	*Fusarium oxysporum*	I, radial growth	Richardson (1970)
	50	SM	*Fusarium oxysporum*	No inhibition	Houseworth and Tweedy (1972)
	20–80	SS	*Fusarium oxysporum* f. sp. *vasinfectum*	S, CO_2 production after 6 days	Rodriguez-Kabana and Curl (1970)
	40–80	SM	*Fusarium oxysporum* f. sp. *vasinfectum*	I, growth during 6 days	Rodriguez-Kabana and Curl (1970)
	50	SM	*Gibberella zeae*	I, radial growth	Houseworth and Tweedy (1972)
	250	SM	*Phymatotrichum omnivorum*	No effect on growth	Gunasekaran and Ahuja (1975)
	5–17	SM	*Rhizoctonia solani*	No inhibition	Richardson (1970)
	17–40	SM	*Rhizoctonia solani*	I, radial growth	Richardson (1970)
	8	SS	*Sclerotium rolfsii*	S, CO_2 production	Rodriguez-Kabana et al. (1968)
	20–80	SS	*Sclerotium rolfsii*	I, CO_2 production	Rodriguez-Kabana et al. (1968)
	50	SM	*Sclerotium rolfsii*	I, mycelial growth, sclerotia production	Bozarth and Tweedy (1971)
	100	SM	*Sclerotium rolfsii*	I, mycelial growth, sclerotia production	Ercegovitch et al. (1973)
Prometryne	20–80	SS	*Trichoderma viride*	S, CO_2 production	Rodriguez-Kabana et al. (1968)
	20	SS	*Fusarium oxysporum* f. sp. *vasinfectum*	S, CO_2 production	Chopra et al. (1970a)
38-substituted s-triazines	100	SM	*Sclerotium rolfsii*	I, mycelial growth or sclerotia production (by 20 products)	Ercegovitch et al. (1973)
Ureas					
Fluometuron	50	SM	*Diplodia maydis*	I, radial growth	Houseworth and Tweedy (1972)
	50	SM	*Fusarium oxysporum*	No inhibition	Houseworth and Tweedy (1972)
	50	SM	*Gibberella zeae*	I, radial growth	Houseworth and Tweedy (1972)
	10–100	SM	*Sclerotium rolfsii*	I, mycelial growth and sclerotia production	Bozarth and Tweedy (1971)

Compound	Conc.	Medium[a]	Organism	Effect[b]	Reference
	1–10	SM	*Sclerotium rolfsii*	S, sclerotia initial formation	Bozarth (1969)
	25	SM	*Sclerotium rolfsii*	I, sclerotia initial formation	Bozarth (1969)
	2.5–20	SS	*Sclerotium rolfsii*	I, sclerotia production	Curl and Rodriguez-Kabana (1972)
Metobromuron and fluometuron	50	SM	*Sclerotium rolfsii*	I, mycelial growth and sclerotia production	Bozarth and Tweedy (1971)
Diuron	2.5–20	SS	*Sclerotium rolfsii*	I, sclerotia production	Curl and Rodriguez-Kabana (1972)
Monuron	50–250	SM	*Phymatotrichum omnivorium*	I, growth	Gunasekaran and Ahuja (1975)
Other compounds					
Alachlor	50	SM	*Diplodia maydis*	I, radial growth	Houseworth and Tweedy (1972)
	50	SM	*Fusarium oxysporum*	I, radial growth	Houseworth and Tweedy (1972)
	50	SM	*Gibberella zeae*	No inhibition	Houseworth and Tweedy (1972)
Diphenamid	20	SM	*Rhizoctonia solani*	I, linear growth	Katan and Eshel (1974)
	50–100	SM	*Pythium aphanidernatum*	I, mycelial growth	Cole and Batson (1975)
	50–100	SM	*Rhizoctonia solani*	I, mycelial growth	Cole and Batson (1975)
Trifluralin	10	SM	*Aphanomyces euteiches*	I, dry weight	Harvey et al. (1975)
	0.6–5.2	SS	*Fusarium oxysporum* f. sp. *vasinfectum*	S, spore production	Tang et al. (1970)
	50	SM	*Rhizoctonia solani*	I, mycelial growth	Grinstein et al. (1976)
	6.2–12.5	SS	*Sclerotium rolfsii*	S, CO_2 production	Rodriguez-Kabana et al. (1969)
	10–100	SM	*Sclerotium rolfsii*	I, mycelial production	Rodriguez-Kabana et al. (1969)
Paraquat	250	SM	*Phymatotrichum omnivorum*	No inhibition	Gunasekaran and Ahuja (1975)
	10–100	SM	*Septoria tritici, Septoria nodorum*	I, mycelial growth, spore production	Jones and Williams (1971)
(Gramoxone w)	2	SM	*Gaeumannomyces graminis*	I, colony diameter	Grossbard and Harris (1976)
Fluorodifen	50	SM	*Diplodia maydis*	I, radial growth	Houseworth and Tweedy (1972)
	50	SM	*Fusarium oxysporum*	I, radial growth	Houseworth and Tweedy (1972)
	50	SM	*Gibberella zeae*	I, radial growth	Houseworth and Tweedy (1972)
EPTC	10–50	SM	*Sclerotium rolfsii*	I, mycelial production	Rodriguez-Kabana et al. (1970)
	1–10	SM	*Sclerotium rolfsii*	S, mycelial production	Peeples and Curl (1970)
	40	SM	*Sclerotium rolfsii*	I, mycelial production	Peeples and Curl (1970)
Chlorpropham	20–60	SM	*Phymatotrichum omnivorum*	I, mycelial growth	Gunasekaran and Ahuja (1975)

[a] SM—synthetic medium. SS—sterilized soil.
[b] I—inhibition. S—stimulation.

(33, 90, and 330 mg/l, respectively), increases noticeably under the same conditions. This observation agrees with that of Richardson and Miller (1960), who noticed that the fungitoxicity of certain "chlorinated hydrocarbon insecticides" was related to the solubility and steam pressure of the substances.

Influence of organic compounds.   We must also consider the probable influence of organic compounds on the interactions between pesticides and pathogenic microflora. Thus, Pitts *et al.* (1970) studied the growth response of *Sclerotium rolfsii* in soil supplemented with atrazine and glucose. A strong inhibition of fungal growth occurred with a high glucose level. With a low level of glucose, the effects of the herbicides were slight.

Influence of soil sterility.   Few experimentalists study the direct effects of pesticides on the pathogens in a nonsterile soil. However, Chopra *et al.* (1970a), after having shown an inhibiting effect of prometryne on the germination of the spores of *Fusarium oxysporum* f. sp. *vasinfectum* in a sterile soil, noticed no effect with higher doses in a nonsterile soil. In contrast, Tang *et al.* (1970) observed that trifluralin has an effect on the germination of the spores of *Fusarium oxysporum* f. sp. *vasinfectum* in a nonsterile soil. Finally, Percich and Lockwood (1975) reported, in the field, an increase in the number of *Fusarium* spp. after treatments corresponding to 10 and 100 mg of atrazine per kilogram of soil.

Influence of formulation additives and mixtures of pesticides.   One must mention again the fact that the activity of the products can be altered by the presence of formulation ingredients which often constitute a great proportion of the commercial products, or by the presence of other pesticides. Thus, *in vitro* a combination of the active product of thiophanate methyl with the inert ingredients of a benomyl formulation prevents production of sclerotes in *Typhula incarnata*. None of the substances used showed a great activity when used separately. Consequently, it may be concluded that the combination of the two products creates a synergistic effect (Ebenebe and Fehrmann, 1976). Furthermore, Brunnelli *et al.* (1975) have proved, on plants studied *in vitro,* the possibility of antagonistic or synergistic interactions of certain acaricides on the activity of some fungicides used in mixtures.

Effect of pesticides on the virulence of pathogens.   The pathogenicity or aggressiveness of microorganisms may be altered in the presence of pesticides, this phenomenon not necessarily being associated with an alteration in their growth. There are few reports on the subject, partly because it is not always possible to make a distinction between the effects of the products on the pathogen and their effects on the susceptibility of the host plant. Katan and Eshel (1973) found no increase in the virulence of *Rhizoctonia solani* grown on a medium supplemented with diphenamid. Percich and Locwood (1975) sowed peas or corn in soils infested with *Fusarium solani* f. sp. *pisi* and *Fusarium roseum* f. sp. *cerealis*. The preliminary growth of the inoculum in a medium containing some atrazine did not modify the virulence observed later. In contrast, Deep and Young (1965) proved that a nonvirulent strain of *Agrobacterium tumefaciens* becomes virulent in the presence of the fungicide dichlone. One might also mention the research of Hubbeling and Basu Chaudhary (1970), who studied the mutagenic effects of chloramben on *Verticillium dahliae*. With all the variant strains obtained, these researchers observed less intense attacks on the plants. However, they did not rule out the possibility of an occurrence of mutants with a higher level of pathogenicity.

    c. Indirect Effects on Pathogens.   The effects of pesticides on the microflora of the soil can entail some indirect consequences on the activity of the pathogens. Indeed, biological balances in the soil are determined, among other things, by the conditions of the medium, but also by a set of antagonistic or complementary relationships between the species confronting each other in the same ecological habitat.

    It has been accepted that two mechanisms govern the antagonistic relationships between species: on the one hand the possibility of growth of the species and their competition for the substrate, and, on the other hand, the antibiotic activity of the microorganisms.

    Influence of pesticides on substrate utilization.   The importance of competition for substrates has been shown by Lai and Bruehl (1968). Wilkinson and Lucas (1969), using vegetal tissues as a substrate, observed that the presence of paraquat modifies the competition between various fungic strains. For instance, on potato haulm, paraquat favors *Fusarium culmorum* as opposed to *Trichoderma viride*. When sprayed on wheat chaff, it inhibits invasion by *Rhizopus* sp. and favors *Aspergillus niger*. In the same way, treatment of *Raphanus raphanistrum* with MCPA interferes with the competition between *Aspergillus niger* and *Penicillium notatum* and favors *Aspergillus niger*. Wilkinson and Lucas indicated that the presence of residues of pesticides in vegetal remains can be an important factor in the competition between saprophytes. Sometimes treatments can lessen the competitiveness of the pathogen. For instance, after fumigation of infested citrus roots with methyl bromide, *Armillaria mellea* survives when incubated in a sterile soil but not in a nonsterile soil. In nonsterile soil, *Trichoderma* spp. are directly involved in the death of sublethally fumigated *Armillaria* by their invasion of living cells (Ohr *et al.*, 1973).

    Influence of pesticides on the antagonistic relationships between species.   Some studies have shown evidence *in vitro* of the possible influence of pesticides on the antagonistic relationships between strains and, more particularly, on the synthesis of toxins by some of these strains. Kosinkievicz (1968, 1970) found that linuron could inhibit or stimulate the production of antibiotic substances in *Bacillus polymyxa, Bacillus brevis,* and *Bacillus oligonitrophilus*. He also showed the influence of linuron on the production of antibiotics by a strain of *Actinomyces abikoensum* (1968). According to Krezel and Leszczynska

(1970), 5 ppm of chlorpropham or 10 ppm of linuron or of prometryne can cause an inhibition of the antibiotic activity of *Streptomyces griseus* toward *Erwinia carotovora*. Balicka and Krezel (1969) showed that the same compounds at concentrations between 10 and 100 ppm affected the antagonism between strains of *Bacillus* sp. type *mesentericus* and *Pseudomonas phaseoli* by disturbing the metabolism of *Bacillus* sp. These results varied with the strain used. According to Williams and Ayanaba (1975), an increased incidence of *Pythium* stem rot on cowpeas, which occurs with benzimidazole fungicides, seems likely due to the suppression of antagonists and competitors of *Pythium aphanidermatum*.

Other possible indirect effects of pesticides on pathogens have been indicated by Kavanagh (1969, 1974), who notes that the use of pesticides entails a modification of the agricultural practices and, at the same time, an alteration in the biological activity of the soils. Destroyed vegetals can represent an extra source of substrate. Moreover, the destruction of the vegetal cover can modify the microclimate at the soil level. Finally, some indirect effects may come from the destruction of a host plant. For instance, Minton (1972) shows that the "weed control" practiced in a sorghum and cotton rotation makes it possible to control *Amaranthus* spp. (a host of *Verticillium alboatrum*) and, thus, to lessen the incidence of Verticillium wilt in cotton.

B. EFFECT OF PESTICIDES ON MICROORGANISM RESISTANCE

1. The Agronomic Problem

The regular use of some pesticides may induce or promote the appearance of species or strains of organisms resistant to the action of these substances. It is well known that the use of some insecticides may be followed by the appearance of resistant strains of insects. Concerning the microorganisms, the adaptation of human pathogenic bacteria to antibiotics is of particular importance.

In regard to the soil microflora, as far back as 1954, English and Van Halsema published a note on the time required for the emergence of resistant strains of *Xanthomonas* and *Erwinia* in soil after the use of streptomycin and terramycin combinations. Likewise, Horsfall (1956; quoted by Ashida, 1965) suggested that the apparent reduction in the effectiveness of Bordeaux mixture on potato blight might be attributed to the greater tolerance developed by *Phytophthora* after an application of fungicides for over 60 years.

In fact, the possibility of developing strains resistant to various classes of fungicides *in vitro* has been known for a long time. Parry and Wood (1958), by growing mycelium of *Botrytis cinerea* on media containing copper sulfate, thus obtained a strain that resisted a concentration twice as high as that inhibiting the growth of the wild type. Partridge and Rich (1962) obtained the adaptation of *Penicillium notatum*, *Sclerotinia fructicola*, and *Stemphylium sarcinaeforme* to various fungicides including $CuSO_4$, $HgCl_2$, and captan. A review of the adaptation of fungi to metal toxicants was made by Ashida (1965), who listed twenty-four examples of positive adaptation to copper, mercuric salts, or organomercurials and noted the acquisition of tolerance to other metals. In a subsequent review, Toledo (1974) also gave numerous examples concerning organic fungicides. As for dithiocarbamate fungicides, he mentioned among other studies those of Parry and Wood (1959), who demonstrated a progressive adaptation of *Botrytis cinerea* to ferbam, and those of Elsaid and Sinclair (1964), who isolated a strain of *Rhizoctonia solani* resistant to thiram. Toledo reported also the adaptation of strains of *Penicillium*, *Sclerotinia*, *Stemphylium*, *Botrytis*, and *Venturia* to captan, glyodin, or dichlone. He also cited several examples of acquired resistance to PCNB and other halogenonitrobenzenes (notably with *Botrytis* and *Hypomyces* strains).

However, the relatively recent introduction of the systemic fungicides made the problem of resistance more acute. Indeed, after the great interest aroused by these substances, the quick loss of sensitivity in the target pathogens that sometimes occurred *in vivo* gave rise to a real agronomic problem. A review of this question was published by Dekker (1973), who drew up a list of pathogenic strains resistant to benomyl and related compounds, and also to thiophanates, ethirimol, oxathiins, chloroneb, and other systemic products. Fehrmann (1976) has published an exhaustive bibliography on the various aspects of the problem of resistance. He presents an extensive list of investigations carried out under field or laboratory conditions. Table VIII lists some recent studies on this subject. Most of them were carried out *in vitro*, but a relationship between the laboratory studies and the agronomic observations often may be established. Thus, Bollen and Sholten (1971), after stating the ineffectiveness of benomyl against *Botrytis cinerea* on cyclamen, isolated a strain able to grow on artificial media containing much higher rates of fungicides. On the other hand, Hollomon (1975), after characterizing a strain of *Erysiphe graminis* resistant to ethirimol *in vitro*, did not succeed in his attempts to isolate similar resistant strains in the field. Under field conditions the excellent control obtained during the first years of treatment with systemic fungicides was sometimes followed by a reoccurrence of the damages. These observations were made, for example, concerning benzimidazole treatments on *Phialophora cinerescens* (Tramier and Bettachini, 1974) and *Septoria leucanthemi* (Paulus *et al.*, 1976). Staunton and Kavanagh (1975), studying twenty-six tomato and twenty-eight strawberry crops, concluded that 92% and 46%, respectively, of the sampled crops presented strains of *Botrytis cinerea* resistant to benomyl. Resistant strains of *Fulvia fulva* were also detected on tomato. Jones and Walker (1976), studying strains of the apple scab fungus *Venturia inaequalis*, observed the development of tolerant strains within

TABLE VIII
Recent Studies on Resistance of Fungi to Fungicides

Fungicides	Resistant strains	Authors
Benomyl and related compounds	*Botrytis cinerea*	Gessler (1976), Abelentsev (1973) Staunton and Kavanagh (1975)
	Colletotrichum coffeanum	Okioga (1976)
	Colletotrichum lindemuthianum	Meyer (1976)
	Fulvia fulva	Staunton and Kavanagh (1975)
	Fusarium roseum	Smiley and Howard (1976)
	Neurospora crassa	Borck and Braymer (1974)
	Penicillium digitatum	Wild and Rippon (1975)
	Pythium paroecandrum	Gessler (1976)
	Rhizoctonia solani	Kataria and Grover (1974)
	Sclerotinia homoeocarpa	Warren *et al.* (1974)
	Sporobolomyces roseus	Nachmias and Barash (1976)
	Venturia inaequalis	Abelentsev (1973)
	Verticillium albo-atrum	Hall (1975)
	Verticillium dahliae	Talboys and Davies (1976), Hall (1975)
	Verticillium malthousei	Lambert and Wuest (1975, 1976)
	Verticillium nigrescens	Hall (1975)
Dodine	*Venturia inaequalis*	Yoder and Klos (1976)
Ethirimol and dimethirimol	*Erysiphe graminis* f. sp. *hordei*	Hollomon (1976), Shephard *et al.* (1975)
Ethazol (ethridiazol)	*Pythium sylvaticum, P. vexans, P. ultimum, P. debaryanum*	Halos and Huisman (1976)
Carboxin	*Aspergillus nidulans*	Van Tuyl *et al.* (1974)
	Alternaria burnsii	Mathur *et al.* (1972)
	Rhizoctonia bataticola	Mathur *et al.* (1972)
Chloroneb	*Rhizoctonia solani*	Kataria and Grover (1974)
Quintozene	*Rhizoctonia solani*	Kataria and Grover (1974)

three years of regular use of benomyl. With dodine, a fungicide that is not really systemic but may, however, penetrate into the leaves, tolerant strains were detected after about ten years of use (Szkolnik and Gilpatrick, 1969; Jones and Walker, 1976).

Reports on the agronomic implications of the appearance of resistant bacteria strains after treatment with antibiotics were also made by some authors. Among others, Moller *et al.* (1972) describe fire-blight resistance to streptomycin in some orchards.

2. Resistance Phenomena and the Behavior of the Pathogenic Population in the Field

Resistance to pesticides may correspond to some physiological modifications in the cell that do not induce a genetic transformation—for example, the development of adaptable enzymatic systems. Concerning fungicides, Dekker (1973) believes that this type of tolerance is of little practical importance because it seldom reaches a high level and is lost by the fungus when there is no more fungicide in the medium. More often, the adaptation results from genetic mutations. Products with no particular specific action can act as multisite inhibitors (Kaars Sijpesteijn, 1970, quoted by Dekker, 1973). On the other hand, specific products such as the systemic fungicides generally act on a single site. The probability that resistance will develop is therefore more important. Cell resistance to systemic fungicides may be the result of different physiological mechanisms, some of which have been mentioned: binding of the toxic ion, alternative pathway of electron transport bypassing the sensitive site, changes in cell permeability, etc. (Georgopoulos, 1975; Fehrmann, 1976; Nachmias and Barash, 1975). The mutagenicity of some products may also be implicated in the appearance of resistance, although Ferhrmann (1976) showed only a weak mutagenicity of MBC and benomyl.

If the possibility of the appearance of resistant mutants is clearly established *in vitro,* what really happens in the field is still to be settled. Some studies seem to show that pathogenic populations of untreated crops are not tolerant to certain fungicides such as dodine (Gilpatrick and Blowers, 1974) and MBC (Shabi and Ben-Yephet, 1976). On the other hand, other research reveals the presence of naturally resistant strains before the introduction of the fungicide. Thus, Wuest *et al.* (1974) reported that a strain of *Verticillium malthousei* isolated by them in 1958 was tolerant to benomyl. Kuramoto (1976) also isolated strains of *Penicillium* spp. resistant to benzimidazole fungicides in packing houses; these strains came from orchards that had never been treated with these products.

The development of the resistant population may result from the selective pressure of the fungicide treatment, or it may be related to the suppression of antagonists or competitors (Warren *et al.,* 1975). Heterokaryosis or anastomosis should also enhance the development of resistance within a fungus population (Fehrmann, 1976).

The development of resistance to systemic fungicides is often a rapid phenomenon. In experiments on tomatoes, Fletcher and Sholefield (1976) observed rapid progress in the percentage of *Botrytis* isolates tolerant to benomyl in sprayed plants—none in 1972, 31% in 1973, and 70% in 1974.

Levels of tolerance may fluctuate greatly; for example, according to Geeson (1976), isolates of *Botrytis cinerea* from various origins may be classified into three groups according to their sensitivity to MBC: sensitive, tolerant, or fully tolerant strains. The fully tolerant strains can develop in the presence of concentrations about one thousand times as high as the doses acting on the sensitive strains. Gilpatrick and Blowers (1974) reported that ascospores of *Venturia inaequalis* from a dodine-tolerant source were only one-half to one-third as sensitive as those from a nontolerant source. However, because of the low solubility of many compounds, these values are only indicative.

Sometimes the tolerance to a particular compound is associated with a tolerance to other substances. This is true especially for tolerance to benzimidazole fungicides. Isolates of benomyl-tolerant strains of *Monilia fructicola* were also tolerant to thiophanate methyl and to four other benzimidazole fungicides (Jones and Ehret, 1976). Erwin (1973), Shabi and Ben-Yephet (1976), and Meyer (1976) reported similar observations, althoughVan Thuyl *et al.* (1974) in their studies on strains of *Aspergillus nidulans* with ultraviolet-induced resistance, pointed out the rare occurrence of benomyl-resistant strains that are sensitive to thiabendazole.

Lastly, tolerance may sometimes be correlated with an increased sensitivity to another product. This is the case for the increased sensitivity to zineb of a benomyl-tolerant strain of *Verticillium malthousei* (Lambert and Wuest, 1975).

Tolerance *in vitro* sometimes appears as a steady characteristic and sometimes vanishes more or less rapidly in the absence of pesticides. The tolerance of two isolates of benomyl-tolerant strains of *Monilia fructicola* remained unchanged after passages through benomyl-free substrate including four passages through peach (Jones and Ehret, 1976). In contrast, resistance to maneb and PCNB was rapidly lost in strains of *Rhizoctonia solani* grown in the absence of fungicide (Elsaid and Sinclair, 1964). Ruppel (1975) observed that benomyl-tolerant strains of *Cercospora beticola* did not differ from benomyl-sensitive strains, either in growth and sporulation *in vitro* or in sporulation *in vivo*. The tolerance of one strain remained unchanged after three passages through sugar beet. After a mixed inoculation with a sensitive and a tolerant strain, the tolerant strain population declined but did not disappear, as shown after reisolation and enumeration of the two strains. Wicks (1976) also pointed out in his work that a benomyl-tolerant population of *Venturia inaequalis* had persisted in an orchard for two years after the last application of the fungicide. Concerning the agronomic uses of antibiotics, Russel (1975) demonstrated that some mutants of *Pseudomonas phaseolicola* retained this characteristic after a passage through the plant, whereas others did not.

Lastly, the behavior in soil of the resistant population of pathogens seems no different from that of the sensitive population (Dovas *et al.*, 1976). Likewise, the virulence of the pathogens generally seems not to be modified, as reported by Ruppel (1975), Cook and Pereira (1976), and Okioga (1976), although Moller *et al.* (1972) observed a contrasting example.

C. CONCLUSION

Certain pesticides can have an influence on the progress of plant diseases. This influence is complex because the effects of pesticides on the plants and the effects of the same pesticides on the pathogenic microorganisms responsible for the disease are often mixed. In itself, however, the study of the effects of pesticides on the pathogenic microflora is of great interest. It stresses the sensitivity of certain processes linked with the specific activity of a small number of organisms. It shows, moreover, that the action of the pesticides can also induce a change in the biological equilibrium between the strains. Finally, the appearance of strains resistant to some poisons reveals once again the capacity of the microflora to adapt to environmental conditions. The consequences of this last phenomenon, as well as the facts mentioned above, go far beyond the mere problem of the relationship between pesticides and pathogenic microflora.

VI. Effect of Pesticides on the Microflora Responsible for Pesticide Degradation

Most pesticides are degraded in the soil primarily through microbiological processes. The importance of the microflora responsible for this degradation is considerable because its activity determines the persistence and thus the agricultural use of the pesticides.

A pesticide acts in two ways on the microflora. On the one hand, it can influence the microorganisms responsible for its degradation because of its physiological activity. This action may also concern the organisms involved in the degradation of other substances. On the other hand, the pesticide acts also as a substrate for the soil microflora. As a matter of fact, its degradation may supply certain organisms with the carbon and the energy, sometimes the nitrogen, necessary to their growth. In any case, the most important consequences of these two aspects appear essentially in relation to pesticide degradation. The microorganisms that form the microflora involved are extremely diversified (Goring *et al.*, 1975; Menzie 1969). Among them are bacteria as well as actinomycetes and fungi. However, the degradation of a particular product is often due to a rather limited number of organisms. Obviously this may confer a certain weakness on the microflora considered.

A. EFFECT OF PESTICIDES ON THE MICROFLORA RESPONSIBLE FOR THEIR DEGRADATION

1. Inhibiting Effects

"The dose makes the poison." This aphorism applies also to the effects of pesticides on the microflora able to degrade them. Jensen and Petersen (1952) reported that a strain of bacteria was able to grow on an agar medium containing

0.1–0.2% of 2,4-D as sole carbon source. In contrast, no growth took place when the concentration of 2,4-D in the medium reached 0.4%.

According to Weinberger and Bollag (1972), *Rhizoctonia solani* might degrade various phenyl-substituted urea herbicides, including diuron and linuron. But Ebner (1965) had previously reported the toxic activity of these two compounds against *Rhizoctonia solani*. With the same species Catroux *et al.* (1973) gave an example of a strain that degraded diuron in pure culture but was inhibited by it at relatively low concentrations. As far as we know, there is no study proving that such activity exists under soil conditions. The quantity of product corresponding to a normal agronomic use of the pesticide is generally very small. The heterogeneity of the product distribution may, however, result in relatively high concentrations locally in the aqueous phase of the soils. This toxic effect is probably one of the determining factors in the equilibrium between the different microorganisms able to degrade a pesticide in the soil.

2. Stimulating Effects

The addition of certain pesticides to the soil determines the proliferation of microorganisms able to degrade these products. This process, called soil enrichment, was first described by Audus (1949) in his study on the biological detoxication of 2,4-D in soil. This observation was then used by numerous research workers to make the isolation of the involved microorganisms easier. In the same report, however, Audus stressed one of the main consequences of this enrichment: Pesticide degradation in the soil samples was much more rapid when the treatment was repeated a second time a few days later. For this work Audus used the soil perfusion technique previously described by Lees and Quastel (1946). Newman and Thomas (1949) have confirmed Audus's results by observing the same phenomenon in soil samples. Moreover, Newman *et al.* (1952) observed under field conditions that the concentration of 2,4-D was reduced more rapidly in soil where it had decomposed before. In contrast, they failed to demonstrate an effect of a soil pretreatment on the behavior of 2,4,5-T.

A number of studies over the past twenty-five years have shown that the treatment of soil with pesticides can cause a considerable increase in the degrading power of the microflora toward these pesticides. Studies on the mechanisms of this adaptation of the soil to degradation are much rarer.

a. Field Experiments. The agronomic importance of soil adaptation may be considerable. However, application of the results obtained in field experiments is difficult because of the influence of environmental factors on microflora activity in the soil. Some studies carried out under these conditions can be reported here, however.

Some research indicates an increase in the speed of degradation of 2,4-D and MCPA after several applications under field conditions (Anon, 1956; Walker, 1957,) quoted by Kirkland and Fryer, 1966, 1972). Kirkland and Fryer (1966),

Kirkland (1967), and Fryer and Kirkland (1970) have reported the results of long-term experiments with four herbicides: MCPA, simazine, linuron, and triallate. Evidence of the influence of repeated treatments on the breakdown rate was obtained only with MCPA. The MCPA was applied to uncropped plots at 48 ounces per acre twice a year. The time necessary for the phytotoxic effect to disappear was reduced from 3 weeks (after three previous treatments) to 4 days (after ten applications). On the other hand, in plots put to barley and treated with MCPA each year, no significant results could be obtained because of the great variability of this type of experiment.

Torstensson et al. (1975) also studied the effect of repeated applications of 2,4-D and MCPA on their degradation during a long-term experiment on a meadow. After a single application, the authors observed a degradation time of about 10 weeks for 2,4-D and 20 weeks for MCPA. A second application the following year resulted in a shortening of the period to 7 and 10 weeks. After eighteen or nineteen annual applications the time for the disappearance of the toxic effect was only 4 weeks for 2,4-D and 7 weeks for MPCA.

As far as we know, except for the results obtained with the phenoxyacetic herbicides, few of the results observed under agronomic conditions suggest any process of soil adaptation to pesticide degradation. However, Aggour et al. (1977) observed that TCA applications sometimes lead to the persistence of an undesirable phytotoxic effect the following year. They noted the presence of some herbicide traces in a soil treated only one time with TCA. On the other hand, they found no residue in a soil treated several times. McGrath (1976) twice sprayed soil plots in situ with TCA. He then studied the degradation of the compound in the twice-sprayed soil and in a control. No difference was noticed during the first month following the second application. But the total disappearance of TCA was more rapid in the pretreated soil.

In a field survey Savage (1973) stated that nitralin residues were on an average slightly higher in fields with no history of previous use. In a field experiment nitralin disappeared more rapidly from a soil that had received repeated treatments over a five-year period than from an untreated field plot. On the other hand, no significant result was obtained during a similar experiment with trifluralin. Some long-term field experiments can be mentioned here. They concern simazine (Allot, 1969), bromacil, DCPA, diphenamid, diuron, fluometuron, neburon, prometryn, pyrazon, simazine, and trifluralin (Horowitz et al., 1974). In no case was a decrease in the activity of the different compounds reported. The authors even noticed an intensification of the herbicide activity with the number of pyrazon and neburon applications.

b. In Vitro Studies.   Numerous studies are carried out partly or totally under controlled conditions, which give greater consistency in experimental results.

Studies with herbicides.   PHENOXYALKANOIC HERBICIDES.   The phenoxyalkanoic herbicides are among the earliest and most widely studied products. The experimental conditions vary considerably, however. Since the first work of

Audus (1949), the technique of percolation of soil samples with pesticide solutions has often been used. With this method Audus (1951) observed the total disappearance of the lag phase preceding the actual degradation phase of 2,4-D or MCPA after several successive additions of the product (Audus, 1951; Brownbridge, 1956). An increase in the degradation rate of several other phenoxyalkanoic acids (4-CPA, 2-(4-CPP), dichlorprop) applied under the same conditions was also reported by Brownbridge (1956).

Some workers report the use of soil suspensions. Newman and Thomas (1949) inoculated a liquid medium containing 2,4-D with previously treated soils and untreated soils. Cultures from pretreated soils were more active than cultures from untreated soils. Whiteside and Alexander (1960) used a soil suspension in a mineral medium containing 2,4-D or 4-(2,4-DB). Subsequent additions were transformed in shorter periods than were the initial quantities added. This method was also used by Torstensson et al. (1975) to study the degradation of 2,4-D and MCPA in soil samples from their field experiment and for isolating strains able to degrade either compound.

Some studies have been carried out on soil samples incubated under controlled conditions, but sometimes taken from plots treated in situ. Thus, Newmann and Thomas (1949), in an experiment in pots, estimated the "adaptation" of a soil to the degradation of 2,4-D by determining seedling emergence and plant response. Some indications in the same direction have been given by Jensen and Petersen (1952), although the precise conditions of experimentation were not described. Kirkland and Fryer (1966, 1972) reported the results of some investigations complementary to their experimentations under the field conditions described above. The breakdown of MCPA was greatly accelerated in soil samples taken in plots that had received seven or nine previous field treatments with herbicides (3.3 kg/ha) at intervals of about six months. The time necessary for the total degradation of the product was thus divided by a factor of nearly 4 in the samples taken in treated plots. The lag phase that was sometimes revealed in the previously untreated samples did not occur in a soil already treated with MCPA. Finally, Hurle and Rademacher (1970) studied the degradation of ^{14}C-2,4-D, at relatively high concentrations, in samples taken in a soil that had received yearly treatments for weed control in cereal crops for twelve years. The pesticide degradation studied eight months after the last field application seemed more rapid in a sample from pretreated soil. The difference, in comparison with the degradation kinetics observed in a sample with no pretreatment, was due mainly to a reduction in the length of the lag phase.

The results are not always so decisive with certain phenoxyalkanoic acids. As in the experiments of Newman et al. (1952) already mentioned, Byast and Hance (1975) failed to prove a "conditioning effect" of the soils after repeated applications of 2,4,5-T.

PHENYLAMIDE HERBICIDES.   Kaufman and Kearney (1965) observed in a soil percolation system that both CIPC and CEPC were degraded more rapidly after a second application. Suess *et al.* (1974) studied the degradation of ^{14}C-monolinuron labeled on the carbonyl of the urea group or on the ring, or on both. They found an increase in the decomposition rate of monolinuron in soils previously treated with the herbicide. El Dib and Aly (1976) observed the adaptation of the microflora after additions of aniline, 4-chloroaniline, or acetanilide in a sample of river water. In contrast, under the same conditions they observed practically no degradation of IPC, CIPC, linuron, monuron, propanil, or carboxin (a fungicide). When a large inoculum of *Bacillus cereus* was added to the medium, however, the degradation of propanil, carboxin, and linuron began, and the rate of degradation increased after several successive additions of the product. In the same conditions chlorpropham and monuron did not seem to be degraded.

HALOGENATED ALIPHATIC ACIDS. Thiegs was the first (1955) to note that dalapon was degraded more rapidly after several successive additions than after a first application. Leasure (1964) observed in a greenhouse experiment the considerable differences in activity which resulted from application of dalapon to a soil treated for the first time or to a soil treated several times during the preceding weeks.

Concerning TCA, McGrath (1976) confirmed *in vitro* the results he obtained in the fields. He proved that soils treated with TCA (20 ppm) acquired the ability to degrade new additions of the product without any lag phase. The time necessary to obtain total degradation of the product was reduced from about 30 to 5 days. Similar results were obtained by Aggour *et al.* (1977).

OTHER HERBICIDES. Soil conditioning for pesticide degradation was also proved for various other types of products. Thus, Drescher and Otto (1969) studied the degradation of pyrazon in soil samples treated four times with 1000 ppm of pyrazon. The degradation of the first addition lasted about 6 weeks, the degradation of subsequent treatments 10 days. With the same product a reduction of the lag phase preceding the degradation was also observed by Engvild and Jensen (1969).

Soil adaptation to endothal was observed by Horowitz (1966) in an experiment in pots. But the interval between the two treatments was only 2 months.

Riepma (1962), working with soil suspensions, studied the decomposition of two successive additions of amitrole. The breakdown rate of the second addition of the compound was higher than that of the first addition. His results also showed, after the second treatment, the nearly total disappearance of the degradation lag phase.

Lastly, a more rapid breakdown of DNOC was observed by Hurle and Pfefferkorn (1972) as a result of pretreatment of the soil with various concentrations of the same product.

In contrast to the preceding results, one of the products that do not seem to lead to soil adaptation is maleic hydrazide (Helweg, 1975). The degradation of this product was not enhanced by six treatments with herbicide within 3 months.

Studies with insecticides. Repeated applications of the insecticide–acaricide diazinon to paddy water may be ineffective for controlling the brown plant hopper (*Nilaparvata lugens*) (Irri, 1970, quoted by Sethunathan and Pathak, 1972). Sethunathan and Pathak collected soil and water samples from untreated plots and from plots treated with diazinon. Aliquots of the water samples, or of the supernatant of the soil samples shaken with water, were added to an aqueous solution of diazinon. After 10 days of incubation, diazinon degradation was null in untreated samples, but complete in treated ones.

Fung and Uren (1977) studied the decomposition of methomyl, an insecticide used especially to control the tobacco yellow dwarf virus. The transformation of methomyl determined by a perfusion study occurred after a lag phase of about 7–14 days. However, the lag phase was nearly absent when the degradation was studied with a previously enriched soil.

Studies with fungicides. Few investigations concerning the possibility of conditioning a soil to fungicide degradation have been described. Carboxin has already been discussed above. Groves and Chough (1970) observed the failure to control white rot (*Sclerotium cepivorum*) in onion fields after some years of application of the fungicide dichloran (2,6-dichloro-4-nitroaniline). It was demonstrated *in vitro* that pretreatment of the soil with the fungicide increased the degradation rate of the product on later additions. When ^{14}C-dichloran was added to untreated soils, no ^{14}CO$_2$ was evolved. But in soils pretreated with dichloran, 25–50% of the total radioactivity evolved as ^{14}CO$_2$.

c. Biological Phenomena Involved. The process of conditioning soils to pesticide degradation is undoubtedly linked with the microbial activity. This increased capacity of the soils to decompose a pesticide results primarily in a suppression or a reduction of the lag phase normally preceding the actual degradation of numerous products in the soil. More rarely, the degradation rate observed during the active phase seems to be modified.

The degradation lag phase. Conditioning soils to pesticide degradation seems to correspond to the juxtaposition or superposition of a group of phenomena, some corresponding to the appearance of a "degrading power" in a microflora (enzymatic adaptation, selection, mutation, and so on), and others to the development of this character (mainly the microorganisms proliferation). Audus (1960) noted that the lag phase preceding the actual degradation of 2,4-D in perfusion experiments was not modified by the previous adaptation of the soil in the presence of a bacteriostatic agent. From this observation he deduced the preponderant-role of microorganisms proliferation in the enrichment process. Brownbridge (1956) and Audus (1960) used a perfusate from a previously enriched soil to induce a great reduction in the lag phase of a nonpretreated soil. Engvild and Jensen (1969) inoculated a nonsterile soil containing pyrazon with a previously isolated bacteria strain, which was able to grow by using this herbicide as its sole source of carbon. They observed a rapid degradation of pyrazon after a lag phase of a few days. On the other hand, no degradation took place in a noninoculated soil during the experiment. Audus (1960) isolated a soil bacterium able to grow on a 2,4-D agar medium. When added to a solution of pesticide perfused on a fresh soil sample, this organism induced an immediate degradation of the compound. It should be stressed that in nearly all the examples of positive conditioning effects previously mentioned the substances involved are able to withstand proliferation of organisms. The following products are especially able to act as sole carbon source for certain species: pyrazon (Eperspaecher *et al.*, 1976; 2,4-D and MCPA (Jensen and Petersen, 1952; Steenson and Walker, (1956), propham, chlorpropham and CEPC (Clark and Wright, 1970; Kaufman and Kearney, 1965), DNOC (Gundersen and Jensen, 1956; Jensen and Lautrup-Larsen, 1967), dalapon and TCA (Kaufman, 1964; Foy, 1969), diazinon (Menzie, 1969), and monuron (Geissbuhler, 1969). The specificity of the pesticide substrate for the microorganism may also be important (Jensen and Petersen, 1952); Fröhner *et al.*, 1970). Results concerning other products are less decisive. Siedschlag and Camper (1976) isolated strains during soil enrichment studies with nitralin but detected no $^{14}CO_2$ evolution from ring-labeled herbicide with any isolate tested. On the other hand, Carter and Camper (1975) observed isolated strains that exhibited an increase in the number of viable cells when grown in a medium with nitralin or trifluralin as sole source of carbon and nitrogen.

Some research workers have tried to count the microorganisms able to degrade certain pesticides in soil. Burge (1969) used the method of "most probable number" to count the microorganisms degrading dalapon in soil samples before and after an incubation with 50 ppm of this product. He observed after the treatment a large increase (hundredfold) in the number of microorganisms degrading the herbicide. The same method was applied, also by Burge (1972), to follow the evolution of the degrading population of propanil after a treatment with high doses of this product. The number of bacteria able to degrade propanil appeared to increase markedly in most of the soils studied. In contrast, in experiments with 2,4-D and MCPA, Torstensson *et al.* (1975) did not succeed in revealing significant modifications in the number of microorganisms able to use these products for their growth even after nineteen yearly applications. The quantities of products applied each year were much less, however—about 1 ppm, compared with the 50 ppm of dalapon applied by Burge. The increase in the number of microorganisms given by Burge (1 to 5 × 10^6) corresponds to a yield of approximately 5–20% in the transformation of the molecular carbon in cell carbon. Such a yield estimated for the experiments with 2,4-D and MCPA would mean a yearly increase in the number of microorganisms of about 10^5 not perceived by the counting method used.

The length of the lag phase is not an absolute criterion that allows one to detect systematically the possible effects of soil adaptation to degradation. It varies in relation to most of the environmental factors acting on the microflora activity. It also varies with the type of product (Audus, 1952) and with the doses given. Hurle (1973) studied the effect of different doses of 2,4-D and MPCA on the length of the lag phase. With high concentrations, the length of the lag phase increased almost linearly. With low concentrations, however, no distinct lag phase was detected.

   The active degradation phase.   Whatever the degree of adaptation of the soils and the lag phase observed may be, the rate of the active phase of product degradation is often only partially modified. Audus (1952) estimates that this phase represents the soil's maximum capacity for herbicide degradation under given conditions and is probably limited by the total number of fully adapted organisms that it can support. According to Engvild and Jenssen (1969), it took more than a month to degrade 80% of the pyrazon added to a soil sample with no pretreatment (not including the lag phase preceding the degradation). On the other hand, it takes 10 days or so to obtain an equivalent degradation in a pretreated soil. Hurle and Pfefferkorn (1972), in their study on the effect of pretreatment on decomposition of DNOC in soil, observed that the degradation rate increased with the increase in the doses given for pretreatment. But with the lowest doses, the rates observed were unexpectedly less than that obtained during the active degradation phase in an untreated sample. Some hypotheses can be set forth: One of them would be that only a part of the microorganisms preadapted to the degradation would survive between the two treatments, whereas some of the most effective species would have disappeared.

   Soil adaptation and type of metabolism.   In most of the examples previously described, the conditioning of the soil to pesticide degradation is linked with a proliferation of the microorganisms able to carry out the initial metabolic attack on the molecule. This means that the microorganisms involved associate molecule degradation with energy recovery, and the resulting multiplication leads to an increase in the supply of enzyme. In certain microorganisms, however, the association between degradation and energy recovery may fail. The microorganisms capable of carrying out certain reaction sequences cannot proliferate. In practice, the development of these microorganisms depends, therefore, on a secondary energy source, which explains the different more or less restrictive terms used to describe these mechanisms—detoxication, co-oxidation, or, according to Horvath (1972), co-metabolism. It has not been clearly established whether pesticide degradation by co-metabolic processes may lead to a soil adaptation phenomenon. Some results indicate that the possibility of adaptation sometimes does exist. This is especially true for the results referred to above concerning some substituted phenylureas (El Dib and Aly, 1976). In fact, Soulas and Reudet (1977) showed that the degradation of phenylurea herbicides could begin with one or several co-metabolic stages, followed by a probable "metabolic" process. It is quite likely that some apparent adaptation processes are due to the increase in the capacity of the microflora to degrade certain possibly toxic intermediary products of the degradation.

Another aspect must be noted here. The same compound can be degraded at once by different processes, which may imply adaptation or proliferation of the microorganisms and sometimes not imply these phenomena. On the other hand, the proportion of strains in the microflora able to degrade a pesticide by a co-metabolic process is sometimes considerable (Juengst and Alexander, 1976). Therefore, microflora may play an important role when small quantities of pesticides are applied. In contrast, with larger doses the microflora would become the limiting factor of the degradation, and the microorganisms using the pesticide as substrate could proliferate and become predominant. The consequences of the previous assumption on the possibility of soil adaptation in agronomic conditions require further investigation. It would also be of interest to link soil adaptation with the metabolic capacity of some large groups of microorganisms in the soil. Bollag (1972, 1974) especially considers that bacteria can often completely degrade a particular pesticide, and the enzyme systems involved are generally adaptive. In contrast, the main activity of fungi seems related to constitutive and less specific enzyme systems.

The persistence of soil adaptation. The persistence of the process of the soil's adaptation to degradation has been the subject of very few studies, in spite of its obvious agronomic importance. Audus (1952) states that a long-term perfusion of soils adapted to degradation, with water, induces the reappearance of a lag phase in the degradation. Audus (1960) also asserts, however, that soil samples enriched in 2,4-D and kept under good conditions may retain their degrading power for a year at least. Indications that such a persistence of adaptation also occurred under field conditions were reported by Newman et al. (1952). McGrath (1976) took soil samples from field plots pretreated with TCA. He observed *in vitro* that the conditioning effect was retained at least partially during 32 months.

The reasons for the persistence of the soil's degrading power toward certain products are not clear. In the absence of the pesticide, the microflora cannot survive except at the expense of other types of substrate. It must be noted, however, that traces of pesticide probably persist in soils several years after their application and may induce enzyme synthesis. Moreover, Jensen and Lautrup-Larsen (1967) noted that the bacteria retained their ability to metabolize DNOC through numerous passages on DNOC-free media, although the formation of inactive mutants was observed. Clark and Wright (1970) also reported that an *Arthrobacter* strain maintained its ability to utilize IPC as its sole carbon source when it was cultivated on a medium without pesticide. In contrast, an *Achromobacter* strain lost this capacity. Moreover, adaptation of the microflora to the degradation of pesticides applied at higher rates is frequently obtained with a very small initial dose of the product (Audus, 1949; Thornton, 1955). Waid

(1972) suggested the possible implication of an infectious transfer in soil, from donor cells to recipient cells, of genetic factors that control the formation of enzymes that decompose recalcitrant molecules. The process would be similar to the transfer of the resistance of bacterial cells toward antibiotics.

Finally, it must be stressed that the enrichment of a soil in microorganisms able to degrade a pesticide is not always a consequence of the use of this same pesticide. We shall see in Section VII,B that this enrichment can result from the previous use of another pesticide. Furthermore, the supply of various organic material may induce a multiplication of the active microorganisms or an increase in their activity (Burge, 1969; McClure, 1970). This is especially important for co-metabolic processes. In this regard, we may refer to some observations on the degradation of 2,6-dichlorobenzamide. An increase in the degradation of 2,6-dichlorobenzamide in soil obtained only after repeated additions of co-substrate persisted for several weeks after the suspension of these additions (Fournier, 1975). Finally, some attempts have been made to introduce a fungal or bacterial inoculum directly into the soil. Among the trials carried out in nonsterile soils is that of Suess (1970). A large inoculum of *Bacillus sphaericus* increased the rate of monolinuron inactivation for more than 4 weeks after the inoculation.

B. INFLUENCE OF PESTICIDE COMBINATIONS

A mixing of various pesticides may occur in soil under many circumstances. Combinations of herbicides or other products are commonly applied to obtain an increase in the activity spectrum of the substances. The combined use of different compounds on any one culture, the crop succession, the persistence in soil of the degradation products—all these factors may promote the occurrence of secondary effects which are due mainly to modifications of the activity and persistence of the substances.

1. Modifications of the Activity

The modifications of the activity of pesticides used in combination are known essentially through the modifications of the effects on the protected or target organisms, owing to the agronomic problems involved.

The subject of pesticide interaction in higher plants was extensively reviewed by Putnam and Penner (1974). Synergistic or antagonistic effects may result from alterations in the uptake, translocation, or metabolism of pesticides. Such alterations probably appear also with microorganisms. We have listed in Table IX some recent studies on this subject. Special attention was paid to the study of herbicide mixtures. But herbicides may also interact with other types of pesticide. Thus, Freeman and Finlayson (1977) studied the interactions of 25 her-

TABLE IX
Plant Effect of Some Pesticide Combinations

Pesticide[a]		Interactions between herbicides	Authors
(Ant.)	2,4-D	Removes inhibition of the RNA synthesis by EPTC	Beste and Schreiber (1972)
(Syn.)	Ethofumisate	Increases the foliar penetration of desmedipham	Eshel et al. (1976)
(Syn.)	EPTC	Increases uptake of atrazine	Wyse et al. (1976a)
(Ant.)	Trifluralin, nitralin	Inhibit root growth and reduce absorption of triazines and linuron	O'Donovan and Prendeville (1976)
(Syn.)	Diallate	Increases foliar uptake of 2,4-D, 2,4-D, atrazine, TCA, diquat	Davis and Dusbabek (1973)

Effect of insecticides on herbicide activity

(Syn.)	Carbofuran	Inhibits the aryl acylamidase that detoxifies propanil	Smith and Tugwell (1975)
(Syn.)	Diazinon, carbaryl	Inhibits the aryl acylamidase that detoxifies propanil	El-Refai and Mowafy (1973)
(Syn.)	Carbofuran	Reduces the metabolism of alachlor	Hamill and Penner (1973a)
(Syn.)	Carbofuran	Reduces the metabolism of chlorbromuron	Hamill and Penner (1973b)
(Syn.)	Carbofuran	Decreases the metabolism of butylate	Hamill and Penner (1973c)
(Syn.)	Organophosphates	Decrease the metabolism of dicamba, chlorpropham, linuron, pronamide, pyrazon	Chang et al. (1971a)
(Syn.)	Carbamates	Decrease the metabolism of propanil, pronamide, pyrazon	Chang et al. (1971a)
(Syn.)	Phorate	Increases the absorption of prometryne	Parks et al. (1972)
(Ant.)	Disulfoton	Decreases the phytotoxic effect of fluometuron	Savage and Ivy (1973)
(Syn.)	Carbaryl	Reduces the metabolism of linuron	Del Rosario and Putnam (1973)

Effect of herbicides on metabolism of insecticides by plant

Propanil, linuron		Inhibits the metabolism of dyfonate and malathion	Chang et al. (1971b)
Linuron		Inhibits the metabolism of carbaryl	Chang et al. (1971b)
Propanil		Inhibits the conversion of one metabolite of carbaryl	Chang et al. (1971b)
Chlorpropham		Stimulates the metabolism of carbaryl	Chang et al. (1971b)
EPTC		Increases translocation and/or metabolism of phorate	Schultz et al. (1976)

[a] (Ant.)—antagonistic interaction. (Syn.)—synergistic interaction.

bicides and 3 insecticides. Of the 215 herbicide–insecticide combinations tested, 29 resulted in phytotoxic interactions.

In the same way, synergistic or antagonistic interactions between insecticides and other synthetic chemicals may alter the activity of these compounds against insects (De Giovanni *et al.*, 1971; Lichtenstein *et al.*, 1973; Lichtenstein, 1975). Lastly, we may also mention a study on the toxicity of pesticide combinations toward some animal higher organisms (Statham and Lech, 1976).

With regard to pathogenic microflora, some effects of pesticide mixtures have already been discussed (Section V,A,2). Nene and Dwivedi (1972) studied the effect of insecticides and herbicides on the fungitoxicity of thiram for *Helmintosporium maydis*. Foschi *et al.* (1970) carried out research on the behavior of the fungicides dodine and TMTD, used in mixture with various acaricides and insecticides. They observed a general diminution in the fungicide activity. According to Richardson (1960), however, combinations of aldrin, dieldrin, or lindane with thiram, captan, and chloranil ensured a better protection against *Rhizoctonia solani* and *Pythium ultimum* diseases than did the fungicides alone. Other experiments of Foschi and Svampa (1972) or of Svampa *et al.* (1974) clearly showed that the insecticides interfered with the effects of some fungicides. Usually the authors observed an enhancement of this activity, but some examples of antagonistic behavior were also reported.

To conclude, we find that the influence of pesticide combinations on the major physiological groups of soil microflora has remained almost unstudied up to now.

2. Modifications of Persistence

Interactions between pesticides may result in modifications of their persistence. Such interactions occur at the soil surface or in the soil.

a. Direct Reactions between Pesticides. Direct reactions between pesticides have rarely been noted. According to Ivie and Casida (1971), however, the insecticide rotenone can act as a photosensitizer to accelerate the photodegradation of some types of insecticides, including organophosphates, methylcarbamates, pyrethroids, and chlorinated cyclodiene compounds. In the same way the esterification of the polyvalent soil fumigant metam-sodium (sodium N-methyldithiocarbamate) by nematocides containing halogenated hydrocarbons prevents its conversion into the active N-methylisothiocyanate (Miller and Lukens, 1966). Certain residues of degradation can also react; thus, N-methylisothiocyanate reacts with metam-sodium to yield N,N'-dimethylthiuram disulfide, or with methylamine, another product of metam-sodium degradation (and also a photolytic product of paraquat degradation), to give N,N'-dimethylthiourea (Tuner and Corden, 1963; Wright and Cain, 1969). Such processes can also result from activity of microorganisms. Kearney *et al.*

(1969) have shown that the metabolism of chlorinated aniline mixtures could result in the formation of asymmetrical chloroazobenzenes in soil.

b. Stimulation of Degrading Activity of Microflora. The interactions between pesticides are usually expressed in terms of their influence on the breakdown of these pesticides by microorganisms. Some phenomena, by stimulating microflora activity, may reduce the persistence of the products.

Effects of co-metabolism. During tests with isolated strains, Horvath and Alexander (1970) obtained the co-metabolism of 22 substituted benzoates, including herbicides, after growing cells in an enrichment medium containing chloro- or aminobenzoate. In soil incubations, Fournier (1975) obtained an accelerated degradation of 2,6-dichlorobenzamide (the main product of degradation of the herbicide dichlobenil) by addition of nonsubstituted benzamide. As far as we know, however, there are no studies that clearly show the stimulation of the degradation in soil of a pesticide in the presence of another pesticide acting as co-substrate. In fact, the quantities of co-substrate thus available are normally very limited.

On the other hand, the biocidal activity of certain pesticides against microflora might produce a release of the organic substrate, favorable to an enhancement of co-metabolism.

Effects of enzymatic cross-induction phenomena. Phenomena of enzymatic induction sometimes cause a decrease in the persistence of combined substances. Audus (1951) percolated 2,4-D-enriched soil with MCPA. He observed a much shorter time of detoxication for this herbicide than was found in an unenriched soil. An analogous result was obtained by percolating an MCPA-enriched soil with 2,4-D. The enriched soil degraded more rapidly 2,4-D than an untreated soil. Two points were noticed by Audus: Firstly, the lag phase was suppressed in each case, but secondly, the degradation of the pesticides was slower in the cross-induced soils than in direct induction experiments. Audus suggested that either different microorganisms were involved, or different adaptive enzyme systems developed in the same organism. In another experiment Audus showed that MCPA, but not 2,4-D, can also temporarily induce 2,4,5-T detoxication. From a practical point of view this observation permitted Kaufman (1972, 1975) to consider the possibility of adding small quantities of MCPA to commercial formulations containing 2,4,5-T to accelerate the degradation of the latter. Patterns of cross-adaptation in soils treated with various phenoxyalkanoic acids have been investigated by Brownbridge (1956).

Kirkland and Fryer (1972) described an MCPA-enriched soil that was also able to degrade MCPB with equal facility. In contrast, the degradations of dichlorprop, mecoprop, silvex, and dicamba were not affected by the previous treatments with MCPA. Kaufman (1975), however, studying the degradation of dicamba in soils previously treated with dicamba itself or with 2,4-D, concluded that a rapid degradation occurred only in 2,4-D-enriched soils.

Hurle (1973) also observed a slightly quicker degradation of DNOC in a soil treated five months before with dinoseb acetate, but it is not clear whether there is really a cross-adaptation phenomenon. The stimulation of co-metabolism may also happen in connection with the biocidal action of dinoseb acetate against microflora.

Concerning carbamate herbicides, we may also refer here to the work of Kaufman and Kearney (1965), who isolated strains able to use chlorpropham (CIPC) or CEPC as their sole carbon source. CEPC-effective isolates degraded chlorpropham whereas chlorpropham-effective isolates degraded CEPC, but much more slowly. All isolates degraded propham more rapidly than either chlorpropham or CEPC. Finally, McClure (1972) observed an accelerated degradation of propham, but also of chlorpropham and swep, after an application on soil samples of a suspension of microorganisms grown with propham as sole carbon source.

The interpretation of the results obtained by cross-adaptation often remains very difficult by reason of the complex phenomena that interfere: enzymatic adaptation, proliferation of the microorganism caused by the excessive quantities of product sometimes used, co-metabolism, etc.

   c. *Inhibition of the Degrading Activity of Microflora.*   Other processes can increase the persistence of pesticides in combination by reducing the activity of the degrading microflora. Again, various phenomena can interfere. Kaufman (1975) mentions the biocidal properties of pesticides toward soil microorganisms which may prevent their degradation, the inhibition of adaptive enzymes of effective strains, and, finally, the inhibition of the microorganisms proliferation.

   Herbicide interactions.   As early as 1966 Kaufman observed that phytotoxic levels of dalapon persisted in fields 128 days after its application in combination with 5 ppm of amitrole. According to studies *in vitro*, the inhibition of dalapon degradation seemed to be in relation to the toxicity of amitrole for effective microorganisms. Moreover, Hilton (1969) showed that the inhibition by amitrole of the dehydratase enzyme in the biosynthetic pathway is the major factor controlling growth in heterotrophic microorganisms. Many other interactions between the herbicides have since been reported. Thus, according to Hamroll and Jumar (1973), the duration of the half-life of proximpham ($O(N$-phenylcarbamoyl)propanonoxim) was increased twofold when 10–20% of a secondary herbicide was added. In particular, fenuron, diuron, lenacil (3-cyclohexyl-5,6-trimethylene uracil), CIPC, and IPC were effective. Likewise, monuron or the thiolcarbamate herbicide diallate, used in mixtures with linuron, increased its persistence in soil (Poschenrieder *et al.*, 1975). A slight inhibition of the linuron degradation was observed with only 10^{-6} mole of inhibiting substances per kilogram of soil. At higher rates of inhibitor (hundredfold), the inhibition reached 35% with diallate and 56% with monuron, after two months of incubation. In unpublished results quoted by Wallnöffer *et al.*, (1977), similar observations were made concerning a decrease in the degradation rate of ^{14}C-monolinuron in diallate- and monuron-treated soils. Working with an isolated strain of *Bacillus sphaericus,* these authors also observed an inhibition of the linuron degradation in the presence of monuron.

In most studies previously mentioned, competitive inhibition of esterases or amidases implicated in the hydrolysis of phenylcarbamate or acylanilide herbicides appears likely. The same phenomenon could explain some observations on the degradation of ^{14}C-labeled 2,6-dichlorobenzamide in soil incubation (Fournier, 1974). As indicated by determination of the $^{14}CO_2$ release, 2,6-dichlorobenzamide applied alone was slowly degraded, and no metabolite accumulated. In contrast, the simultaneous addition of various analogues resulted in a decrease in $^{14}CO_2$ release and in an accumulation of labeled 2,6-dichlorobenzoic acid.

For another chlorobenzoic herbicide, chloramben, Talbert *et al.*, (1970) described the inhibition of the methyl ester hydrolysis in the presence of dinoseb. During recent soil incubation experiments we have also noticed (Repiquet and Fournier, 1977) a temporary decrease in the degradation rate of ^{14}C-2,4-D in soil samples, with simultaneous applications of 0.5–5 ppm of dinoseb. Moreover, Hurle (1975) reported the inhibiting power of another nitro compound, DNOC, on 2,4-D degradation. Concerning the results of Repiquet and Hurle, it would be of interest to determine if they correspond to the antibiotic and especially the antifungic, activity of the nitro compounds.

The products of degradation can also interfere in the breakdown of the parent molecule. McClure (1974) studied the degradation of various anilide herbicides in a mixed population of microorganisms adapted to propham. He observed that the activity of microorganisms was a function first of the molecular structure, and then probably of the toxicity of the metabolites produced. In the same way Blake and Kaufman (1975) showed that halogenated anilines at high concentrations were able to inhibit amidases. Interactions between the pesticide and its degradation products are suggested by Oleinikov *et al.* (1976), who observed in soil experiments that the microorganisms that appeared to be implicated in the breakdown of 3,4-DCA were inhibited by both 3,4-DCA and propanil.

Interactions between herbicides and other pesticides. Some interactions between herbicides and fungicides enhance the persistence of the herbicide. Thus, Wagenbreth and Kluge (1971) noticed a diminution of prometryn degradation with thiram. After 100 days 10% of the initial concentration of prometryn remained in the soil when it was used alone, and 60% when it was combined with the fungicide. According to these authors thiram also inhibited simazine breakdown, and polyvalent metham-sodium acted in the same way with simazine and prometryn. Thiram was also found by Wallnöfer *et al.* (1977) to have a slightly inhibiting effect on the degradation of monolinuron in the soil.

Previous investigations of Walker (1970) have shown that the highly stable fungicide quintozene applied at high rates caused an increase in the persistence of chlorpropham in the soil. The half-life of the herbicide was increased almost twofold. In contrast, in similar experiments the half-life of sulfallate was only slightly increased. In soil incubation studies with field rates of ^{14}C-isoproturon, Repiquet and Fournier (1977) noticed a marked inhibition of the herbicide degradation when maneb or benomyl was added, both at rates of 50 ppm. This inhibition rose above 50% of that in the control during the first two weeks of experimentation.

Interactions between herbicides and insecticides have been widely studied, especially the effects of combinations of phenylamide herbicides and methylcarbamate insecticides.

Kaufmann et al. (1970, 1971) first demonstrated that various insecticide methylcarbamates, including 1-naphthyl-, O-isopropoxyphenyl-, 4-dimethyl-amino)m-totyl-, 6-chloro-3,4-xylyl-, and 4-dimethylamino-3,5-xylyl-N-methylcarbamate, increased the persistence of CIPC in the soil. The most effective was 1-naphthyl-N-methylcarbamate, or carbaryl, which, applied at the level of 1 ppm, brought about an extended lag phase in the degradation of chlorpropham. Likewise, the addition of PCMC (p-chlorophenylmethylcarbamate) at a concentration of 5 ppm in soil delayed the degradation of propanil and greatly reduced the formation of 3,3',4',4'-tetrachloroazobenzene (TCAB) (Kaufman et al., 1971). Burge (1972) reported that propanil applied to moist soils was rapidly hydrolyzed by an acylamidase directly extractable from the soil, but was stable in soils treated with PCMC. Repiquet and Fournier (1977) showed the inhibition of a phenylurea herbicide, isoproturon, by high rates of PCMC. Poschenrieder et al. (1975) observed a decrease in the degradation of labeled linuron when metmercapturon, carbaryl, propoxur but also aldicarb, or parathion were added. Other unpublished data of Poschenrieder were given by Wallnöfer (1977), who reported that the degradation of monolinuron in soil incubation was significantly reduced in the presence of the five insecticides mentioned above. Furthermore, the release of O_2 by *Chlorella* was normally inhibited by extracts of soil treated with solan for a few days. A combination of the herbicide with metmercapturon, however, resulted in an extended inhibiting effect. Lastly, an increase in the persistence of chlorpropham in the soil caused by heptachlor or a combination of DDT and captan (Kaufman, 1972, 1975) has been reported.

Studies with phenylcarbamate- or acylanilide-hydrolyzing enzymes isolated from soil bacteria and fungi have demonstrated that methylcarbamates are competitive inhibitors of these enzymes (Kaufman et al., 1970, 1971). Engelhardt and Wallnöfer (1975), Poschenrieder et al. (1975), and Wallnöfer et al. (1977) have confirmed earlier results by studying the degradation of linuron, monolinuron, and solan in mixture with various insecticides in a pure culture of

Bacillus sphaericus. The methylthiocarbamate metmercapturon was the most active inhibitor, acting at rates of 10^{-8} M, 10^{-7} M, and 10^{-6} M, respectively, with linuron, monolinuron, and solan, each at a concentration of 5.10^{-5} M in the medium. The competitive inhibition of the arylacylamidase of *Bacillus sphaericus* by metmercapturon during linuron degradation was demonstrated. Carbaryl appeared to be an effective inhibitor of linuron and monolinuron degradation but to a lesser degree than metmercapturon. An inhibition of solan degradation was also noticed. Another methylcarbamate, propoxur, showed some activity. The breakdown of the three products was also inhibited by the systemic insecticide aldicarb; in the same way linuron degradation was strongly reduced when parathion or fenthion was added to the medium. Kaufman (1975), in his review on the persistence and biodegradation of pesticide combinations, reported that parathion and diazinon were able to inhibit certain enzymes catalyzing the phenylcarbamate of acylanilide hydrolysis. Finally, the PCMC inhibition of barban degradation by *Penicillium* sp. has been described by Wright and Forey (1972).

Interactions between insecticides. Only a few interactions have been found in the degradation of insecticides by microorganisms. Anderson and Lichtenstein (1972) showed that parathion dyfonate and to a lesser extent lindane brought about a reduction of DDT degradation in pure cultures of *Mucor alternans,* although no drastic inhibition of the fungal growth occurred. Tabet and Lichtenstein (1976) reported that *Trichoderma viride* was able to degrade to a considerable extent ^{14}C-photodieldrin in water-soluble compounds. However, in fungal cultures treated with compounds structurally similar to photodieldrin (aldrin and dieldrin), more photodieldrin remained. An enzyme-inhibiting competitive effect due to the similarity of the molecules was suggested by the authors. Chlordane also acted, but to a lesser degree. Less photodieldrin was recovered, with endrin, DDT, and DDE, although this was not reflected in the production of water-soluble metabolites.

Other combinations. The problem of mixture degradation is not limited only to pesticide interactions. Some effects on microorganisms due to formulation products have already been discussed. Effects on degrading microflora should be considered. Attention must also be drawn to the possible interactions of the pesticides with other micropollutants in the soil. Thus, the addition of the detergents ABS (alkyl benzene sulfonate) and LAS (linear alkyl benzene sulfonate) to insecticide-treated soils increased the persistence of the organophosphorous insecticides parathion and diazinon (Lichtenstein, 1966). Moreover, attempts to look for pesticide synergistic substances (Casida 1970; Felton *et al.*, 1970), as well as antidotal products (Chang *et al.*, 1973a,b, 1974; Holm and Szabo, 1974), will probably continue in the future. The basic phenomena involved in the modification of plant, insect, or microorganism response to pesticide mixtures are often similar. The interference of these mixtures with the degradation of the pesticide in the soil must therefore be taken into account. Finally, certain problems relating to the presence of active impurities in the chemicals should not be neglected (Umetsu *et al.*, 1977).

C. CONCLUSION

Several examples described in this chapter reveal the twofold influence of pesticides on the microorganisms responsible for their degradation in the soil. This influence is exerted through the biological characteristics of the substances or through the proliferation of microorganisms able to use them as more or less specific substrates. Whereas a thorough knowledge of the metabolic processes of degradation can be obtained through studies carried out with isolated microorganisms, the behavior of the degrading microflora *in situ* is almost unknown. It is only determined empirically through observations that have some obvious practical value. A more systematic study might lead to a better control of the pesticide in the soil and also to a more general knowledge of microorganism ecology.

VII. Conclusions

The attention of most research workers has been captured by the complexity of the effect of pesticides on the soil microflora. In spite of all the assays published, no uniform results have been obtained, no generalizations can be made, and no conclusions drawn.

Incontestably, the use of pesticides in agriculture leads to changes in the soil microflora; proofs of this have been given throughout this chapter, especially for the pathogenic microflora and for decomposing microflora. In the major biological cycles, nitrification and cellulolysis seem to be the phases that are most affected by pesticides. In the case of the nitrifying bacteria, their frailty, added to their specialization, results in a greater sensitivity. Moreover, the effects of pesticides on the biological life of soils seem to appear more often under precarious environmental conditions that impede the growth of microorganisms. At normal field rates and with short-term applications, the effects are generally more limited. Very often, they are not apparent and do not stimulate much research. Nevertheless, we must remain cautious and not draw conclusions too quickly.

Research must be continued but only in certain areas, where the most sensitive microorganisms appear to intervene. Moreover, methods of evaluating the effects of pesticides must be refined and improved. A better quantification—perhaps by means of new biological methods, such as measurement of the biomass and radiorespirometry—would provide more precise information. It would then be possible to broaden our knowledge about some points that are still unresolved. Research workers should make a particular effort to determine whether the effects of pesticides appear later, say at middle term; whether they become cumulative after repeated treatments; whether they increase or disappear after a series of applications; and whether resistant strains appear in the saprophytic organisms, as they are already known to do in the pathogenic microflora. The microbiologists must adopt standardized experimental protocols and methods of analysis, both for experiments *in vitro* and for field assays. Only then will it be possible to compare the different data obtained in various laboratories and draw conclusions concerning the eventual consequences of the application of pesticides on the soil microflora.

ACKNOWLEDGMENTS

We are very grateful to three colleagues in our group, Pesticides-Soils, of Institut National de la Recherche Agronomique (INRA, National Institute of Agricultural Research)—Dr. Calvet, Mr. G. Catroux, and Mr. G. Soulas— for their useful comments and ideas obtained from several discussions. We are also greatly indebted to Mrs. Emeše Cruiziat, Mrs. Marie-Odile Grandgirard, and Mr. J. M. Pouchain for their kind help in correction of the English text.

REFERENCES

Abelentsev, V.I. 1973. *Khim. Sel. Khoz.* **11**, 32–36. (*Rev. Plant Pathol.* 1975, **54**, 313.)
Abueva, A.A., and Bagaev, V.B. 1975. *Izv. Timiryaz. Selskokhoz. Akad.* **2**, 127–130.
Aggour, M., Bartels, M., and Heitefuss, R. 1977. *Z. Pflanzenkr. Pflanzenschutz.* **8**, 209–213.
Agnihotri, V.P. 1971. *Can. J. Microbiol.* **17**, 377–383.
Agnihotri, V.P. 1974. *Indian J. Exp. Biol.* **12**, 85–88.
Akotkar, S.N., and Deshmukh, V.A. 1974. *Punjabro Krishi Vidyapeeth Res. J.* 3(1), 68–71.
Aliev, Sh.A., Septner, V.A., and Stonov, L.D. 1973. *Khim. Sredstva Zashch. Rast.* **3**, 110–117.
Allott, D.J. 1969. *Weed Res.* **9**(4), 279–287.
Altman, J. 1973. *Int. Congr. Phytopathol., 2nd, Minneapolis* No. 1130.
Amantaev, E., Ilyaletdinov, A., and Kudshev, T. 1963. *Agrobiologiya* **3**, 462–464.
Anderson, J.P.E., and Domsch, K.H. 1973. *Arch. Mikrobiol.* **93**, 113–127.
Anderson, J.P.E., and Lichtenstein, E.P. 1972. *Can. J. Microbiol.* **18**, 553–560.
Apltauer, J., and Skopalikova, O. 1966. *Ust. Ved. Inf. MZLH Rastl. Vyroba* **38**, 47–55.
Arvik, J.H., Hysak, D.L., and Zimdahl, R.L. 1973. *Weed Sci.* **21**(3), 173–180.
Ashida, J. 1965. *Annu. Rev. Phytopathol.* **3**, 153–174.
Atlavinyte, O., Daciulyte, J., and Lugauskas, A. 1974. *Din. Mikrobiol. Protsessov Pochve Obuslovlivayushchie Ec. Faktory, Mater. Simp.* **2**, 137–140.
Audus, L.J. 1949. *Plant Soil* **2**, 31–36.
Audus, L.J. 1951. *Plant Soil* **3**, 170–192.
Audus, L.J. 1952. *J. Sci. Food Agric.* **3**, 268–274.
Audus, L.J. 1960. "Herbicides and the Soil," pp. 1–19. Blackwell, Oxford.
Audus, L.J. 1970. *Colloq. Int. Gand Pestic. Microflore Sol* pp. 465–492.
Ausmus, B.S. 1973. *Bull. Ecol. Res. Commun. (Stockholm)* **17**, 223–234.
Avezdhanova, G.P., Tkachenko, M.P., and Yunusov, M.R. 1976. *Khim. Sel. Khoz.* **14**(4), 35–36.

Backman, P.A., Rodriguez-Kabana, R., and Williams, J.C. 1975. *Phytopathology* **65**, 773-776.
Bakalivanov, D., and Nikolova, G. 1969. *Int. Congr. Soil Sci., 1st, Sofia* pp. 221-226.
Balezina, L.S., and Tretyakova, A.N. 1974. *Din. Mikrobiol. Protsessov Pochve Obuslov-livayushchie Ec. Faktory, Mater. Simp.* **2**, 115-117.
Balicka, N., and Krezel, Z. 1969. *Weed Res.* **9**, 37-42.
Balicka, N., and Sobieszczanski, J. 1969. *Acta Microbiol. Pol.* **18**, 7-10.
Balicka, N., Kosinkiewicz, B., and Wegrzyn, T. 1970. *Colloq. Int. Gand Pestic. Microflore Sol* pp. 647-654.
Bardiya, M.C., and Gaur, A.C. 1970. *Zentralbl. Bakteriol. Parasitenkd. Infectionskr. Hyg.* **124**(6), 552-555.
Bartha, R., Lanzilotta, R.P., and Pramer, D. 1967. *Appl. Microbiol.* **15**(1), 67-75.
Batson, W.E., and Cole, A.W. 1975. *Proc. Am. Phytopathol. Soc.* **2**, 132.
Beckmann, E. O., and Pestemer, W. 1975. *Landwirtsch. Forsch.* **28**(1), 24-33.
Belohov, A.V. 1974. *Khim. Sel. Khoz.* **8**, 48-50.
Bertrand, D., and De Wolf, A. 1972. *C.R. Acad. Agric. Fr.* **58**(17), 1469-1473.
Beste, C.E., and Schreiber, M.M. 1972. *Weed Sci.* **20**, 4-7.
Billes, G. 1971. *Rev. Ecol. Biol. Sol* **8**(2), 235-241.
Blake, J., and Kaufman, D.D. 1975. *Pestic. Biochem. Physiol.* **5**, 305-313.
Bliev, Th.K. 1973. *Pochvovedenie* **7**, 61-68.
Bliev, Yu.R. 1972. *Khim. Sel. Khoz.* **11**, 50-53.
Bollag, J.M. 1972. *Crit. Rev. Microbiol.* **2**, 35-58.
Bollag, J.M. 1974. *Adv. Appl. Microbiol.* **18**, 75-130.
Bollen, G.J., and Sholten, G. 1971. *Neth. J. Plant Pathol.* **77**, 83-90.
Borck, K., and Braymer, H.D. 1974. *J. Gen. Microbiol.* **85**, 51-56.
Bozarth, G.A. 1969. *Diss. Abstr.* **30**, 1984-1985.
Bozarth, G.A., and Tweedy, B.G. 1971. *Phytopathology* **61**, 1140-1142.
Brakel, J. 1963. *Ann. Inst. Pasteur* **105**, 143-149.
Breazeale, F.W., and Camper, N.D. 1972. *Appl. Microbiol.* **23**(2), 431-432.
Brenner, F.J., and Corson, S.D.W. 1974. *Proc. Pa. Acad. Sci.* **48**, 65-67.
Brownbridge, N. 1956. *Ph.D. Thesis, University of London.*
Brunelli, A., Swanpa, G., and Pancaldi, D. 1975. *Inf. Fitopatol.* **25**, 9-16.
Buczacki, S.T. 1973. *Ann. Appl. Biol.* **75**, 25-30.
Bunt, J.S., and Rovira, A.D. 1955. *J. Soil Sci.* **6**, 119-128.
Burge, W.D. 1969. *Appl. Microbiol.* **17**, 545-550.
Burge, W.D. 1972. *Soil Biol. Biochem.* **4**, 379-386.
Byast, T.H., and Hance, R.J. 1975. *Bull. Environ. Contam. Toxicol.* **14**, 71-75.
Campbell, J.E., and Mears, A.D. 1974. *Aust. J. Exp. Agric. Anim. Husb.* **14**(67), 262-265.
Campbell, L. 1956. *Phytopathology* **46**, 635.
Camper, N.D., Moherek, E.A., and Huffman, J. 1973. *Weed Res.* **13**(2), 231-233.
Carter, G.E., Jr., and Camper, N.D. 1975. *Weed Sci.* **23**, 71-74.
Casida, J.E. 1970. *J. Agr. Food Chem.* **18**, 753-772.
Catroux, G., and Fournier, J.C. 1971. *C.R. Acad. Sci. Ser. D.* **273**(7), 716-718.
Catroux, G., Fournier, J.C., and Soulas, G. 1973. *Proc. Eur. Weed Res. Symp. Herbic. Soils Versailles* pp. 141-150.
Cervelli, S., Aringhieri, R., and Nannipieri, P. 1974. *Agric. Ital. (Pisa)* **74**,(1), 38-44.
Cesari, A., and Rapparini, G. 1969. *Atti Giorn. Fitopatol. Cagliari* 631-636.
Chandler, J.M., and Santelman, P.W. 1968. *Weed Sci.* **16**, 453.
Chandra, P. 1964. *Weed Res.* **4**, 54-63.
Chandra, P. 1966. *C.R. Colloq. Dyna. Biocénose Sol Braunschweig-Völkenrode* pp. 320-330.
Chandra, P., and Bollen, W.B. 1961. *Soil Sci.* **92**, 387-393.

Chang, F.Y., Smith, L.W., and Stephenson, G.R. 1971a. *J. Agric. Food Chem.* **19**, 1183-1186.
Chang, F.Y., Stephenson, G.R., and Smith, L.W. 1971b. *J. Agric. Food Chem.* **19**, 1187-1190.
Chang, F.Y., Bandeen, J.D., and Stephenson, G.R. 1973a. *Weed Res.* **13**, 399-406.
Chang, F.Y., Stephenson, G.R., and Bandeen, J.D. 1973b. *Weed Sci.* **21**, 292-295.
Chang, F.Y., Stephenson, G.R., and Bandeen, J.D. 1974. *J. Agric. Food Chem.* **22**, 245-248.
Chappell, W.E., and Miller, L.J. 1956. *Plant Dis. Rep.* **40**, 52-56.
Chopra, B.K., Curl, E.A., and Rodriguez-Kabana, R. 1970a. *Phytopathology* **60**, 717-722.
Chopra, S.L., Niranjan, Das, and Bhagwan, Das, 1970b. *J. Indian Soc. Soil Sci.* **18**(4), 437-446.
Chulakov, Sh.A., and Zharasov, Sh. V. 1975. *Rocz. Glebozn.* **26**(2), 179-184.
Clark, C.G., and Wright, S.J.L. 1970. *Soil Biol. Biochem.* **2**, 19-26.
Cole, A.W., and Batson, W.E. 1975. *Phytopathology* **65**, 431-434.
Cook, R.T.A., and Pereira, J.L. 1976. *Ann. Appl. Biol.* **83**, 365-379.
Corden, M.E., and Young, R.A. 1965. *Soil Sci.* **99**, 272-277.
Correa-Salazar, A. 1976. M.Sc. Thesis, Univ. Dijon.
Cowley, G.T., and Lichtenstein, E.P. 1970. *J. Gen. Microbiol.* **62**, 27-34.
Cullimore, D.R., and MacCann, A.E. 1977. *Plant Soil* **46**, 499-510.
Darveshov, Z. 1973. *Nauchn. Tr. Tashk. Gos. Univ.* **439**, 333-336.
Davidson, J.G., and Clay, D.V. 1972. *Span* **15**(2), 68-71.
Davis, D.G., and Dimond, A.E. 1953. *Phytopathology* **43**, 137.
Davis, D.G., and Dusbabek, K.E. 1973. *Weed Sci.* **21**, 16-18.
Davis, D.E., Pillai, C.G.P., and Truelove, B. 1976. *Weed Sci.* **24**(6), 587-593.
Deep, I.W., and Young, R.A. 1965. *Phytopathology* **55**, 212-216.
De Giovanni, G., Brunelli, A., and Stanzani R. 1971. *Atti Giorn. Fitopatol. Bologna* 527-529.
Dekker, J. 1973. *OEPP EPPO* **10**, 47-57.
Del Rosario, D.A., and Putnam, A.R. 1973. *Weed Sci.* **21**, 465-473.
Deshmukh, V.A., and Shrikhande, J.G. 1974. *J. Indian Soc. Soil Sci.* **22**(1), 36-42.
Dommergues, Y. 1960. *Agron. Trop.* **15**, 61-70.
Dommergues, Y. 1968. *Ann. Inst. Pasteur* **115**, 627-656.
Dommergues, Y., and Mangenot, F. 1970. In "Ecologie Microbienne du Sol," p. 18. Masson, Paris.
Domsch, K.H., Anderson, J.P.E., and Ahlers, R. 1973. *Appl. Microbiol.* **25**(5), 819-824.
Dovas, C., Skylakakis, G., and Georgopoulos, S.G. 1976. *Phytopathology* **66**, 1452-1456.
Doxtaker, K.G. 1968. *Bacteriol. Proc.* **A**, 21.
Drescher, N., and Otto, S. 1969. *Z. Pflanzenkr. Pflanzenschutz.* **76**, 27-33.
Drouineau, G., and Lefevre, G. 1949. *Ann. Agron.* **19**, 518-536.
Drouineau, G., Gouny, P., and Lahaye, Th. 1947. *C.R. Acad. Agric. Fr.* **33**, 203-204.
Dubey, H.D. 1969. *Soil Sci. Soc. Am. Proc.* **33**(6), 893-896.
Dubey, H.D., and Rodriguez, R.L. 1970. *Proc. Am. Soc. Soil Sci.* **34** 435-439.
Ebenebe, C., and Fehrmann, H. 1976. *Rev. Plant Pathol.* **55**, 777-778.
Ebner, L. 1965. *Z. Pflanzenkr. Pflanzenschutz.* **72**, 344-350.
Edgington, L.W., and MacEven, F.L. 1975. *Proc. Am. Phytopathol. Soc.* **2**, 96.
Eisenhardt, A.R. 1975. *Tidsskr. Planteavlo* **79**(2), 254-258.
El-Dib, M.A., and Aly, O.A. 1976. *Water Res.* **10**, 1055-1059.
Elkan, G.H., and Moore, W.E. 1960. *Can. J. Microbiol.* **6**(3), 339-347.
El-Refai, A.R., and Mowafy, M. 1973. *Weed Sci.* **21**, 246-253.
Elsaid, H.M., and Sinclair, J.B. 1964. *Phytopathology* **54**, 518-522.
Engelhardt, G., and Wallnöfer, P. 1975. *Weed Abstr.* **25**, 387.
English, A.R., and Van Halsema, G. 1954. *Plant Dis. Rep.* **38**, 429-431.
Engvild, K.C., and Jensen, H.L. 1969. *Soil Biol. Biochem.* **1**, 295-300.
Enkina, O.V., and Vasilev, D.S. 1974. *Khim. Sel. Khoz* **12**(10), 778-780.
Eno, Ch.F. 1962. *Proc. Fl. Soc. Soil Crop Sci.* **22**, 49-56.

Eno, Ch.F., and Everett, P.H. 1958. *Proc. Am. Soc. Soil Sci.* **22**, 235–238.

Eperspächer, J., Haug, S., Blobel, F., and Sauber, K. 1976. *Zentralbl. Bakteriol. Parasitenkd. Infektionskr. Hyg.* **162**, 145–148.

Ercegovich, C.D., Chrzanowski, R.L., Cole, H., Herendeen, N., and Witkonton, S. 1973. *Can. J. Microbiol.* **19**, 329–334.

Erwin, D.C. 1973. *Annu. Rev. Phytopathol.* **11**, 389–422.

Eshel, Y., and Katan, J. 1972. *Weed Sci.* **20**, 468–471.

Eshel, Y., Zimdahl, R.L., and Schweizer, E.E. 1976. *Weed Sci.* **24**, 619–627.

Fehrmann, H. 1976. *Phytopathol. Z.* **86**, 144–185.

Felton, J.C., Jenner, D.W., and Kirby, P. 1970. *J. Agric. Food Chem.* **18**, 671–673.

Fisyunov, A.V., Vobsob'ev, F.E., and Zhemela, G.P. 1973. *Khim. Sel. Khoz* **3**, 50–53.

Fletcher, J.T., and Sholefield, S.M. 1976. *Ann. Appl. Biol.* **82**, 529–536.

Focht, D.D., and Josseph, D.D. 1974. *J. Environ. Qual.* **3**(4), 327–328.

Forsberg, J.L. 1955. *Plant Dis. Rep.* **39**, 106–114.

Foschi, S., and Svampa, G. 1972. *Inf. Fitopatol.* **15**, 13–20.

Foschi, S., Bentivogli, P.G., and Ponti, I. 1970. *Inf. Fitopatol.* **7**, 3–9.

Foster, M.G. 1975. *Bull. Environ. Contam. Toxicol.* **14**(3), 353–360.

Fournier, J.C. 1974. *Chemosphere* **2**, 77–82.

Fournier, J.C. 1975. *Chemosphere* **1**, 35–40.

Foy, C.L. 1969. *In* "Degradation of Herbicides," (P.C. Kearney and D.D. Kaufman, eds.), pp. 201–253. Dekker, New York.

Freeman, J.A., and Finlayson, D.G. 1977. *J. Am. Soc. Hortic. Sci.* **102**, 29–31.

Freney, J.R. 1965. *Aust. J. Agric. Res.* **16**, 257–263.

Fröhner, C., Oltmanns, O., and Lingens, F. 1970. *Arch. Microbiol.* **74**, 82–89.

Fryer, J.D., and Kirkland, K. 1970. *Weed Res.* **10**, 133–158.

Fung, K.K.H., and Uren, N.C. 1977. *J. Agric. Food Chem.* **25**, 966–969.

Fusi, P., and Franci, M. 1971. *Agrochimica* **15**(6), 557–563.

Fusi, P., and Franci, M. 1972. *Agrochimica* **16**(4–5), 377–386.

Garcia, M.M., and Jordan, D.C. 1969. *Plant Soil* **2**, 317–339.

Garretson, A.L., and San Clemente, C.L. 1968. *J. Econ. Entomol.* **61**, 285–288.

Geeson, J.D. 1976. *Trans. Br. Mycol. Soc.* **66**, 123–129.

Geissbuhler, H. 1969. *In* "Degradation of Herbicides" (P.C. Kearney and D.D. Kaufman, eds.), pp. 79–111. Dekker, New York.

Georgopoulos, S.G. 1975. *Ann. Appl. Biol.* **81**, 444.

Gessler, C. 1976. *Phytopathol. Z.* **85**, 35–38 (*Rev. Plant Pathol.* 1976, **55**.)

Giardina, M.C., Tomati, U., and Pietrosanti, W. 1970. *Colloq. Int. Gand Pestic. Microflore Sol* pp. 615–626.

Gilpatrick, J.D., and Blowers, D.R. 1974. *Phytopathology* **64**, 649–652.

Goring, C.A.I., Laskowski, D.A., Hamaker, J.W., and Meikle, R.W. 1975. *Environ. Sci. Res.* **6**, 135–172.

Goss, O.M., and Shipton, W.A. 1965. *J. Agric. West. Austr.* **6**(11), 659–661.

Grinstein, A., Katan, J., and Eshel, Y. 1976. *Phytopathology* **66**, 517–522.

Grossbard, E. 1970a. *Colloq. Int. Gand Pestic. Microflore Sol* pp. 515–530.

Grossbard, E. 1970b. *Colloq. Int. Gand Pestic. Microflore Sol* pp. 531–542.

Grossbard, E. 1971. *Weed Res.* **11**(4), 263–275.

Grossbard, E. 1973. *Bull. Ecol. Res. Commun. (Stockholm)* **17**, 275–276.

Grossbard, E. 1974. *Chem. Ind.* **15**, 611–614.

Grossbard, E. 1975. *Rocz. Glebozn.* **26**(2), 117–130.

Grossbard, E., and Harris, D. 1976 *Meded. Fac. Landbouwwet Rijksuniv. Gent* **41**(2), 693–702.

Grossbard, E., and Marsh, J.A.P. 1974. *Pestic. Sci.* **5**, 609–623.

Grossbard, E., and Wingfield, G.I. 1975. In "Some Methods for Microbiological Assay" (R.G. Board and D.W. Lovelock, eds.), pp. 236-256. Academic Press, New York.

Grover, R. 1972. Weed Res. 12(1), 112-114.

Groves, K., and Chough, K.S. 1970. J. Agric. Food Chem. 18, 1127-1128.

Guillemat, J., Charpentier, M., Tardieux, P., and Pochon, J. 1960. Ann. Epiphyt. 11(3), 261-296.

Guiscafré-Arrilaga, J. 1949. Phytopathology 39, 8.

Gunasekaran, M., and Ahuja, A.S. 1975. Trans. Br. Mycol. Soc. 64, 324-327.

Gundersen, K., and Jensen, H.L. 1956. Acta Agric. Scand. 6, 100-114.

Haglund, W.A., Gabrielson, R.L., and Tomkins, D.R. 1972. Phytopathology 62, 287-289.

Hall, R. 1975. Can. J. Bot. 53, 452-455. (Rev. Plant Pathol. 1975, 54, 808.

Halos, P.M., and Huisman, O.C. 1976. Phytopathology 66, 152-157.

Hamill, A.S., and Penner, D. 1973a. Weed Sci. 21, 330-335.

Hamill, A.S., and Penner, D. 1973b. Weed Sci. 21, 335-338.

Hamill, A.S., and Penner, D. 1973c. Weed Sci. 21, 339-349.

Hamroll, B., and Jumar, A. 1973. Chem. Technol. 7, 423-424.

Harvey, R.G., Hagedorn, D.J., and De Loughery, R.L. 1975. Crop Sci. 15, 67-71.

Hauke-Pacewiczowa, T. 1970. Colloq. Int. Gand Pestic. Microflore Sol pp. 497-503.

Heitefuss, R., and Bodendörfer, H. 1968. Z. Pflanzenkr. Pflanzenschutz 75, 641-659.

Heitefuss, R., and Brandes, W. 1970. Nachrichtenbl. Dtsch. Pflanzenschutzdienstes (Stuttgart) 22, 40.

Helling, C.S. 1971. Proc. Am. Soc. Soil Sci. 8, 1091-1099.

Helweg, A. 1975. Weed Res. 15, 53-58.

Hilton, J.L. 1969. J. Agric. Food Chem. 17, 182-183.

Hofer, I., von, Beck, Th., and Wallnöfer, P. 1971. Z. Pflanzenkr. Pflanzenschutz 78(7), 398-405.

Hollomon, D.W. 1975. Proc. Br. Insectic. Fungic. Conf., 8th, Brighton (Rev. Plant Pathol. 55, 772.)

Holm, R.E., and Szabo, S.S. 1974. Weed Res. 14, 119-122.

Horner, C.E. 1965. Plant Dis. Rep. 49, 393-395.

Horowitz, M. 1966. Weed Res. 6, 168-171.

Horowitz, M., and Blumenfeld, T. 1973. Phytoparasitica 1(2), 101-110.

Horowitz, M., and Herzlinger, G. 1974. Weed Res. 14, 252-259.

Horowitz, M., Blumenfeld, T., Herzlinger, G., and Hulin, N. 1974a. Weed Res. 14, 97-109.

Horowitz, M., Hulin, N., and Blumenfeld, T. 1974b. Weed Res. 14, 213-220.

Horvath, R.S. 1972. Bacteriol. Rev. 36, 146-155.

Horvath, R.S., and Alexander, M. 1970. Appl. Microbiol. 20, 254-258.

Houseworth, L.D., and Tweedy, B.G. 1972. Phytopathology 62, 765.

Houseworth, L.D., and Tweedy, B.G. 1973. Plant Soil 38(3), 493-500.

Hubbeling, N., and Basu Chaudhary, K.C. 1970. Colloq. Int. Gand Pestic. Microflore Sol pp. 627-635.

Huge, P.L. 1970a. Colloq. Int. Gand Pestic. Microflore Sol pp. 811-827.

Huge, P.L. 1970b. Rap. 'Act. Gembloux pp. 76-80.

Hurle, K. 1973. Proc. Eur. Weed Res. Symp. Herbic. Soil E.W.R.C. Versailles pp. 151-162.

Hurle, K. 1975. Z. Pflanzenkr. Pflanzenschutz 7, 101-108.

Hurle, K., and Pfefferkorn, V. 1972. Proc. Br. Weed Control Conf., 11th 2, 806-810.

Hurle, K., and Rademacher, B. 1970. Weed Res. 10, 159-164.

Husarova, M. 1972. Rostl. Vyroba 18(9), 959-970.

Idris, M. 1973. Pak. J. Sci. Res. 25(3-4), 272-276.

Ivie, G.W., and Casida, J.E. 1971. J. Agric. Food Chem. 19, 410-416.

Jacobsen, B.J., and Hopen, H.J. 1975. Phytopathology 2, 26.

Jaques, R.P., Robinson, J.B., and Chase, F.E. 1959. Can J. Soil Sci. 39, 235-243.

Jenkinson, D.B., and Powlson, D.S. 1976. *Soil Biol. Biochem.* **8**(3), 209–213.

Jensen, H.L., and Lautrup-Larsen, G. 1967. *Acta Agric. Scand.* **17**, 115–126.

Jensen, H.L., and Petersen, I.H. 1952. *Acta Agric. Scand.* **2**, 215–231.

Johnson, E.J., and Colmer, A.R. 1958. *Plant Physiol.* **33**, 99–101.

Jones, A.L., and Ehret, G.R. 1976. *Plant Dis. Rep.* **60**, 765–769.

Jones, A.L., and Walker, R.J. 1976. *Plant Dis. Rep.* **60**, 40–44.

Jones, D.G., and Williams, J.R. 1971. *Trans. Br. Mycol. Soc.* **57**, 351–357.

Jones, L.W. 1952. *Soil Sci.* **73**, 237–241.

Jones, L.W. 1956. *Utah State Agric. Coll. Bull.* 390.

Juengst, F.W., Jr., and Alexander, M. 1976. *J. Agric. Food Chem.* **24**, 111–115.

Kaiser, P., and Reber, H. 1970. *Colloq. Int. Gand Pestic. Microflore Sol* pp. 689–705.

Karanth, N.G.K., Chitra, C., and Vasantharajan, V.N. 1975. *Ind. J. Exp. Biol.* **13**,(1), 52–54.

Karki, A.B., and Kaiser, P. 1974. *Rev. Ecol. Biol. Sol.* **11**(4), 477–498.

Karki, A.B., Coupin, L., Kaiser, P., and Moussin, M. 1973. *Rev. Ecol. Biol. Sol* **10**(1), 3–11.

Kaszubiak, H. 1970. *Colloq. Int. Gand Pestic. Microflore Sol* pp. 543–550.

Katan, K., and Eshel, Y. 1973. *Residue Rev.* **45**, 145–177.

Katan, J., and Eshel, Y. 1974. *Phytopathology* **64**, 1186–1192.

Kataria, H.R., and Grover, R.K. 1974. *Z. Pflanzenkr. Pflanzenschutz* **81**, 472–478. (*Rev. Plant Pathol.* 1975, **54**, 498.)

Kaufman, D.D. 1964. *Can. J. Microbiol.* **10**, 843–852.

Kaufman, D.D. 1966. *Weeds* **14**, 130–134.

Kaufman, D.D. 1972. *Pestic. Chem.* **6**, 175–204.

Kaufmann, D.D. 1975. *Rocz. Glebozn.* **26**, 55–64.

Kaufman, D.D., and Kearney, P.C. 1965. *Appl. Microbiol.* **13**, 443–446.

Kaufman, D.D., Kearney, P.C., Von Endt, D.W., and Miller, D.E. 1970. *J. Agric. Food Chem.* **18**, 513–519.

Kaufman, D.D., Blake, J., and Miller, D.E. 1971. *J. Agric. Food Chem.* **204**, 55–64.

Kaunat, H. 1965. *Zentralbl. Bakteriol. Parasitenkde. Infektionskr. Hyg.* **119**(2), 158–165.

Kavanagh, T. 1969. *Sci. Proc. R. Dublin Soc. B* **2**, 179–190.

Kavanagh, T. 1974. *Sci. Proc. R. Dublin Soc.* **3**, 251–265.

Kearney, P.C., and Kaufman, D.D. 1965. *Science* **147**, 740–741.

Kearney, P.C., Plimmer, J.R., and Guardia, F.S. 1969. *J. Agric. Food Chem.* **17**, 1418–1419.

Kiperman-Reznik, L.D., and Grimalovski, A.M. 1974. *Minist. Sel. Khoz SSSR Kishinev SkH Inst. Frunze Tr.* **127**, 48–54.

Kir, I.N., Kaplun, S.A., and Rakhmatov, I. 1974. *Tr. Vses. Nauchno-Issled. Inst. Khlopkovod.* **26**, 81–88.

Kirkland, K. 1967. *Weed Res.* **7**, 364–367.

Kirkland, K., and Fryer, J.D. 1966. *Proc. Br. Weed Control Conf., 8th* **2**, 616–621.

Kirkland, K., and Fryer, J.D. 1972. *Weed Res.* **12**, 90–95.

Kirkwood, R.C., and Fletcher, W.W. 1970. *Weed Res.* **10**(1), 3–10.

Klyuchnikov, L.Yu., Petrova, A.N., and Polesko, Yu.A. 1964. *Microbiology* **33**(6), 879–882.

Kolcheva, B., Nikov, M., and Fetvarzhieva, N. 1974. *Tr. Mezhdunar. Kongr. Pochvoved. 10th* **3**, 165–172.

Kolesnikov, V.A., Sodovov, V.I., and Chkhetiani, V.R. 1973. *Khim. Sel. Khoz.* **11**(8), 50–54.

Kosinkievicz, B. 1968. *Commun. Natl. Conf. Gen. Appl. Microbiol. Bucharest* p. 174. (Abstr.)

Kosinkiewicz, B. 1970. *Colloq. Int. Gand. Pestic. Microflore Sol* pp. 673–680.

Kozaczenko, H. 1974. *Zesz. Nauk. Akad. Roln. Techn. Olsztynie Roln.* **8**, 3–58.

Kozaczenko, H., and Sobieraj, W. 1973. *Buil. Warzywniczy,* **14**, 161–178.

Kozlova, E.I., Belousova, A.A., and Vandar'eva, U.S. 1964. *Agrobiologija* **2**, 271–277.

Kozlova, L.M., Bliev, Yu.K., and Solov'eva, E.N. 1974. *Sb. Nauchn. Tr. Leningrad Nauchno Issled. Inst. Lesn. Khoz.* **18**, 33-41.

Kozyrev, M.A., and Laptev, A.A. 1972. *Khim Sel. Khoz.* **8**, 54-57.

Krezel, Z., and Leszczynska, D. 1970. *Colloq. Int. Gand Pestic. Microflore Sol* pp. 655-661.

Kruglov, Yu.V., and Kvyatkovskaya, L.B. 1975. *Rocz. Glebozn.* **26**(2), 145-149.

Kruglov, Yu.V., Pertseva, A.N., and Galkina, G.A. 1975a. *Dokl. Vses. Akad. Skh Nauk Lenina* **2**, 20-22.

Kruglov, Yu.V., Gersh, N.B., Pertseva, A.N., Bei-Bienko, N.V., and Mikhailova, E.I. 1975b. *Rocz. Glebozn.* **26**(2), 159-164.

Krumzdorov, A.M. 1974. *Khim Sel. Khoz.* **8**, 44-46.

Kudzin, Yu.K., Fisyunov, A.B., Chernyauskaya, N.A., and Makarova, A.Ya. 1973. *Dokl. Vses. Akad. Skh. Nauk Lenina* **9**, 13-15.

Kulinska, D. 1967. *Rocz. Nauk. Roln. Ser. A* **93**(2), 229-262.

Kulinska, D., and Romanov, I. 1970. *Colloq. Int. Gand Pestic. Microflore Sol* pp. 551-558.

Kuramoto, T. 1976. *Plant Dis. Rep.* **60**, 168-172.

Kuryndina, T.I. 1965. *Sborn. Nauch. Rab. Inst. Sadov. Michurina* **11**, 113-116.

Lai, M.T., and Semeniuk, G. 1970. *Phytopathology* **60**, 563-564.

Lai, P., and Bruehl, G.W. 1968. *Phytopathology* **58**, 562-566.

Lakshmi-Kumari, M., Biswas, A., Vijayalakshmi, K., Narayana, H.S., and Rao, N.S. 1974. *Proc. Indian Natl. Sci. Akad. Part B* **40**(5), 528-534.

Lambert, D.H., and Wuest, P.J. 1975. *Phytopathology* **65**, 637-638.

Lambert, D.H., and Wuest, P.J. 1976. *Phytopathology* **66**, 1144-1147.

Lauss, F., and Danneberg, O.H. 1975. *Bodenkultur* **26**(1), 22-30.

Lay, M.M., and Ilnicki, R.D. 1974. *Weed Res.* **14**, 289-291.

Leasure, J.K. 1964. *J. Agric. Food Chem.* **12**, 40-43.

Lee, C.C., Harris, R.F., Williams, J.D.H., Armstrong, D.E., and Syers, J.K. 1971. *Proc. Am. Soc. Soil Sci.* **35**, 82-86.

Lees, H., and Quastel, J.H. 1946. *Biochem. J.* **40**, 803-815.

Lefebvre-Drouet, E., and Calvet, R. 1978. *Weed. Res.* **18**, 33-39.

Lembeck, W.J., and Colmer, A.R. 1967. *Appl. Microbiol.* **15**(2), 300-303.

Lethbridge, G., and Burns, R.G. 1976. *Soil Biol. Biochem.* **8**, 99-102.

Lichtenstein, E.P. 1966. *J. Econ. Entomol.* **59**, 985-993.

Lichtenstein, E.P. 1975. *Proc. IAEA Meet. Chem. Residues Food, 2nd, Vienna.* (*Chem. Abstr.* 1976, **85**, 115.)

Lichtenstein, E.P., Liang, T.T. and Anderegg, B.N. 1973. *Science* **181**, 847-489.

Livens, J., Mayaudon, J., Saive, R., Voets, J.P., and Verstraete, W. 1973. *Inst. Belg. Amelior. Betterave* **4**, 135-181.

Lobanov, V.E., and Poddubnaya, L.P. 1967. *Khim. Sel. Khoz.* **5**(10), 32-34.

Lorinczi, F. 1974. *Contrib. Bot. Gradina Bot. Univ. Babes-Bolya Cluj* pp. 160-168.

Lozano-Calle, J.M. 1970. *Colloq. Int. Gand Pestic. Microflore Sol* pp. 599-614.

McClure, G.W. 1970. *Contrib. Boyce Thompson Inst.* **24**, 235-246.

McClure, G.W. 1972. *J. Environ. Qual.* **1**, 176-177.

McClure, G.W. 1974. *Weed Sci.* **22**, 323-329.

MacCrady, M.H. 1915. *J. Infect. Dis.* **17**, 183-212.

MacFayden, A. 1961. *J. Exp. Biol.* **38** 323-341.

MacGarity, J.W., Gilmour, C.M., and Bollen, W.B. 1958. *Can. J. Microbiol.* **4**, 303-316.

McGrath, D. 1976. *Weed Res.* **16**, 131-137.

MacLeod, N.H., Chappelle, E.H., and Crawford, A.M. 1969. *Nature (London)* **223**, 267-268.

Macura, J. 1974. *Geoderma* **12**(4), 311-329.

Makawai, A.A.M., and Ghaffar, A.S. 1970. *U.A.R.J. Microbiol.* **5**(1-2), 109-117.

Malichenko, S.M. 1971. *Mickrobiol. Zh.* **33**, 734-735.

Markert, S., and Kundler, 1975. *Arch. Acker-Pflanz. Bodenkd.* **19**(7), 487-497.

Martin, J.F. 1950. *Soil Sci.* **69**, 215-232.

Martin, J.P., Harding, R.B., Cannell, G.H., and Anderson, L.D. 1959. *Soil Sci.* **87**, 334-338.

Mashtakov, S.M., Gurinouvitch, E.S., Simenko, T.L., and Kabailova, I.V. 1962. *Microbiologija* **31**, 85-89.

Mathur, R.L., Masih, B., and Sankhla, H.C. 1972. *Indian Phytopathol.* **24**, 548-552. (*Rev. Plant Pathol.* 1973, **52**, 122.)

Mayaudon, J. 1973. *Inst. Belg. Amelior. Betterave* **4**, 151-160.

Mendoza, M. 1973. *An. Inst. Nac. Invest. Agrar. Ser. Gen.* **2**, 21-35.

Menzie, C.M. 1969. "Metabolism of Pesticides." Special Scientific Report, Wildlife N. 127. Bureau of Sport Fisheries and Wildlife. Washington, D.C.

Merenyuk, G.V., Tarkov, M.I., Timchenko, L.A., and Tikhovskaya, T.M. 1973. *Aktual. Vopr. Sanit. Mikrobiol.* 14-15.

Meyer, E. 1976. *Mitt. Biol. Bundesanst. Land Forstwirtsch.* **166**, 135. (*Rev. Plant Pathol.* 1976, **55**, 575.)

Meynell, G.G., and Meynell, E. 1965. *In* "Theory and Practice in Experimental Bacteriology," pp. 1-29. Cambridge Univ. Press, London and New York.

Mezharaupe, V.A. 1967. *Mikroog. Rast. Trud. Inst. Mikrobiol. Akad. Nauk SSSR* 65-83.

Micev, N. 1970. *Contemp. Agric.* **18**(3), 245-249.

Miklaszenski, S. 1975. *Rocz. Glebozn.* **26**(2), 241-245.

Milkowska, A., and Grozelak, A. 1966. *Sylwan* **110**(11), 13-22.

Millar, W.N., and Casida, L.E. 1970. *Can. J. Microbiol.* **16**, 299-304.

Miller, P.M., and Ahrens, J.F. 1964. *Phytopathology* **54**, 901.

Miller, P.M., and Lukens, R.J. 1966. *Phytopathology* **56**, 967.

Minton, E.B. 1972. *Phytopathology* **62**, 582-583.

Moller, W.J., Beutel, J.A., Reil, W.O., and Zoller, B.G. 1972. *Phytopathology* **62**, 779.

Nachmias, A., and Barash, I. 1976. *J. Gen. Microbiol.* **94**, 167-172.

Namdeo, K.N., and Dube, J.N. 1973. *Indian J. Exp. Biol.* **2**, 548-550.

Nayak, M.L., Singh, R.P., and Verma, J.P. 1976. *Z. Pflanzenkr. Pflanzenschutz.* **83**, 407-415.

Nene, Y.L., and Dwivedi, T.S. 1972. *Pestic. Bombay* **6**, 13-17. (*Rev. Plant Pathol.* **52**, 507.)

Nepomiluev, V.F., Bebin, S.I., und Kuzyakina, T.T. 1966. *Izv. Timiryazevsk Sel. Khoz. Akad.* **4**, 88-94.

Neubauer, R., and Avizohar-Hershenson, Z. 1973. *Phytopathology* **63**, 651-652.

Neven, H., Vlassak, K., and Heremans, K.A.H. 1975. *Int. Symp. Gand Phytopharmacol. Phytiatr. 27th* pp. 1221-1230.

Newman, A.S., and Thomas, J.R. 1949. *Proc. Am. Soc. Soil Sci.* **14**, 160-164.

Newman, A.S., Thomas, J.R., and Walker, R.L. 1952. *Proc. Am. Soc. Soil Sci.* **16**, 21-24.

Nilsson, H.E. 1973. *Swed. J. Agric. Res.* **3**, 115-118.

Novak, B. 1973. *Zentralbl. Bakteriol. Parasitenkd. Infektionskr. Hyg. Abt. II* **128**(3-4), 316-335.

Novogrudskaya, E.E., and Isaeva, L.I. 1965. *Agrobiologija* **4**, 572-582.

O'Donovan J.T., and Prendeville, G.N. 1976. *Weed Res.* **16**, 331-336.

Ohr, H.D., Munnecke, D.E., and Bricker, J.L. 1973. *Phytopathology* **63**, 965-973.

Okioga, D.M. 1976. *Ann. Appl. Biol.* **84**, 21-30.

Oleinikov, R.R., Strekozov, B.P., and Garnicheva, Z.D. 1975. *Mikrobiologija* **44**(1), 147-150.

Oleinikov, R.R., Strekozov, B.P., and Garnicheva, Z.D. 1976. *Weed Abstr.* **25**, 203.

Olsen, O.A. 1966. *Can. Plant Dis. Surv.* **46**, 1-4.

Pajewska, M. 1972. *Poznan. Tow. Przyj. Nauk Pr. Kom. Nauk Roln. Kom. Nauk Lesn.* **33,** 245–249.

Pantera, H. 1970. *Colloq. Int. Gand Pestic. Microflore Sol* pp. 847–854.

Pantera, H. 1974. *Pam. Pulawski* **60,** 231–240.

Panterowa, H., Zurawski, H., and Gonetova, I. 1975. *Rocz. Glebozn.* **26**(2), 247–256.

Parks, J.P., Truelove, B., Buchanan, G.A. 1972. *Weed Sci.* **20,** 89–92.

Paromenskaya, L.N. 1975. *Rocz. Glebozn.* **26**(2), 131–135.

Parry, K.E., and Wood, R.K.S. 1958. *Ann. Appl. Biol.* **46,** 446–456.

Parry, K.E., and Wood, R.K.S. 1959. *Ann. Appl. Biol.* **47,** 10–16.

Partridge, A.D., and Rich, A.E. 1962. *Phytopathology* **52,** 1000–1004.

Pathak, A.N., Shankar, H., and Awasthi, K.S. 1960–1961. *J. Indian Soc. Soil Sci.* **8**(9), 197–200.

Paulus, A.O., Nelson, J., and Besemer, S. 1976. *Plant Dis. Rep.* **60,** 695–697.

Peeples, J.L., and Curl, E.A. 1970. *Phytopathology* **60,** 586.

Peeters, J.F., Van Rossen, A.R., Heremans, K.A., and Delcambe, L. 1975. *J. Agric. Food Chem.* **23,**(3), 404–406.

Percich, J.A. 1975. *Diss. Abstr. Int. B.* **36**(3), 1049.

Percich, J.A., and Lockwood, J.L. 1975. *Phytopathology* **65,** 154–159.

Pillay, A.R., and Tchan, Y.T. 1972. *Plant Soil* **36,** 571–594.

Pitts, G., Curl, E.A., and Rodriguez-Kabana, R. 1970. *Phytopathology* **60,** 587.

Pochon, J., and De Barjac, H. 1958. *In* "Traité de Microbiologie des Sols. Applications Agronomiques," pp. 45–46. Dunod, Paris.

Pochon, J., and Tardieux, P. 1962. *In* "Techniques d'Analyse en Microbiologie du Sol," pp. 49–80. La Tourelle, Paris.

Pociene, S., Tarvidas, J., and Tamasauskaite, A. 1974. *Liet. TSR Aukst. Mokyklu Mokslo Darb. Biol.* **13,** 21–27.

Ponchet, J., and Tramier, R. 1971. *Ann. Phytopathol.* **3**(3), 401–406.

Pop, L., Gingioveanu, I., Matei, I., and Preda, A. 1968. *Probl. Agric.* **11,** 49–56.

Poschenrieder, G., Wallnöfer, P., and Beck, T. 1975. *Z. Pflanzenkr. Pflanzenschutz* **82,** 398–405.

Pugh, G.J.F., Williams, J.I., and Wainwright, M. 1973. *Proc. Int. Colloq. Prog. Soil Zool., 5th* pp. 489–496.

Purushothaman, D., Marimuthu, T., and Kesavan, R. 1974. *Indian J. Exp. Biol.* **12**(6), 580–581.

Putnam, A.R., and Penner, D. 1974. *Residue Rev.* **50,** 73–110.

Ragab, M. Th. 1974. *Can. J. Plant Sci.* **54**(4), 713–716.

Rakhimov, A.A., and Rybina, V.F. 1963. *Usb. Biol. Zh.* **7,** 74–76.

Rankov, V., and Velev, B. 1975. *Rocz. Glebozn.* **26**(2), 233–239.

Raynal, G., and Ferrari, F. 1973. **22,** 259–272.

Reddy, D.D.R., Rao, J.V., and Kumar, D.K. 1975. *Plant Dis. Rep.* **59**(7), 592–595.

Reichlova, E. 1975. *Rostl. Vyroba* **21**(6), 607–615.

Repiquet, C., and Fournier, J.C. 1977. *Conf. Desherbage (COLUMA), 9th, Paris* **2,** 801–810.

Ricci, P. 1974. *Ann. Phytopathol.* **6,** 441–443.

Richardson, L.T. 1957. *Can. J. Plant Sci.* **37,** 196–204.

Richardson, L.T. 1959. *Can. J. Plant Sci.* **39,** 30–38.

Richardson, L.T. 1960. *Abstr. Mycol.* **39,** 521.

Richardson, L.T. 1970. *Can. J. Plant Sci.* **50,** 594–596.

Richardson, L.T., and Miller, D.M. 1960. *Can. J. Bot.* **38,** 163–175.

Riepma, P. 1962. *Weed Res.* **2,** 41–50.

Ritter, M., Simon-Sylvestre, G., and Bonnel, L. 1970. *Int. Congr. Prot. Plantes, 7th, Paris,* pp. 384–386.

Roa, D.P. 1959. *M.Sc. Thesis, Oregon State Univ.,* Corvallis.

Rodriguez-Kabana, R., and Curl, E.A. 1970. *Phytopathology* **60**, 65-69.
Rodriguez-Kabana, R., and Curl, E.A. 1972. *Fitopatologia* **6**, 7-17.
Rodriguez-Kabana, R., Curl, E.A., and Funderburk, H.H. 1968. *Can. J. Microbiol.* **14**, 1283-1288.
Rodriguez-Kabana, R., Curl, E.A., and Funderburk, H.H., Jr. 1969. *Phytopathology* **59**, 228-232.
Rodriguez-Kabana, R., Curl, E.A., and Peeples, J.L. 1970. *Phytopathology* **60**, 431-436.
Rodriguez-Kabana, R., Backman, P.A., and King, P.S. 1976a. *Plant Dis. Rep.* **60**, 255-259.
Rodriguez-Kabana, R., Backman, P.A., Karr, G.W., Jr., and King, P.S. 1976b. *Plant Dis. Rep.* **60**, 521-524.
Ross, H.F. 1952. *Proc. Fla. Soc. Soil Crop Sci* pp. 58-61.
Ruffin, J. 1974. *Bull. Ga. Acad. Sci.* **32**(3-4), 76-82.
Ruppel, E.G. 1975. *Phytopathology* **65**, 785-789.
Russel, P.E. 1975. *J. Appl. Bacteriol.* **39**, 175-180.
Russell, E.W. 1973. *In* "Soil Conditions and Plant Growth, 10th ed. p. 163. Lougman, London.
Saive, R. 1974. *Ann. Gembloux* **80**, 55-70.
Savage, K.E. 1973. *Weed Sci.* **21**, 285-288.
Savage, K.E., and Ivy, H.W. 1973. *Weed Sci.* **21**, 275-284.
Schultz, K.R., Fuhremann, T.W., and Lichtenstein, E.P. 1976. *J. Agric. Food Chem.* **24**, 296-299.
Sethunathan, N., and Pathak, M.D. 1972. *J. Agric. Food Chem.* **20**, 586.
Shabi, E., and Ben-Yephet, Y. 1976. *Plant Dis. Rep.* **60**, 451-454.
Sharma, L.N., and Saxena, S.N. 1974. *J. Indian Soc. Soil Sci.* **22**, 168-171.
Shaw, W.M., and Robinson, B. 1960. *Soil Sci.* **90**(5), 320-323.
Shephard, M.C., Bent, K.J., Woolner, M., and Cole, A.M. 1975. *Proc. Br. Insectic. Fungic. Conf., 8th, Brighton. Res. Rep.* (1-2) (*Rev. Plant Pathol.* 1976, **55**, 772.)
Shklyar, M.Z., Voevodin, A.V., and Beshanov, A.V. 1961. *Agrobiologija* **2**, 222-225.
Siedschlag, E.W., Jr, and Camper, N.D. 1976. *Weed Res.* **16**, 295-300.
Siegel, M.R. 1975. *Phytopathology* **65**(2), 219-220.
Simon, J.C., Jamet, P., Lemaire, J.M., and Jouan, B. 1973. *Sci. Agron. Rennes.* pp. 231-244.
Simon-Sylvestre, G. 1974. *Agrochimica* **18**(4), 334-343.
Simon-Sylvestre, G., and Chabannes, J. 1974. *Symp. Maize Novi-Sad* pp. 549-561.
Simon-Sylvestre, G., and Chabannes, J. 1975. *Ann. Agron.* **26**(1), 75-98.
Smiley, R.W., and Howard, R.J. 1976. *Plant Dis. Rep.* **60**, 91-94.
Smith, A.E. 1972. *Soil Sci.* **113**(1), 36-41.
Smith, M.S., and Weeraratna, C.S. 1975. *Pestic. Sci.* **5**, 721-729.
Smith, R.J., and Tugwell, N.P. 1975. *Weed Sci.* **23**, 176-178.
Sobieszczanski, J. 1970. *Colloq. Int. Gand Pestic. Microflore Sol* pp. 681-688.
Sobieszczanski, J., Rola, J., and Zurek, M. 1975. *Rocz. Glebozn.* **26**(2), 165-178.
Sokolov, N.S., and Knye, L.L. 1973. *Khim. Sel. Khoz.* **9**, 43-48.
Soulas, G., and Reudet, M.A. 1977. *Z. Pflanzenkr. Pflanzenschutz* **8**, 227-235.
Stanlake, G.J., and Clark, J.B. 1975. *Appl. Microbiol.* **30**(2), 335-336.
Statham, C.N., and Lech, J.J. 1976. *Toxicol. Appl. Pharmacol.* **36**, 281-296.
Staunton, W.P., and Kavanagh, T. 1975. *Proc. Br. Insectic. Fungic. Conf., 8th, Brighton Res. Rep.* (1-2) (*Rev. Plant Pathol.* 1976, **55**, 771.)
Steenbjerg, F., Larsen, I., Jensen, I., and Bille, S. 1972. *Plant Soil* **36**(2), 475-496.
Steenson, T.I., and Walker, N. 1956. *Plant Soil* **8**, 17-31.
Suess, A. 1970. *Z. Pflanzenkr. Pflanzenschutz.* **77**, 568-576.
Suess, A., Eben, C., Siegmund, H., and Pater, T. 1974. *In* "Comparative Studies of Food and Environmental Contamination," pp 373-384. International Atomic Energy Agency, Vienna.
Suriawiria, V. 1974. *Proc. Inst. Teknol. Bandung* **8**(1), 1-6.
Svampa, G., Brunelli, A., and Tosatti, E.M. 1974. *Inf. Fitopatol.* **12**, 5-13.

Swaroop, S. 1951. *Indian J. Med. Res.* **39**, 107-134.

Szegi, J. 1970. *Colloq. Int. Gand Pestic. Microflore Sol* pp. 559-561.

Szegi, J., Gulyas, F., Manninger, E., and Zamory-Bakondi, E. 1974. *Tr. Mezhdunar. Kongr. Pochvoved 10th* **3**, 179-184.

Szember, A., Gostkowska, K., and Furczak, J. 1973. *Pol. J. Soil Sci.* **6**(2), 141-147.

Szkolnik, M., and Gilpatrick, J.D. 1969. *Plant Dis. Rep.* **53**, 861-868.

Tabet, J.C.K., and Lichtenstein, E.P. 1976. *Can. J. Microbiol.* **22**, 1345-1356.

Taha, S.M., Mahmoud, S.A.Z., Abdel-Hafez, A.M., and Hamed, A.S. 1972. *Egypt. J. Microbial.* **7**(1-2), 53-61.

Talbert, R.E., Runyan, R.L., and Baker, H.R. 1970. *Weed Sci.* **18**, 10.

Talboys, P.W., and Davies, M.K. 1976. *Ann. Appl. Biol.* **82**, 41-50.

Tang, A., Curl, E.A., and Rodriguez-Kabana, R. 1970. *Phytopathology* **60**, 1082-1086.

Teuteberg, A. 1968. *Z. Pflanzenkr. Pflanzenschutz* **75**(2), 72-86.

Thiegs, B.J. 1955. *Down Earth* **11**, 2.

Thornton, H.G., 1955. *Rep. Rothamsted Exp. Stn.* pp. 66-67.

Toledo, A.D. 1974. *Biologico* **11**, 163-170.

Tolkachev, N.Z. 1974. *Din. Mikrobiol-Protsessov Pochve Obuslovlivayuschie Ec. Faktory, Mater. Simp.* **2**, 130-132.

Torstensson, L. 1974. *Swed. J. Agric. Res.* **4**, 151-160.

Torstensson, L. 1975. *Swed. J. Agric. Res.* **5**(4), 177-183.

Torstensson, N.T.L., Stark, J., and Göransson, B. 1975. *Weed Res.* **15**, 159-164.

Tramier, R., and Bettachini, A. 1974. *Ann. Phytopathol.* **6**, 231-236.

Trzecki, S., and Kowalski, E. 1974. *Zesz. Nauk Akad. Roln. Warsz. Roln.* **15**, 93-101.

Tu, C.M. 1970. *Appl. Microbiol.* **19**, 479-484.

Tu, C.M. 1972. *Appl. Microbiol.* **23**(2), 398-401.

Tu, C.M. 1973. *Can. J. Plant Sci.* **53**(2), 401-405.

Tu, C.M. 1975. *Arch. Microbiol.* **105**(2), 131-134.

Tu, C.M., and Bollen, W.B. 1968. *Weed Res.* **8**, 28-37.

Tuner, N.J., and Corden, M.E. 1963. *Physiol. Plant* **53**, 1388-1394.

Tyunyaeva, G.N., Minenkov, A.K., and Penkov, L.A. 1974. *Agrokhimija* **6**, 110-114.

Ujevic, I., and Kovacikova, E. 1975. *Polnohospodarst.* **21**(1), 31-37.

Ulasevich, E.J., and Drach, Y.O. 1971. *Mikrobiol. Zh.* **33**, 735-736.

Umetsu, N., Grose, F.H., Allahyari, R., Abu-El-Haj, S., and Fukuto, T.R. 1977. *J. Agric. Food. Chem.* **25**, 946.

Urusbaev, K. 1975. *Vestn. Skh. Nauki Kaz.* **18**(8), 45-48.

Valaskowa, E. 1968. *Pflanzenschutz* **38**, 135-146.

Van Der Zweep, W. 1970. *Proc. Br. Weed Control Conf., 10th* pp. 917-919.

Van Faassen, A.G. 1974. *Soil Biol. Biochem.* **6**(2), 231-233.

Van Thuyl, J.M., Davidse, L.C., and Dekker, J. 1974. *Neth. J. Plant Pathol.* **80**, 165-168.

Velev, B., and Rankov, V. 1975. *Rocz. Glebozn.* **26**(2), 223-232.

Verstraete, W., and Voets, J.P. 1974. *Meded. Fac. Landbouwwet. Rijksuniv. Gent* **39**(2), 1263-1277.

Vlassak, K., and Livens, J. 1975. *Sci. Total Environ.* **3**(4), 363-372.

Voets, J.P., and Vandamme, E. 1970. *Colloq. Int. Gand Pestic. Microflore Sol* pp. 563-580.

Voets, J.P., Meerschman, P., and Verstraete, W. 1974. *Soil Biol. Biochem.* **6**, 149-152.

Voinova, Z., Petkova, P., and Kostov, O. 1975. *Rocz. Glebozn.* **26**(2), 207-211.

Wagenbreth, D., and Kluge, E. 1971. *Arch. Pflanzenschutz* **7**, 451-459.

Waid, J.S. 1972. *Residue Rev.* **44**, 65-71.

Wainwright, M., and Pugh, G.J.F. 1973. *Soil Biol. Biochem.* **5**, 557-584.

Wainwright, M., and Pugh, G.J.F. 1974. *Soil Biol. Biochem.* **6**(4), 263-267.

Wainwright, M., and Pugh, G.J.F. 1975. *Plant Soil* **43**(3), 561-572.
Walker, A. 1970. *Hortic. Res.* **10**, 45-49.
Walker, A. 1973. *Weed Res.* **13**(4), 416-421.
Walker, R.L., and Newman, A.S. 1956. *Appl. Microbiol.* **4**, 201-206.
Wallnöfer, P., Poschenrieder, G., and Engelhardt, G. 1977. *Z. Pflanzenkr. Pflanzenschutz* **8**, 199-207.
Wanic, D., and Kawecki, Z. 1971. *Rocz. Nauk Roln. Ser. A* **97**(4), 115-126.
Wanic, D., and Kawecki, Z. 1972. *Rocz. Nauk Roln. Ser. A* **98**(1), 175-187.
Wargo, P.M. 1972. *Phytopathology* **62**, 1278-1283.
Warren, C.G., Sanders, P., and Cole, H. 1974. *Phytopathology* **64**, 1139-1142.
Warren, C.G., Sanders, P., and Cole, H. 1975. *Phytopathology* **65**, 836.
Weeks, R.E., and Hedrick, H.G. 1971. *Bacteriol. Proc.* **A**, 78.
Wegrzyn, T. 1975. *Rocz. Glebozn.* **26**(2), 91-97.
Weinberger, M., and Bollag, J.M. 1972. *Appl. Microbiol.* **24**, 750-754.
Welvaert, W. 1974. *Agro-Ecosystems* **1**(2), 157-168.
Whiteside, J.S., and Alexander, M. 1960. *Weeds* **8**, 204-213.
Wicks, T. 1976. *Plant Dis. Rep.* **60**, 818-819.
Wiese, I.H., and Basson, N.C.J. 1966. *Agric. Res.* **1**, 58-60.
Wild, B.L., and Rippon, L.E. 1975. *Phytopathology* **65**, 1176-1177.
Wilkinson, V., and Lucas, R.L. 1969. *New Phytol.* **68**, 701-708.
Williams, R.J., and Ayanaba, A. 1975. *Phytopathology* **65**, 217-218.
Wolf, D.C., Bakalivanov, D., and Martin, J.P. 1975. *Rocz. Glebozn.* **26**(2), 35-48.
Wright, S.J.L. 1972. *Chemosphere* **1**(1), 11-14.
Wright, K.A., and Cain, R.B. 1969. *Soil Biol. Biochem.* **1**, 5-14.
Wright, S.T.L., and Foster, A. 1972. *Soil Biol. Biochem.* **4**, 207-213.
Wuest, P.J., Cole, H., and Sanders, P.L. 1974. *Phytopathology* **64**, 331-334.
Wyse, D.L., Meggitt, W.F., and Penner, D. 1976a. *Weed Sci.* **24**, 5-10.
Wyse, D.L., Meggitt, W.F., and Penner, D. 1976b. *Weed Sci.* **24**, 11-15.
Yakovlev, V.G., and Stenina, N.P. 1974. *Sb. Nauchn. Tr-Leningr. Nauchno-Issled. Inst. Lesn. Khoz.* **18**, 111-117.
Yoder, K.S., and Klos, E.J. 1976. *Phytopathology* **66**, 918-923.
Yurkevich, I.V., and Tolkachev, N.Z. 1972. *Khim. Sel. Khoz.* **9**, 56-58.
Zharasov, Sh. U., Tsukerman, G.M., and Chulakov, Sh.A. 1972. *Khim Sel. Khoz.* **8**, 57-59.
Zubets, T.P. 1973. *Nauchn. Tr. Sev-Zapadn. Nauchno-Issled Inst. Sel. Khoz.* **24**, 103-109.

ADVANCES IN AGRONOMY, VOL. 31

FACTORS AFFECTING ROOT EXUDATION II: 1970-1978

M. G. Hale and L. D. Moore

Department of Plant Pathology and Physiology
Virginia Polytechnic Institute and State University
Blacksburg, Virginia

I. Introduction

The root in its soil environment is part of a complex, interacting system. One of the ways the root interacts is through the release of organic matter, which may vary from simple organic molecules to cells and tissues that are sloughed in the process of growth. The roles of chemicals in organismic interactions have been classified into the general categories of nutrients, foods, and allelochemics (Whittaker and Feeney, 1971). In an ecological sense the intricate pattern of exchanges of materials relates the organisms of a community to the environment and to each other. Such intricate interactions occur in the rhizosphere of plants and involve root exudates. Root exudates are also involved in chelation and solubilization of nutrients, in soil aggregation, and in effects on the pH of the soil solution.

Soil microbiologists were the first to draw attention to the differences in environment for microorganisms close to the root surface and at some distance from the root. Hiltner (1904) is given credit for coining the term "rhizosphere," which is the sphere of influence of the root on its environment for a distance of at least a millimeter or two from the root surface. Agronomists have since become aware of "soil sickness" problems in the rotation of crops in which it is postulated that there is a chemical effect of residual compounds from previous crops (Muller, 1966).

The intriguing idea that populations of rhizosphere organisms can be controlled by foliar applications of chemicals, which are either translocated to the roots and exuded into the rhizosphere intact, or which change the pattern of exudation and in this manner affect the resistant of the plant, has excited plant pathologists and soil microbiologists.

In the last two decades a large body of literature has been developed on root and other plant part exudates. Understandably, a large percentage of the experiments have been conducted under controlled environmental conditions, usually with axenic (aseptic) seedlings in nutrient solutions. Those experiments conducted under natural environmental and soil conditions have usually involved the assay of microorganism populations in rhizosphere and nonrhizosphere soils. The use of $^{14}CO_2$ to label plant metabolites and to trace their movement from leaves to roots and into the rhizosphere has begun to yield valuable information. New quantitative estimates of carbon lost via the roots indicate that the quantities are far greater than have been estimated previously. Such experiments have led to refinements of the term "exudate" to denote the source of the organic compounds and whether they are water-soluble or insoluble; diffusible or nondiffusible; gaseous or volatile. One begins to hope that meaningful quantitative determinations of root exudates can be made *in situ* and that evaluation can be made of the effect of exudation on the carbon balance of the plant as well as on the ecology of the rhizosphere. The data accumulated have been substantive enough that loss of organic compounds or organic matter into the soil from roots as they grow cannot be ignored in interpretation or design of experiments.

Brief reviews of exudation include those of Tinker and Sanders (1975) and Lespinat and Berlier (1975) on factors affecting exudation. Bowen and Rovira (1976) have reviewed exudation in relation to root diseases. Hale *et al.* (1978) reviewed the principles and concepts of root exudation. The relationship of exudates to soil microorganisms has been comprehensively covered in the two volumes edited by Dommergues and Krupa (1978; Krupa and Dommergues, 1978). Since the review of Hale *et al.* (1971), the factors affecting root exudation have not been reviewed comprehensively.

The present review covers contributions to our knowledge and understanding of the factors affecting root exudation that have been published during the period

1970-1978. Since there has been increasing interest in seed exudation and reevaluation of seedling root exudates in comparison with seed exudates, the factors affecting seed exudation are included.

II. Plant Factors

A. GROWTH

1. Root Caps and Root Hair Secretion

Unevenly distributed layers of granular and fibrillar material (mucilage) cover the outer surface of roots and root hairs. Mucilage was found on the roots of sixteen plant species examined by Greaves and Darbyshire (1972); it seemed to fill the space between the root and soil particles. Under nonaxenic conditions, some of the mucilage may be of microbial origin. The nature of the mucilage or slime is complex, but it seems to be composed primarily of polysaccharides. Floyd and Ohlrogge (1971) collected droplets from the nodal roots of corn (Table I) every 2 to 6 hours. Analysis of the hydrolyzate revealed the presence of galactose, arabinose, xylose, fucose, and uronic acid in the ratio of 7:8:5:11:3. Some acid phosphatase and adenosine triphosphatase were also present. Slime from the primary root tips of corn seedlings contained, in addition, the sugars mannose, glucose, and rhamnose (Jones and Morre, 1973). In the unhydrolyzed slime, galactose occurred in the largest quantities.

Paul et al. (1975) estimated the slime production rate of excised root tips of corn 1-5 cm long to be 0.79 μg hr^{-1} per root tip in a medium containing 40 mM sucrose, 0.5 Hoagland's nutrient, and boric acid. Fucose made up 39%, galactose 30%, and arabinose plus xylose 22% of slime sugars. Fucose occurred in the slime of all cultivars studied. Since fucose occurred only in the slime polysaccharide, its presence could be used as a marker for slime production (Bowles and Northcote, 1972). Fucose apparently occurred only in the membrane fraction of maize seedling root tips (Bowles and Northcote, 1972).

The hypothesis of Jones and Morre (1967) that the Golgi apparatus in corn is involved in the secretion of root cap slime was confirmed in detailed biochemical studies by Paul and Jones (1975 a,b, 1976). The slime originated in a layer one to three cells deep at the root cap surface. It accumulated between the plasmalemma and the cell wall and was eventually lost to the exterior of the cell wall.

Leppard (1974) used the scanning electron microscope to discover microfibrils in the rhizoplane of the wheat root tip between the root cap and the root hair zone. The microfibrils were determined to be polygalacturonic acids and because of their physical structure could form microhabitats for microorganisms. Leppard and Ramamorthy (1975) have suggested a role for the microfibrils in ion uptake.

TABLE I
Common Name, Scientific Name, and Reference for Plants Mentioned by Common Name in the Text

Common name	Scientific binomial	Citations
Alfalfa	*Medicago sativa* L.	Hamlen *et al.* (1972, 1973)
		Rao (1976)
Barley	*Hordeum vulgare* L.	Barber and Gunn (1974)
		Barber and Lee (1974)
		Barber and Martin (1976)
Bean	*Phaseolus vulgaris* L.	Manning *et al.* (1971)
		Vancura and Stanek (1975)
		Vancura and Stotzky (1976)
		Wyse *et al.* (1976)
Bermuda grass	*Cynodon dactylon* (L.) Pers.	Singh and Singh (1971)
Broom rape	*Orobanche ramosa* L.	Ballard *et al.* (1978)
		Hameed *et al.* (1973)
Cabbage	*Brassica oleracea* L.	Vancura and Stotzky (1976)
Corn	*Zea mays* L.	Barber and Gunn (1974)
		Barlow (1974)
		Bowles and Northcote (1972)
		Clowes and Woolston (1978)
		Floyd and Ohlrogge (1971)
		Hussain and Vancura (1970)
		Jones and Morre (1973)
		Kohl and Matthaei (1971)
		Paul and Jones (1975a,b, 1976)
		Paul *et al.* (1975)
		Vancura *et al.* (1977)
		Vancura and Stotzky (1976)
Cotton	*Gossypium herbeum* L.	Vancura and Stotzky (1976)
	Gossypium hirsutum L.	Booth (1974)
Cucumber	*Cucumis sativa* L.	Vancura and Stotzky (1976)
Flax	*Linum usitatissiumum* L.	Ballard *et al.* (1978)
		Hameed *et al.* (1973)
Gum	*Eucalyptus calaphylla* R. Br.	Malajczak and McComb (1977)
	Eucalyptus marginata Donn ex. Sm.	Malajczak and McComb (1977)
Hyacinth bean	*Dolichos lablab* L.	Bhat *et al.* (1971)
Lupine	*Lupinus angustifolia* L.	Young *et al.* (1977)
Margosa	*Azadirachta indico* Juss.	Alam *et al.* (1975)
Marigold	*Tagetes erecta* L.	Alam *et al.* (1975)
		Hameed (1971)
		Hameed *et al.* (1973)
Pea	*Pisum sativum* L.	Beute and Lockwood (1968)
		Brannstrom (1977)
		Christenson and Hadwigh (1973)
		Short and Lacy (1976)
		Tietz (1975)
		Van Egaraat (1975a,b)
		Vancura and Stotzky (1976)
Peanut	*Arachis hypogaea* L.	Griffin *et al.* (1976)
		Hale and Griffin (1974)

96

TABLE I—*Continued*

Common name	Scientific binomial	Citations
		Hale *et al.* (1977)
		Thompson (1978)
Pine, lodgepole	*Pinus contorta* Dougl. ex Loud.	Reid and Mexal (1977)
Ponderosa	*Pinus ponderosa* Laws	Reid (1974)
		Harley (1969)
		Vancura and Stotzky (1976)
Scots	*Pinus sylvestris* L.	Krupa and Nylund (1972)
		Krupa and Fries (1971)
Rattlebox	*Crotalaria medicaginea* Lank	Sullia (1973)
Red clover	*Trifolium pratense* L.	Bonish (1973)
Red pepper	*Capsicum annum* L.	Alagianagalingen and
		Ramakrishnan (1972)
Rice	*Oryza sativa* L.	Asanuma *et al.* (1978)
		Yoshida and Takashi (1974)
Sorghum	*Sorghum vulgare* Pers.	Balasubramanian and
		Rangaswami (1973)
		Ballard *et al.* (1978)
		Hameed *et al.* (1973)
		Werker and Kislev (1978)
Soybean	*Glycine max* L.	Lee and Lockwood (1977)
		Shapovalov (1972)
		Tingey and Blum (1973)
Squash	*Cucurbita pepo* L.	Magyarosy and Hancock (1974)
		Vancura and Stotzky (1976)
Sugar maple	*Acer saccharum* Marsh.	Smith (1970, 1972)
Sunnhemp	*Crotalaria juncea* L.	Balasubramanian and
		Rangaswami (1973)
Tobacco	*Nicotiana tabacum* L.	Ballard *et al.* (1978)
		Joyner (1975)
Tomato	*Lycopersicon esculentum* Mill.	Ballard *et al.* (1978)
		Hameed *et al.* (1973)
		Vancura and Stotzky (1976)
		Wang and Bergeson (1974)
Urid	*Phaseolis mungo* L.	Rao (1976)
Western wheat grass	*Agropyron smithii* Rydb.	Bokhari and Singh (1974)
Wheat	*Triticum aestivum* L. & Thell.	Ayers and Thornton (1968)
		Barber and Martin (1976)
		Bowen and Rovira (1973)
		Holden (1975)
		Jalali (1976)
		Jalali and Suryanarayana (1971,
		1972, 1974)
		Leppard (1974)
		McDougall (1970)
		McDougall and Rovira (1970)
		Martin (1977a,b)
		Rovira and Ridge (1973)
		Srivastava and Mishra (1971)
		Vancura *et al.* (1977)
		Vagnerova and Macura (1974)

In sorghum, a fibrillar, mucilaginous layer was found on the surface of the root hairs (Werker and Kislev, 1978). A pectic material seemed to arise from the endoplasmic reticulum, but the fibrillar material outside the cell walls arose from the Golgi bodies and mitochondria.

2. Sloughage

As plant roots grow both in length and in diameter, some of the outer tissues are sloughed and decompose either by autolysis or through the activities of microorganisms. Sauerbeck and Johnson (1976) suggested that the total rhizosphere deposition amounts to three to four times as much organic substance as is found in the roots at harvest and postulated that such deposition leads to intensive turnover of organic matter in the rhizosphere. Martin (1977a) stated that there was a significant formation of soil organic matter during active growth of wheat roots and concluded that the contribution directly from roots without the intervention of microorganisms has been underestimated. He estimated that as much as 8% of the shoot carbon ended up in rhizosphere organic matter and that most of the carbon lost from the plant was the result of autolysis of cortical tissue. Three stages were outlined (Martin 1977b) for root decomposition of wheat: (a) continuing release of low-molecular-weight constituents from degenerate epidermal and cortical tissue and sloughed root cap material starting before tillering and continuing to the flowering stage; (b) invasion of epidermal and cortical tissue by soil microflora, causing breakdown of cell walls; and (c) decomposition of epidermal tissue following death. The term "root exudate" is misleading, as it has been used and should be reserved for those water-soluble and diffusible compounds lost from roots (Martin, 1977b). Additional terms should be used specifically to designate sources of organic compounds such as root lysate, mucigel, cell wall residues, and intact plant cells. Similar thoughts led Warembourg and Morrall (1978) to conclude that use of $^{14}CO_2$ to label metabolites and measure carbon loss from roots may be a more reliable way of estimating exudation than collecting exudates and sloughed material over a long period of time. However, Griffin et al. (1976) were able to measure quantitatively the amounts of sloughed cortical tissue and root caps from axenic peanut roots growing in nutrient solution. Microscopic observations of roots indicated that many sloughed cells and tissue fragments remained loosely attached to the roots even after a sonication treatment, so that collection of the sloughed material resulted in an underestimation of the amount. Carbon, hydrogen, and nitrogen content of the sloughed material was calculated and found to be 15.2%, 3.0%, and 1.3%, respectively. For 1.5 mg of sloughed material per gram of root dry weight per week, the loss amounted to 510 μg of carbon, 86 μg of hydrogen, and 66 μg of nitrogen. Bowen and Rovira (1973) have indicated that the insoluble mucilaginous exudate of wheat roots, including sloughed root cap cells, accounted for 0.8–1.6% of the root carbon and 80% of the total carbon released into soil.

Regeneration of sloughed corn root caps appears to result from the increased activity of the quiescent center in the apex of the root (Barlow, 1974). Barlow raised questions concerning the mechanism by which regeneration was controlled and the sort of messenger involved in initiating activity of the quiescent center after root cap sloughing and attributed it to stresses imposed by the root cap cells. Clowes and Woolston (1978) estimated the sloughage into water of root cap cells of primary seedling roots of corn at different densities of roots in the medium. When root density ranged from 50 to 250 roots per liter, sloughed root cap cells ranged from 7000 to 3000 cells per root per day. Lower root densities increased the sloughage, as did more frequent changes of water surrounding the roots. One wonders if a dense population of roots and less frequent water changes resulted in less agitation and dislodgement of sloughing cells rather than any effect of root density on the sloughing process per se.

3. Type of Root System

Monocotylendenous plants, dicotyledenous plants, and gymnosperms have been used in exudation studies. Most work has been done on primary seedling roots or adventitious roots (Hale *et al.*, 1978). Smith (1970) invented techniques by which he could study exudation from lateral root tips of sugar maple. Reid (1974) examined exudation from roots of 9- and 12-month-old Ponderosa pine and 7-year-old lodgepole pine (Reid and Mexal, 1977). The mother roots of lodgepole pine exuded four times as much ^{14}C compounds as did the mycorrhizal roots, even though the mycorrhizal roots accumulated more ^{14}C from the photosynthesizing shoots. In terms of older suberized roots of other species contributing to the rhizosphere organic matter, no information is available.

In a complex natural situation it is conceivable that the type of root system could be a factor dependent on the variations of environment in which the parts of the system were growing. For example, different parts of the root system could be subjected to different moisture stresses and thus different kinds of exudation. It is conceivable also that the sloughing pattern is different in those roots in which an active cambium gives rise to secondary tissues than it is in those roots or parts of roots that do not have secondary tissues.

B. INJURY

Mechanical injury occurs during the normal process of lateral root growth as the laterals force or digest their way from the site of initiation through the cortical cells to the root surface (Esau, 1953). McDougall and Rovira (1970) reported that the lateral root zone of wheat roots was a region of exudation, but the exudation came primarily from the emerging lateral root tips and not from the rupture of cells and tissues of the older root. Furthermore, microscopic observa-

TABLE II
Sources and Possible Causes of Injury Resulting in Increased Exudation[a]

Source	Cause
Microflora and microfauna	Permeability changes, dissolution by enzymes, lysis, puncture, releases of toxins and growth substances
Cultural practices	Pesticides, cultivation, fertilizers, water stress, mineral nutrient deficiency or toxicity
Growth	Lateral root eruption, abrasion by soil particles, sloughing off tissues and root caps, pressure from cambial activities, permeability changes with age and stage of development, regrowth of injured roots
Environmental stress	Water stress, O_2 concentration, CO_2 concentration, temperature extremes, pH, salt concentration

[a] Reprinted with permission from M. G. Hale et al. (1978).

tion of the emerging root tips and careful handling of roots during the experiments suggested that exudation by the root apices was not because of injury. On the other hand, Van Egeraat (1975a) reported that exudation of ninhydrin-positive compounds from pea roots was greater at sites of lateral root emergence, probably because of injury. Van Egeraat showed that homoserine was abundantly present in roots but not in root exudates except where injury occurred either by emerging lateral roots or by damaging roots with a needle. Previously Ayers and Thornton (1968) had demonstrated that wheat roots growing in sand consistently exuded greater amounts of ninhydrin-reacting compounds than did roots growing in nutrient solutions. They attributed the difference to abrasive injury caused by the sand particles. Hale and Griffin (1974) used peanut fruits in various stages of development to demonstrate increased exudation as a result of mechanical injury. Peanut fruits were scarified over a quarter of the surface of each fruit, and the amount of sugars released into an ambient solution was measured. In a 24-hour period over 100 times as much sucrose was exuded from injured immature fruits and 24 times as much from injured mature fruits as from the comparable uninjured fruits. Some sources and causes of injury that might lead to increases in exudation are listed in Table II.

C. ONTOGENETIC DEVELOPMENT

The age and stage of development of plants are factors that affect the amount and kind of exudates, as noted in previous reviews (Rovira, 1969; Hale et al., 1971, 1978). Some of the more recent contributions to our knowledge center around distinguishing between exudates emanating from seeds and seedlings.

As dry seeds imbibe water, they release gaseous and volatile compounds identified as ethanol, methanol, formaldehyde, acetaldehyde, formic acid, ethylene, and propylene (Vancura and Stotzky, 1976). Dried or autoclaved seeds did not evolve the compounds, but pulverized wetted seed material did. Imbibing seeds of bean, corn, and cotton released larger quantities than did imbibing seeds 'of cucumber, squash, and cabbage (Table III). Most of the exudation occurred in the first 2 days during germination and preceded the appearance of the radicle. Evolution of volatiles was independent of alternating light and darkness but appeared to be inversely related to seed size (Stotzky and Schenk, 1976). It had been shown in earlier work that amounts of nonvolatile exudates from seeds are directly correlated with seed size (Vancura and Hanzlikova, 1972).

The age of stored seeds can significantly affect exudation as they imbibe water. Short and Lacy (1976) examined miragreen pea seeds and found 10 times as much exudate from 8-year-old seeds as from 1-year-old seeds. Highest rates of exudation occurred from the micropyle, but significant amounts exuded through the seed coat from the cotyledons.

The contribution of cotyledons to root exudation from seedlings has been studied by Vancura and Stanek (1975). The effects of cotyledon and primary leaf removal from bean plants led them to conclude that compounds stored in the cotyledons moved to the roots and that the exudates reflected this movement. The amount of exudation of cucumber plants up to the twenty-fourth day was always lower when cotyledons were removed and was always higher when

TABLE III
Summary of Volatile and Gaseous Metabolites Released by Germinating Seeds[a-d]

Seed	Methanol	Ethanol	Formaldehyde	Formic acid	Ethylene	Propylene
Bean	+	+	+	0	+	+
Cabbage	0	+		+		
Corn	+	+		+	+	+
Cotton	+	+		0	+	+
Cucumber	0	+	0	0		
Pea	+	+	+	+	+	+
Pinus caribea	+	+	0	0		
Pinus palustris	+	+	+	+		
Pinus ponderosa	+	+		+		
Pinus taeda	+	+		+		
Radish	0	+		+		
Red alder	0	+		0		
Squash	0	+		+		
Tomato	+	+		+		

[a] Reprinted with permission from Vancura and Stotzky (1976).
[b] + indicates compound present, 0 not present.
[c] Ethanol present for all seeds.
[d] Ethylene and propylene found in all seeds examined (four species).

TABLE IV
Scientific Names of Bacteria, Fungi, and Nematodes Mentioned in the Text

Scientific binomial	Citations
Bacteria	
Pseudomas putida (Trevisan) Migula	Vancura *et al.* (1977)
Rhizobium spp.	Allen (1973)
	Currier and Strobel (1976)
	Rao (1976)
Rhizobium leguminosarum Frank em. Baldwin & Fred	Van Egaraat (1975)
Xanthomonas phaseoli var. *fuscans* (Burkh) Starr & Burkh	Vancura and Stanek (1975)
Fungi	
Alternaria humicola Oudemans	Sullia (1973)
Aspergillus niger van Tiegham	Sullia (1973)
Beijerinckia sp.	Bhat *et al.* (1970)
Boletus variegatus Fr.	Krupa and Fries (1971)
	Krupa and Nyland (1972)
Endogone sp.	Jalali and Domsch (1975)
Fomes annosus (Fr.) Cke.(*Fomatopsis annosus*)	Krupa and Nyland (1972)
Fusarium solani f. sp. *cucurbitae* Snyd. & Hans.	Magyarosy and Hancock (1974)
Fusarium solani f. sp. *pisi* (F. R. Jones) Snyd & Hans.	Beute and Lockwood (1968)
Helminthosporium sativum Pam., King. & Bakke	Jalali and Suryanarayana (1971)
Penicillium citrinum Thom	Hameed (1971)
Penicillium herquei Bainier & Sartory	Sullia (1973)
Penicillium simplicissimum (Oud.) Thom	Hameed (1971)
Phytophthora cinnamomi Rands	Malajczuk and McComb (1977)
Puccinia graminis f. sp. *tritici* Eriks. and E. Henn.	Srivastava and Mishra (1971)
Thielaviopsis basicola (Berk. & Be.) Ferr.	Lee and Lockwood (1977)
Trichoderma harzianum Rifai	Joyner (1975)
Trichoderma lignorum Rifai	Sullia (1973)
Trichoderma viride Pers. ex. Fr.	Brannstrom (1977)
Verticillium albo-atrum Reinke & Berth.	Booth (1974)
Nematodes	
Helicotylenchus indicus Siddiqi	Alam *et al.* (1975)
Hoplolaimus indicus Sher.	Alam *et al.* (1975)
Meloidogyne incognita (Kofoid & White) Chitwood	Alam *et al.* (1975)
	Hamlen *et al.* (1973)
	Wang and Bergeson (1974)
Rotylenchulus reniformis Linford & Oleviera	Alam *et al.* (1975)
Tylenchorhynchus brassicae Siddiqi	Alam *et al.* (1975)
Tylenchus filiformis Butschi	Alam *et al.* (1975)

TABLE V
Influence of Age on Effectiveness of Root Exudeates
on Percentage of Germination of *Orabanche Ramosa*
Seeds[a]

Plant	5 weeks	7 weeks
Flax	5.0	7.5
Sorghum	5.2	8.2
Tomato	1.8	2.2
Marigold	0.0	0.0

[a] Reprinted with permission from Hameed *et al.*
(1973).

primary leaves were removed. As reserves in the cotyledons are depleted, changes in exudation over time are important with respect to pathogen attack (discussed in Section V,B,2,b). For example, increases in exudation of isoleucine and glutamic acid occurred beginning on the twenty-fourth day. Disappearance of the pathogen *Xanthomonas phaseoli* (Table IV) from the bean root rhizosphere was attributed to the disappearance of glutamic acid in the root exudates.

As alfalfa plants aged, there were changes in the carbohydrate content of the exudates (Hamlen *et al.*, 1972). Initially, significant increases in the pentoses, arabinose and xylose, were observed, but such increases were not observed in exudates of plants 8 weeks old or older. Also, glucose, fructose, and mannose increased in concentration from the fourth to the sixth week, but fructose and mannose disappeared by the tenth week.

The age of the plant affects the exudation of substances that influence germination of *Orabanche* seeds (Hameed *et al.*, 1973) (Table V). Further studies of effects of exudates on *Orabanche* seed germination (Bailard *et al.*, 1978) showed that active exudates could be leached from the rooting medium of flax 3 days after transplanting, but only after 15 days from transplanting for tomato, tobacco, and sorghum. The maximum effect on germination occurred after 20 to 25 days. This difference in time of appearance of active exudates from the various species after transplanting could be a factor in resistance to parasitism by *Orabanche*.

III. Effects of Environmental Factors

A. INDIRECT EFFECTS

Photosynthates may appear as root exudates in the rhizosphere in a variety of organic forms, which indicates that the translocated material is usually acted

TABLE VI
The Influence of Defoliation on the Quantities of Compounds Exuded by Roots
of Sugar Maples[a,b]

Compound	1969[c]		1970[d]	
	Control	Defoliated	Control	Defoliated
1. Carbohydrates				
Fructose	2.6 ± 0.2	4.3 ± 0.1[e]	4.3 ± 1.2	6.1 ± 0.9
Glucose	0	0	Trace	Trace
Sucrose	7.3 ± 0.1	2.9 ± 0.2[e]	7.9 ± 0.9	6.0 ± 0.3
2. Amino acids/amides				
Alanine	0	0	1.8 ± 0.4	1.7 ± 0.1
Cystine	0.2 ± 0.1	0.5 ± 0.1	0	0
Glutamine	2.3 ± 0.3	3.4 ± 0.5	3.6 ± 0.7	4.3 ± 1.1
Glycine	0.5 ± 0.1	0.3 ± 0.1	1.9 ± 0.3	1.1 ± 0.3
Homoserine	1.1 ± 0.2	0.3 ± 0.1[e]	2.4 ± 0.6	1.8 ± 0.3
Lysine	0.8 ± 0.1	1.8 ± 0.2[e]	0	0.5 ± 0.2[e]
Methionine	0	0	1.5 ± 0.1	1.4 ± 0.4
Phenylalanine	0	0	2.7 ± 0.6	3.1 ± 0.5
Threonine	Trace	Trace	3.4 ± 0.3	0[e]
Tyrosine	0	0	0.9 ± 0.7	1.1 ± 0.2
3. Organic acids				
Acetic	49.7 ± 10.1	24.3 ± 9.4	63.2 ± 11.1	58.1 ± 13.3
Malonic	0	0	Trace	0

[a] Reprinted with permission for Smith (1971).
[b] Data are micrograms × 10^{-1} of each material released during 14 days per milligram oven-dry root.
[c] Mean and standard error of three replicate determinations using one composite exudate sample from 19 and 20 roots of control and defoliated tree, respectively.
[d] Mean and standard error of three replicate determinations using one composite exudate sample from 17 and 23 roots of control and defoliated tree, respectively.
[e] Control and defoliated figures significantly different at 95% level.

upon metabolically before it appears in exudates. Those factors that affect rates of photosynthesis and translocation will have an indirect effect on exudation. For a more thorough discussion of sources and mechanisms of exudation, see Hale *et al.* (1978).

A few investigations have appeared since the previous review (Hale *et al.* 1971) relating to the environmental effects of temperature and light on shoots with consequent changes in root exudates. For example, Shapovalov (1972) explained the effect of temperature on exudation of scopoletin from soybean and oat roots by setting apart three stages based on the Q_{10} of the exudation rate. In the stage 20-24°C, the process appeared to be one of diffusion from free space; from 23 to 30°C, he claimed the process activated diffusion, probably across the plasmalemma; and in the range of 40-60°C, exudation probably increased sharply as a result of denaturation of protein.

The complexity of temperature effects is compounded because of changes in rates of photosynthesis and in rates of translocation of photosynthates to the roots, in rates of enzymatic reactions that synthesize or degrade photosynthate, and in changes in membrane permeability which may occur. For these reasons and others it is difficult to establish a pattern for effects of environment on exudation.

Two interesting investigations need to be mentioned. Smith (1972) defoliated sugar maple trees and measured the changes in exudation. Differences between defoliated and nondefoliated trees were quantitative. Defoliated trees released greater quantities of fructose, cystine, glutamine, lysine, phenylanine, and tyrosine, whereas foliated trees exuded greater amounts of sucrose, glycine, homoserine, methionine, threonine, and acetic acid (Table VI). Many of the differences were not statistically significant, but considering the difficulty of obtaining the amount of quantitative data presented as well as the conditions under which it was obtained, the results are quite interesting.

Using more controlled conditions and a different plant, Bokhari and Singh (1974) examined the effect of clipping and temperature on exudation from western wheat grass. Severe clipping and high temperature stimulated root exudation, with more being exuded in the initial stages of growth than in the later stages of growth. Over a period of 80 days, 1 g dry weight of roots exuded 4.5–6.5 mg of reducing sugars. In terms of carbon balance in a sward system, grazing would apparently cause a greater carbon loss through the roots than would nongrazing (Table VII).

TABLE VII
Root Exudation Expressed as Total Nonstructural Carbohydrate (TNC) Equivalents[a]

Clipping	0–10	10–20	20–30	30–40	40–50	50–60	60–70	70–80	Total
				Days					
	12 hours, day temperature 13°C, night temperature 7°C								
Control	0.595[b]	0.465	0.500	0.432	0.412	0.426	0.407	0.400	4.483
Moderate	0.631	0.560	0.551	0.515	0.473	0.487	0.487	0.476	5.089
Severe	0.595	0.519	0.534	0.515	0.513	0.539	0.538	0.524	5.098
	12 hours, day temperature 24°C, night temperature 13°C								
Control	0.553	0.470	0.381	0.474	0.488	0.511	0.500	0.469	4.640
Moderate	0.519	0.469	0.458	0.486	0.487	0.512	0.500	0.489	4.614
Severe	0.583	0.550	0.530	0.513	0.527	0.538	0.548	0.546	5.001
	12 hours, day temperature 29.5°C, night temperature 18°C								
Control	0.868	0.700	0.637	0.560	0.567	0.576	0.585	0.558	6.092
Moderate	0.794	0.645	0.634	0.603	0.625	0.597	0.608	0.589	6.140
Severe	0.875	0.727	0.729	0.678	0.656	0.628	0.635	0.615	6.543

[a] Reprinted with permission from Bokhari and Singh (1974).
[b] Data are milligrams per gram dry weight of roots.

B. WATER STRESS

Experiments involving direct quantitative measures of the effects of water stress on exudation have been few since Vancura (1964) showed that by allowing a root system to develop water stress and then irrigating he could cause an increase in root exudation. The roots of plants growing in the field are continually exposed to alternately water-stressed and release-of-stress conditions. Methods must be devised to measure the effects of the cycles and their relationship to microbial colonization of roots.

In elegant studies of water stress on exudation from Ponderosa pine (Reid, 1974) and lodgepole pine (Reid and Mexal, 1977), polyethylene glycol (PEG 4000) was used to establish gradients of stress. Relationships between exudation, water stress, and mycorrhizae were investigated. Decreasing water potentials in the rooting medium caused a reduction in the amount of $^{14}CO_2$ absorbed by the leaves, probably as a result of stomatal closure. Decreased absorption might also account for the decrease in translocation of ^{14}C to the roots at the lower water potentials, and it might also account for decreased exudation. For Ponderosa pine no label appeared in exudates at water potentials below -2.6 bars. The amounts of ^{14}C exuded peaked at 3 days after exposure to $^{14}CO_2$ for both Ponderosa and lodgepole pine. Of the three treatments, 0, -2, and -4 bars (Reid and Mexal, 1977), the average cumulative exudation over the first 6 days was greatest at -4 bars and least at -2 bars, but when exudation was expressed as a proportion of the total ^{14}C translocated to the roots, exudation was greatest at 0 bar and somewhat less at -2 and -4 bars. These results may be misleading because of the low oxygen concentration in the rooting medium.

Effects of water stress on exudation are not clearly defined in the literature, but it is apparent that there is an effect on both the amounts and kinds of exudates and that this factor must be considered in interpretation of results from exudation studies.

C. HYDROGEN ION CONCENTRATION

A change in pH affected the exudation of ^{14}C applied as $^{14}CO_2$ to the atmosphere surrounding the shoots of 8-day-old wheat plants. At pH 5.9, ^{14}C-containing compounds exuded accounted for 20,306 dpm. At pH 6.4 the count was reduced to 7057, and at pH 7.0 to 8595 (McDougall, 1970). Rovira and Ridge (1973) found that addition of acetate buffer at pH 5 greatly increased exudation. They attributed the increase to the acetate and not to a pH effect.

In examining exudation of cellulose by red clover roots, Bonish (1973) found that salts decreased exudation of the enzyme to negligible amounts at pH below 5.5, but exudation increased as the pH rose. The general effects of pH on exudation need further study.

D. ANAEROBIOSIS

Because of its effect on the basic metabolism of the root, oxygen deficiency can cause changes in the kinds of compounds exuded. Kohl and Matthaei (1971) found that, under partial anaerobiosis, lactate accumulated in roots of corn at the expense of malate. Lactate was also found in the incubation medium of excised root tips of corn. Under aerobic conditions lactate was not released into the medium. Ethanol, a product of anaerobic metabolism in plants, was found by Young et al. (1977) to occur in the rhizosphere of seedlings of *Lupinus angustifolia* subjected to water logging for 36 hours.

E. MECHANICAL FORCES

To simulate soil pressure on roots and to avoid the difficulties of extracting exuded compounds from soil, Barber and Gunn (1974) used glass ballotine. Compared to unrestricted roots of barley and maize, those growing between the glass ballotine exuded more amino acids and carbohydrates. The increase was from 5% (in unrestricted roots) to 9% (in restricted roots) of dry matter increment of roots.

F. ENVIRONMENTAL POLLUTION

The relationship of air pollution to microbial ecology has been reviewed by Babich and Stotzky (1974) and Smith (1976). They reported numerous effects of air pollutants on microbial reproductive potential and morphology. There were, however, no reports concerning the effects of air pollution on root exudation per se. Manning et al. (1971) did report that pinto bean plants exposed to 0.1–0.15 μl of ozone for 8 hours a day for 28 days had poor root growth, and *Rhizobium* nodules developed only on the nonfumigated bean plants. Similar results were recorded when soybean plants were exposed to 75 pphm (parts per hundred million) of ozone for 1 hour (Tingey and Blum, 1973). No one has yet attempted to evaluate the effects of air pollution on root exudation, although air pollutants such as ozone and sulfur dioxide readily alter the metabolism of higher plants.

IV. Foliar Application of Chemicals

A. EXUDATION OF FOLIARLY APPLIED PESTICIDES

Although foliar application of chemicals is common practice for protection of shoots from pathogen and insect attack, little information is available concerning the effects of such applications on protecting the health of plant roots, even though earlier studies have demonstrated that rhizosphere populations and root exudation patterns have been changed by the application of pesticides. The earlier work has been reviewed by Hale *et al.* (1978). Various foliarly applied pesticides and nutrients affect exudates and rhizosphere populations. Root exudation of growth regulators amounts to 10–15% of the amount applied to the foliage by whatever means (Foy *et al.*, 1971). Unfortunately, no work has been done on exudation of applied pesticides since the 1971 review (Hale *et al.*, 1971).

B. EFFECTS OF FOLIARLY APPLIED CHEMICALS ON EXUDATION OF ENDOGENOUS COMPOUNDS

In subsequent investigations (Balasubramanian and Rangaswami, 1973) foliar applications of 0.1% $NaNO_3$, 0.1% Na_2PO_4, 25 mg of 2,4-dichlorophenoxyacetic acid (2,4-D) per liter, and 200 mg of Dithane 278 per liter (Table VIII) were studied to determine their effects on the exudation of amino acids and sugars from roots of sorghum and sunnhemp. $NaNO_3$ decreased the amounts of amino acids exuded by sorghum but increased the amounts of amino acids exuded by sunnhemp. Na_2PO_4 decreased but 2,4-D increased amino acid exudation. For sorghum, fungal populations in the rhizosphere increased with applications to the foliage of 2,4-D and $NaNO_3$; bacteria increased with applications of 2,4-D, and actinomycetes increased with all applications. For sunnhemp, all three groups of organisms increased with applications of 2,4-D and $NaNO_3$. The 2,4-D effects were correlated with increased populations of microorganisms in the rhizosphere, whereas the application of the other compounds did not lead to such a correlation.

Hale *et al.* (1977) found that applications of 100 mg of 2,4-D per liter increased cholesterol exudation from peanut roots, and both 2,4-D and 200 mg of gibberellic acid per liter decreased fatty acid exudation.

Reported effects of herbicides on exudation and root rot interaction in Sanilac navy bean (Wyse *et al.*, 1976) showed EPTC and dinoseb to increase exudation of electrolytes, amino acids, and sugars from root and hypocotyls and to increase root rot 42–84%. However, Jalali (1976) applied six growth regulators and herbicides to wheat and found that chloramphenicol, and to a lesser extent 2,4-D, reduced rhizosphere populations by suppressing exudation of ribose, maltose, and raffinose, which were exuded abundantly from root-rot-infected roots (Table IX). Lee and Lockwood (1977) applied chloramben, which increased exudation and reduced plant height and stand of soybeans in media infested with *Thielaviopsis basicola*. Compared with the controls, chloramben at 2 $\mu g/ml$ caused roots to exude 540% amino acids, 205% electrolytes, 80% carbohydrate, 123% fatty acids, and 132% nucleic acids. The exudates caused more en-

TABLE VIII
Names of Chemicals Mentioned in Text by Common Name or Abbreviation

Common name or abbreviation	Chemical name
2,4-D	2,4-Dichlorophenoxyacetic acid
GA	Gibberellic acid
EPTC	S-Ethyl dipropylthiocarbamate
Dinoseb	2-sec-Butyl-4,6-dinitrophenol
Chloramphenicol	D(-)-Threo-2,2-dichloro-N-[β-hydrox-α-(hydroxymethyl)-p-dinitrophenmethyl] acetemide
Alachlor	2-Chloro-2',6' diethyl-N-(methoxymethyl)acetanilide
Kinetin	6-Furfurylamino purine
ABA	Abscissic acid
IAA	Indole-3-acetic acid
Dithane Z78	Zinc ethylenebisdithiocarbamate

doconidia to germinate. Alachlor and dinoseb also enhanced root rot, but not as strikingly as chloramben.

Fungitoxicants reportedly restrict the development of mycorrhizal development on wheat roots (Jalali and Domsch, 1975). It was observed that a number of foliarly applied fungicides caused suppression of the total amino acid content of wheat root exudates, although release of some individual acids was enhanced. It is conceivable that many chemicals used in plant protection could have measurable effects on plant metabolism and root exudation.

C. GROWTH REGULATOR EFFECTS

Cytokinins have been found in root exudates (Van Staden, 1976; Itai and Vaadia, 1965), and various forms of nitrogen applied to the leaves of rice plants increased cytokinin exudation (Yoshida and Takashi, 1974). Since microorganisms in the rhizosphere may release cytokinins (Phillips and Torrey, 1970, 1972; Lalove and Hall, 1973; Azcon and Barea, 1975; Vancura et al., 1977) and since cytokinins are synthesized in roots, the effects of cytokinin on mobilization of metabolites and on exudation were investigated by Thompson (1978). Kinetin at 10^{-4} M and 10^{-6} M was applied to the roots of 57-day-old peanut plants. The exudation of fatty acids decreased with application of 10^{-6} M kinetin, but the fatty acid concentration in the roots increased. Kinetin at 10^{-4} M did not have these effects.

Abscissic acid (ABA) and indole-3-acetic acid (IAA) are also exuded (Tietz, 1975); the amount exuded depends on the cultural methods. Exudation of IAA from 5-day-old pea rots was 11.0 mg/kg fresh weight of roots in sand, but only

TABLE IX
Effects of Foliar Sprays on Carbohydrate Exuded from Roots of Wheat Grown under Axenic Conditions[a]

Carbohydrates	Foliar treatments						Control $(\mu g/50 \text{ plants})$
	$(NH_4)_2SO_4$ (350 mM)	Na_2HPO_4 (325 mM)	KCl (255 mM)	$CO(NH_2)_2$ (455 mM)	2,4-D (2.26 mM)	Chloramphenicol (3.41 nM)	
Pentose mono-saccharides							
Ribose	+ 4.3[b]	+1.8	− 30.6	− 13.8	− 7.2	− 23.5	96.99
Xylose	− 15.2	+ 9.1	− 8.1	− 11.7	− 4.8	− 9.8	117.24
Arabinose	−100	+ 3.1	− 9.2	− 2.3	− 8.1	− 10.1	100.10
Hexose mono-saccharides							
Glucose	− 4.5	+ 3.0	− 23.5	− 11.5	− 5.6	− 2.1	147.77
Fructose	+ 3.5	− 1.3	− 9.0	− 5.8	+ 3.4	− 2.2	121.93
Galactose	− 4.1	− 9.0	+ 8.0	− 1.9	− 2.7	+ 12.7	99.93
Rhamnose	+ 4.0	−100	−100	−100	− 10.2	−100	100.03
Disaccharides							
Maltose	− 17.3	− 1.7	−100	−100	− 13.7	− 27.1	92.73
Sucrose	−100	+ 3.27	−100	−100	−100	−100	83.71
Trisaccharides							
Raffinose	− 8.4	− 6.7	−100	−100	−100	−100	87.03

[a] Reprinted with permission from Jalali (1976).
[b] Values are percentage reduction or increase compared with that in controls.

4.1 μg/kg fresh weight in water culture. ABA applied to the foliage translocated in only small amounts to the roots, and the ABA that was exuded was apparently synthesized in the roots. ABA has been shown to increase the hydraulic conductivity of roots (Glinka, 1973), and it may have an effect on outward loss of water and solutes, particularly under water stress conditions during which the abscissic acid/cytokinin ratio increases (Itai and Benzioni, 1976).

The production of biologically active substances by bacteria that predominate in the rhizosphere may play an important role not only in plant growth but also in root exudation. As an example, IAA, gibberellin-like substances, biotin, and pantothenic acid were produced by strains of bacteria isolated from the root surfaces and rhizosphere of maize (Hussain and Vancura, 1970). That plant growth regulators affect cell membrane permeability has been reported by Gregory and Cocking (1966), Etherton (1970), Kennedy and Harvey (1972), and Wood and Paleg (1972).

V. Biotic Factors Affecting Root Exudation

A. RHIZOSPHERE ORGANISMS

Rhizosphere organisms effect higher plants by (a) altering the morphology of the root system, (b) changing the phase equilibria of soil and hence the nutrients so that they are more readily available to plants and are more readily transported, (c) changing the chemical composition of the soil participating in symbiotic processes, and (d) physically blocking the roots surfaces (Nye and Tinker, 1977). To this list should be added the effect on exudation of soil-borne microorganisms.

Soil microorganisms may affect the permeability of root cells by (a) damaging root tissues, (b) altering root metabolism, (c) preferentially utilizing certain exudates, or (d) excreting toxins (Rovira, 1969; Rovira and Davey, 1974). Changes in root exudation caused by microorganisms may indirectly effect general plant health, resistance to disease, or development of other rhizosphere microflora.

Darbyshire and Greaves (1973) reported that the magnitude of the rhizosphere response in terms of microbial number is markedly influenced by biological as well as chemical factors. The rhizosphere can be altered by plant species or variety and by the physiological age and the metabolic state of the plants. Such factors as soil moisture, soil temperature, aeration, and soil fertility have direct effects on the rhizosphere population as well as effects on the plant and root exudation. Similarly, light, relative humidity, and air temperature can also indirectly affect the rhizosphere population.

That plant root exudates serve as nutrient sources for rhizosphere microorganisms is well known (Bowen and Rovira, 1976; Darbyshire and Greaves, 1973; Rovira, 1969). But root exudates can also either stimulate or inhibit the growth of microorganisms. For example, root exudates of *Crotalariae medicaginea* reportedly stimulate the growth of *Penicillium herquei, Aspergillus niger,* and *Alternaria humicola,* but significantly reduce the growth of *Trichoderma lignorum* (Sullia, 1973).

Reid (1974) reported that mycorrhizal Ponderosa pine roots had significantly lower ^{14}C specific radioactivity than did nonmycorrhizal roots when the shoots were exposed to $^{14}CO_2$. Harley (1969), however, proposed that mycorrhizal roots, relative to noninfected roots, acted as metabolic sinks for photosynthetically fixed carbon. Results with lodgepole pine seemed to confirm that mycorrhizal roots are sinks (Reid and Mexal, 1977).

The quantity of total carbon in root exudates of maize and wheat has been shown to increase approximately two to two and one-half times in the presence of microorganisms when compared with axenically cultured plants. The use of carbon compounds in the exudate by *Pseudomas putida* apparently increased the concentration gradient between the root and the nutrient solution, and there was an increase in exudation (Vancura *et al.,* 1977).

In recent years there has been significant interest concerning the roles of rhizosphere microorganisms in plant nutrition (Barber, 1978; Tinker and Sanders, 1975). Although soil-borne bacteria have an uncertain or small effect, mycorrhizal fungi readily improve plant nutrition, usually by increasing the phosphate supply. Mosse (1973) and Tinker (1975) have demonstrated that a position growth response of plants to vesicular–arbuscular mycorrhizae was associated with phosphate nutrition.

Asanuma *et al.* (1978) examined the effects of dilute paddy soil suspensions on the uptake of nitrogen and phosphorus by rice seedlings. More nitrogen was absorbed by the sterile plants at 25 or 50 ppm nitrogen, whereas the inoculated plants absorbed more nitrogen when it was supplied at 100 or 200 ppm. The amount of phosphorus absorbed by the rice seedlings was affected by the concentration supplied in the nutrient solution and by the presence of microorganisms.

Microorganisms do not always have a positive effect on ion uptake. While the uptake of manganese, iron, and zinc by barley grown in solution culture was stimulated by the presence of microorganisms (Barber and Lee, 1974), the uptake of both phosphate and sulfate by pea plants was limited by *Trichoderma viride* (Brannstrom, 1977). Iron transport in the pea plants was apparently retarded.

Enzyme activity in the roots of higher plants may be altered by rhizosphere microorganisms. Vagnerova and Macura (1974) found that protease activity was nil in the roots of axenic wheat. Roots colonized by microorganisms, however, had appreciable protease activity. The activity was detected exclusively in the presence of roots but not in the medium where the organisms were grown without plants.

B. COLONIZING ORGANISMS

In previous sections, environmental factors, chemical compounds, and the soil microflora as factors affecting root exudation were discussed. Some organisms, particularly plant pathogens, infect the plant and cause changes in root exudation.

1. Shoot Colonizers

In the past several years there have been several reports concerning the influence of viruses and leaf surface microorganisms (epiphytes) on root exudation. Not only did infection of pea by bean yellow mosaic virus cause an increase in root exudates, it also caused an increase in root rot incited by *Fusarium solani* (Beute and Lockwood, 1968). There was an appreciable increase in exudation of electrolytes, nucleotides, carbohydrates, and amino compounds from the virus-infected plants; this increase led to an increase in the inoculum potential of the pathogen. Increased cell permeability, a factor associated with many plant diseases, may have accounted for the increased root exudation from the virus infected plants.

Higher populations of rhizosphere microflora were recorded around roots of red pepper infected with tobacco mosaic virus than were found in comparable healthy plants (Alagianagalingan and Ramakrishnan, 1972). The older, virus-infected plants exuded greater quantities of amino acids than did comparable healthy plants. Singh and Singh (1971) found more different kinds of soil microorganisms associated with virus-infected Bermuda grass but greater numbers of fungi associated with healthy plants. They concluded that healthy plant root exudation influenced the microflora differently than did diseased plant root exudation. Magyarosy and Hancock (1974) studied the association of virus-induced changes of laimosphere microflora. Microflora populations were two to seven times as high in soils surrounding hypocotyls of squash plants infected with squash mosaic virus, because exudation was 4% more than for healthy plants. The increase in laimosphere microflora, furthermore, provided protection against infection by *Fusarium solani*.

Reports concerning the influence of leaf surface microorganisms, particularly saprophytic forms (epiphytes), on root exudation are very limited. Bhat *et al.* (1971) inoculated leaves of hyacinth bean grown axenically in flasks with a *Beijerinckia* sp. and recorded a substantially greater amount of amino acids present in exudates from inoculated as compared with uninoculated or completely axenic plants (Table X). Plants grown under axenic conditions and lacking leaf epiphytes exuded far fewer amino acids.

Although it has not been reported extensively in the literature, it is apparent that fungi and bacteria that are foliar pathogens of higher plants may exert significant effects on the root exudation of the host plant. Many of these organisms, as well as the plant pathogenic viruses, have significant effects on the metabolism of host plants causing changes in the metabolism of carbohydrates, amino acids, proteins, lipids, nucleic acids, and natural growth regulators (Heitefuss and Williams, 1976). Such changes should have direct effects on the source to sink pools in colonized tissues. As one of their first actions, most plant pathogenic organisms alter the cell permeability of the suscept tissue. Such an increase in permeability often leads to the leakage of electrolytes. The influence of the colonization of aerial plant parts by pathogenic and nonpathogenic organisms on root exudation and rhizosphere activity certainly deserves more attention.

An example illustrating the effect of an obligate parasite on rhizosphere microflora of wheat is the study of Srivastava and Mishra (1971). The pathogen, *Puccinia graminis* var. *tritici*, caused an increase in the number of fungi per gram of dry root in soils surrounding susceptible as compared with resistant plants, while the numbers of species recorded exhibited no regular trend. Srivas-

TABLE X

Root Exudation of *Dolichos lablab* Grown in Sand or Water Culture and Inoculated with the Epiphyte *Biejerinckia*, Uninoculated, or Full Axenic[a,b]

Amino acid	Sand culture		Water culture		Fully axenic
	Inoculated	Uninoculated[c]	Inoculated[d]	Uninoculated[c]	
Cysteine	0.375	0.3	0.42	0.08	—
Histidine	0.375	0.3	0.52	0.07	Trace
Aspartic acid	0.375	0.375	Trace	0.08	—
Lysine	—	Trace	Trace	—	Trace
Glycine	0.350	0.220	—	—	—
Threonine, alanine	0.560	0.320	0.700	0.200	0.06
Methionine, valine	0.140	—	0.180	—	Trace
Leucine	0.500	0.220	0.520	0.100	0.08
Unidentified, R_f 0.04	0.400	0.200	—	—	—
Unidentified, R_f 0.11	—	0.200	0.070	0.05	0.06
Unidentified, R_f 0.41	0.600	0.300	0.600	—	0.06
Totals	3.675	2.435	3.01	0.59	0.26

[a] Reprinted, with permission, from Bhat *et al.* (1971).
[b] Values given are micromoles per plant.
[c] Only the roots were maintained under sterile conditions; the shoot system was exposed.
[d] In one instance traces of arginine and glutamic acid were also detected.

tava and Mishra contended that the variation in the rhizosphere population was possibly caused by differences in the physiology of the plants as a result of infection. Again, one can relate changes in cell permeability and in metabolic pools to the actions of an organism with subsequent changes in the rhizosphere.

2. Root Colonizers

Although the vast majority of investigations involving soil-borne microorganisms and root exudation have centered around the effects of exudation on the microbial population and root colonization, in the past several years a number of studies have been conducted to determine the effects of soil-borne fungi, bacteria, and nematodes on root exudation. Investigations involving fungi have included both soil saprophytes and plant pathogens.

a. Soil-Borne Saprophytes. When wheat seedlings grown in a nutrient solution received $^{14}CO_2$, Rovira and Ridge (1973) found that the radioactivity present in the solution in root exudates was reduced by the presence of microorganisms. In a second experiment, however, the amount of ^{14}C released into the solution was not affected by the presence of organisms. This discrepancy illustrated that microbial populations varying in numbers and composition in the rooting medium might metabolize the exudates to different extents and possibly cause varying effects on root cell permeability (Rovira and Ridge, 1973).

Barber and Martin (1976) have investigated the quantities of ^{14}C-labeled carbon dioxide produced in the soil by wheat and barley plants and by microbial activity from degradation of photosynthetically labeled organic matter. The microorganisms caused a significant increase in release of photosynthetically fixed carbon equivalent to 18–25% of the dry matter increments of the plants. Barber and Martin suggested that the increase in carbon dioxide released by cropped as compared with fallow soil could largely be ascribed to the immediate utilization by microorganisms of organic substances released by roots. The presence of soil microorganisms was shown to increase significantly the $^{14}CO_2$ released from the rhizosphere, but it had no effect on the ^{14}C content of the soil (Martin, 1977b). The explanation proposed by Martin is based on a report (Holden, 1975) that more than 70% of the cortical cells of seminal roots from 3- to 4-week-old wheat plants were dead. Apparently, root lysis was increased by soil microorganisms penetrating the plant cell wall.

Only recently has the role in plant nutrition, root exudation, and root disease etiology of specific saprophytic microfloral components of the rhizosphere been investigated. With a 40–60% degradation on the roots of marigold plants infected with *Penicillium simplicissimum* under gnotobiotic conditions, Hameed (1971) found significant changes in the root exudates. Compared with the exudates from 72-day-old axenic plants, exudates of infected plants contained significantly higher levels of total water-soluble carbohydrates, reducing sugars, proteins, and total valine-equivalent amino acids (Table XI). Disorganized and sloughed tissues of inoculated roots were observed with a number of conidia adhering to them. A dense mycelial growth formed first on these root tissues and then colonized adjacent cortical tissue. Even though there was degradation of the roots, there was a greater amount of shoot growth of inoculated than of uninoculated marigold plants. Root exudate analyses revealed a change in exudation patterns, which was correlated both with plant development and with fungal growth. After 20 days total organic matter and protein decreased in the exudates of plants inoculated with either *P. simplicissimum* or *P. citrinum*. After 34 days, however, there was an increase in total organic matter. Hameed attributed the original decrease in organic matter and protein to reabsorption by the plant or utilization by the plant or utilization by the two fungi. The subsequent increase in the organic compounds was correlated with an increase in the foliar sugar content of the inoculated plant, and increased fungal colonization.

Joyner (1975) also reported that root infection and colonization affected the exudate pattern of tobacco plants inoculated with *Trichoderma harzianum*. Exudates of axenic roots contained significantly higher levels of reducing sugar than did exudates of colonized roots (Table XII). The ratio of total water-soluble carbohydrates to reducing sugars was higher in exudates from colonized roots (1.76:1) than it was in exudates from axenic roots (0.93:1), and there was a decrease in the reducing sugar content with longer periods of colonization, probably as a result of utilization by the fungus.

b. Soil-Borne Pathogens. It has been well documented that root pathogens cause increases in root exudation (Mitchell, 1976). In most cases the increase is a direct effect of the pathogen, but in some situations fungal metabolites have been reported to alter plant cell membranes (Wheeler and Hanchey, 1968; Wheeler, 1976). Among the metabolites implicated are penicillin, victorin, and pectic enzymes.

TABLE XI

Chemical Composition of Root Exudates of 72-Day-Old Axenic Marigold Plants and *Penicillium simplicissimum*-Colonized Plants at 34 Days after Inoculation[a]

Plants	Total water-soluble organic matter (mg)	Total water-soluble carbohydrates (mg)	Reducing sugars (μg)	Protein (mg)	Total valine-equivalent amino acid (μM)
Axenic plants	462[b]	1.36	224.99	2.77	2.42
P. simplicissimum-inoculated	419	2.80	664.67	4.19	3.02

[a] From Hameed (1971).
[b] Amounts of chemical compounds per plant as an average of 10 plants.

TABLE XII

Chemical Composition of Root Exudates from 79-Day-Old Axenic Tobacco Plants and *Trichoderma harzianum*-Colonized Plants 35 Days after Inoculation[a]

	Axenic plants	Colonized plants
Total water-soluble carbohydrates	4.3 ±0.6[b]	5.8 ± 2.1
Reducing sugars	4.6 ± 0.8	3.3 ± 1.1[c]
Total proline-equivalent amino acids	7.5 ± 2.6	6.7 ± 1.4

[a] From Joyner (1975).
[b] Mean $\pm\ t_{0.5}S_{\bar{y}}$ Milligrams of compound per gram of root tissue (dry weight basis).
[c] Differs significantly at $P = 0.10$.

The metabolism of plant roots can be altered by soil microflora (Christenson and Hadwiger, 1973). Pisatin, an isoflavonoid phytoalexin, was not produced in detectable levels by aseptic pea seedlings, but pea seedlings in nonsterilized soil produced substantial levels.

Root infection of wheat by *Helminthosporium sativum* caused a significant shift in the spectrum of root exudates (Jalali and Suryanarayana, 1971). Greater numbers of sugars were recorded from the healthy root exudates than from the exudates of inoculated plants. Inoculated plants also released greater amounts of ribose, maltose, raffinose, and sucrose, while the release of glucose, fructose, galactose, xylose, and rhamnose was suppressed. Such quantitative changes in carbohydrates indicated a selective utilization of sugars by the pathogen. There was, furthermore, significantly less total carbohydrate exuded from the diseased plants as compared with the healthy plants.

In further studies (Jalali and Suryananrayana, 1972) infected roots exuded greater amounts of glycine, phenylalanine, and tyrosine than did healthy roots. The monoaminodicarboxylic acid group as well as tryptophan and aminobutryic acid were not found in diseased root exudate samples. Jalali and Suryanarayana (1974) reported that, under the stress of infection, the exudation of most of the organic acids identified was suppressed but there was a pronounced increase in the exudation of glycolic acid and succinic acid.

Using two monoxenic culture techniques for growing tomato plants, Wang and Bergeson (1974) studied the effect of nematode infection on root exudation. Root exudates of *Meloidogyne incognita*-infected tomato plants contained 133–836% more sugar than did exudates from healthy plants. In contrast, amino acids were moderately lower in exudates from infected roots than in those from healthy roots. Galled-root exudates contained fewer sugars, amino acids, and organic acids than did healthy-root exudates. Wang and Bergeson suggested that changes in total sugars and amino acids of infected plant xylem sap and root exudates were a probable mechanism by which tomato plants were predisposed to *Fusarium* wilt.

C. ROOT DISEASE ECOLOGY

Root exudation is a primary factor in the determination of population levels of microflora in the rhizosphere (Darbyshire and Greaves, 1973; Rovira, 1965; Sullia, 1973). Certain exudates have been reported to influence the rate of fungal growth (Booth, 1974; Sullia, 1973), fungal sporulation (Kraft, 1974), fungal spore or resting structure germination (Chaturvedi *et al.*, 1974; Coley-Smith and King, 1970; Kraft, 1974), spore attraction (Chang-Ho and Hickman, 1968; Khew and Zentmyer, 1973; Zentmyer, 1968), egg hatch of nematodes (Shepherd, 1968), and soil fungistatis (Griffin, 1969, 1973; Hameed, 1971; Pass and Griffin, 1972; Snyder, 1968).

Correlations between the amount and composition of exudates and the susceptibility of a plant to a particular pathogen have been reported for a number of host–pathogen combinations. For example, Booth (1974) found that choline, which is toxic to *Verticillium albo-atrum*, was 3.5 times as high in *Verticillium*-tolerant cotton as in susceptible cotton. Moreover, L-alanine, which is exuded in greater quantities by the susceptible cultivar, increased the growth of *V. albo-atrum* 320% (dry weight basis) *in vitro*. The polygalacturonase activity of the pathogen was also stimulated by L-alanine and depressed by choline in culture tests. However, there may be no correlation between the susceptibility of a variety and its exudation pattern, as illustrated by the work of Malajczuk and McComb (1977). Seedlings of root-rot-susceptible *Eucalyptus marginata* produced greater concentrations of sugars and amino acids in exudates than did root-rot-resistant *E. calaphylla*. However, zoospores of *Phytophthora cinnamomi*, the pathogen involved, were attracted to both *Eucalyptus* species, and germination of chlamydospores as well as mycelial growth was increased in the presence of root exudates of both species. Through the action of the root exudates as nonspecific nutrient sources, the two major phases of the *P. cinnamomi* life cycle in the soil were affected. The exudates supported germination of survival propagules (chlamydospores) and growth of the infecting propagules (zoospores and mycelium).

The relationship of nematode egg and larval hatch and activity to plant root exudation has been illustrated by the reports of Hamlen *et al.* (1973) and Alam *et al.* (1975). There was no appreciable effect of alfalfa root exudates on the hatching of eggs of *Meloidogyne incognita* over that of distilled water (Hamlen *et al.*, 1973). However, the neutral carbohydrate fraction of root exudates of alfalfa seedlings was more conducive to egg hatch than were comparable fractions obtained from mature plants. Flowering resulted in a neutral carbohydrate exudate fraction that allowed increased hatching when compared with exudates from nonflowering plants of the same age.

Marigold has been grown with several crops or during intervening periods between the crops by the Indian farmer from time immemorial (Alam *et al.*, 1975). Root exudates from marigold seedlings as well as 1-month-old margosa plants were found to be toxic to nematodes and larval hatch of *Meloidogyne incognita*. In addition, the exudates were toxic to the nematodes *Hoplolaimus indicus, Helicotylenchus indicus, Rotylenchulus reniformis, Tylenchorhynchus brassicae*, and *Tylenchus filiformis*.

D. RHIZOBIA AND MYCORRHIZAE

In the past few years there have been several reviews concerning rhizobia, mycorrhizae, and higher plants (Allen, 1973; Schmidt, 1978; Tinker, 1975). Not only do rhizobia and mycorrhizae have direct effects on the root exudation patterns of higher plants, but root exudates may attract specific microorganisms to higher plants.

The bacteria (*Rhizobium* spp.), which nodulate legumes, have specific strains that colonize only certain species of plants or varieties within species (Allen, 1973). These *Rhizobium* spp. exhibit chemotaxis to plant root exudates (Currier and Strobel, 1976). The bacteria are attracted to root exudates of both legume and non-legume plants but show a differential response in that different rhizobia are attracted to different plants. Currier and Strobel report that, although individual strains of *Rhizobium* cause nodules on a number of species of legumes, production of effective nodules is restricted. The chemotaxis is not required for nodulation, nor is the chemotaxis absolutely specific. The chemotaxis may involve simple molecules such as sugars and amino acids, or it may involve more complete compounds such as polypeptides.

It has been suggested, for example, that homoserine may play an important role in the establishment of the rhizosphere microflora of pea plants (Van Egeraat, 1975b). Homoserine released during the formation of lateral roots of pea might selectively stimulate the growth of *Rhizobium leguminosarum*. Since homoserine is an amino acid not associated with most plant species, its importance in the *Rhizobium* sp. activity of pea plants warrants further studies as a possible unique host plant exudate–fungal relationship.

Exudates from nodulated root systems are often different from those in non-nodulated systems. Upon inoculation with *Rhizobium* there is an increase in nonreducing sugars, *ortho*-dihydroxy phenols, amino-N, polygalacturonase, and pectin methylesterase, and a decrease in reducing sugars and total phenols in root exudates of alfalfa and urid. The root exudates of plants inoculated with homologous rhizobia differ quantitatively from those plants inoculated with heterologous rhizobia as well as noninoculated plants (Rao, 1976).

Exudates from mycorrhizal root systems may play a significant role in disease resistance. For example, the ectomycorrhizal (*Boletus variegatus*) root system of Scots pine produces a number of monoterpenes and sesquiterpenes that inhibit the extensive growth of *Phytophthora cinnamomi* and *Fomes annosus* (Krupa and Nylund, 1972). Such compounds were not usually associated with non-mycorrhizal root systems (Krupa and Fries, 1971) and are two- to eightfold higher in mycorrhizal roots.

VI. Summary

Many of the investigations that identify exudates and factors affecting exudation have involved seedling plants in axenic culture. Results have been useful in demonstrating the multifaceted aspects of exudation of a variety of compounds and in quantifying exudation for defined conditions. Information on exudation of plants in the field comes from studies of changing populations of microorganisms and from $^{14}CO_2$ incorporation and subsequent exudation of elaborated metabolites into soil. Microorganisms in the rhizosphere play an important role in changing root exudation patterns, and our knowledge about the factors that affect exudation and plant interactions with soil-borne organisms is increasing rapidly. No where is this more evident than in the area of root disease ecology.

A number of factors that affect exudation should provide profitable research opportunities. Among these are effects of air pollution, foliar epiphytes, and pesticide applications. The factors involved in the mineral nutrient–root exudate–microorganism system are not well understood, and the role of exudates in disease resistance needs more clarification. The impact on agronomic practices in mineral nutrition and disease control may be greater than has been assumed.

REFERENCES

Alagianagalingam, M.N., and Ramakrishnan, K. 1972. *Indian J. Microbiol.* **12,** 23–26.
Alam, M.M., Masood, A., and Hussain, S.I. 1975. *J. Exp. Biol.* **13,** 412–414.
Allen, O.M. 1973. *In* "Forages, the Science of Grassland Agriculture" (M.E. Heath, D.S. Mitcalfe, and R.E. Barnes, eds.), pp. 98–104. Iowa State Univ. Press, Ames.
Asanuma, S., Tanaka, H., and Yatazawa, M. 1978. *Soil Sci. Plant Nutr.* **24,** 207–220.
Ayers, W.A., and Thornton, R.H. 1968. *Plant Soil* **28,** 193–207.
Azcon, R., and Barea, J.M. 1975. *Plant Soil* **43,** 609–619.
Babich, H., and Stotzky, G. 1974. *Crit. Rev. Environ. Control* **4,** 353–420.
Balasubramanian, A., and Rangaswami, G. 1973. *Folia Microbiol.* **18,** 492–498.
Ballard, B.J., Hameed, K.L., Hale, M.G., and Foy, C.L. 1968, *Plant Physiol. Suppl.* **61,** 316.

Barber, D.A. 1978. *In* "Interactions Between Non-pathogenic Soil Microorganisms and Plants" (Y.R. Dommergues and S.V. Krupa, eds.), pp. 131–162. Elsevier, Amsterdam.

Barber, D.A., and Gunn, K.B. 1974. *New Phytol.* **73,** 39–45.

Barber, D.A., and Lee, R.B. 1974. *New Phytol.* **73,** 87–106.

Barber, D.A., and Martin, J.K. 1976. *New Phytol.* **76,** 69–80.

Barlow, P. 1974. *New Phytol.* **73,** 937–954.

Beute, M.K., and Lockwood, J.L. 1968. *Phytopathology* **58,** 1643–1651.

Bhat, J.V., Limay, K.S., and Vasantharajan, V.N. 1971. *In* "Ecology of Leaf Surface Microorganisms" (T.F. Preece and C.H. Dickinson, eds.), pp. 581–595. Academic Press, New York.

Bokhari, U.G., and Singh, J.S. 1974. *Crop Sci.* **14,** 790–794.

Bonish, P.M. 1973. *Plant Soil* **38,** 307–314.

Booth, J.A. 1974. *Can. J. Bot.* **52,** 2219–2224.

Bowen, G.D., and Rovira, A.D. 1973. *Bull. Ecol. Res. Commun. (Stockholm)* **17,** 443–450.

Bowen, G.D., and Rovira, A.D. 1976. *Annu. Rev. Phytopathol.* **14,** 121–144.

Bowles, D.G., and Northcote, D.H. 1972. *Biochem. J.* **130,** 1133–1145.

Brannstrom, G. 1977. *Z. Pflanzenphysiol.* **83,** 341–346.

Chang-Ho, Y., and Hichman, G.J. 1968. *In* "Root Diseases and Soil-borne Pathogens" (T.A. Tousson, R.V. Bega, and P.E. Nelson, eds.), pp. 103–108. Univ. Calif. Press, Los Angeles.

Chaturvedi, S.N., Siradhana, B.S., and Muralia, R.N. 1974. *Plant Soil* **39,** 49–56.

Christenson, J.A., and Hadwiger, L.A. 1973. *Phytopathology* **63,** 784–790.

Clowes, F.A.L., and Woolston, R.E. 1978. *Ann. Bot.* **42,** 83–89.

Coley-Smith, J.R., and King, J.E. 1970. *In* "Root Diseases and Soil-borne Pathogens" (T.A. Tousson, R.V. Bega, and P.E. Nelson, eds.), pp. 130–133. Univ. Calif. Press, Los Angeles.

Currier, W.W., and Strobel, G.A. 1975. *Plant Physiol.* **57,** 820–823.

Darbyshire, J.F., and Greaves, M.P. 1973. *Pestic. Sci.* **4,** 349–360.

Dommergues, Y.R., and Krupa, S.V. 1978. "Interactions Between Non-pathogenic Soil Microorganisms and Plants." Elsevier, Amsterdam.

Esau, K. 1953. "Plant Anatomy." Wiley, New York.

Etherton, B. 1970. *Plant Physiol.* **45,** 527–528.

Floyd, R.A., and Ohlrogge, A.J. 1971. *Plant Soil* **34,** 596–606.

Foy, C.L., Hurt, W., and Hale, M.G. 1971. *Conf. Proc. Biochem. Interact. Plants Natl. Acad. Sci. Washington* pp. 75–85.

Glinka, Z. 1973. *Plant Physiol.* **51,** 217–219.

Greaves, M.P., and Darbyshire, J.F. 1972. *Soil Biol. Biochem.* **4,** 443–449.

Gregory, D.W., and Cocking, E.C. 1966. *J. Exp. Bot.* **17** 68–77.

Griffin, G.J. 1969. *Phytopathology* **59,** 1214–1218.

Griffin, G.J. 1973. *Can. J. Microbiol.* **19,** 999–1005.

Griffin, G.J., Hale, M.G., and Shay, F.J. 1976. *Soil Biol. Biochem..* **8,** 29–32.

Hale, M.G., and G.J. Griffin. 1974. *Soil Biol. Biochem.* **8,** 225–227.

Hale, M.G., Foy, C.L., and Shay, F.J. 1971. *Adv. Agron.* **23,** 89–109.

Hale, M.G., Moore, L.D., and Orcutt, D.M. 1977. *Plant Physiol. Suppl.* **59,** 30.

Hale, M.G., Moore, L.D., and Griffin, G.J. 1978. *In* "Interactions Between Non-pathogenic Soil Microorganisms and Plants" (Y.R. Dommergues and S.V. Krupa, eds.), pp. 163–203. Elsevier, Amsterdam.

Hameed, K.M. 1971. Ph.D. Dissertation, Virginia Polytechnic Institute and State Univ., Blacksburg.

Hameed, K.M., Saghir, A.R., and Foy, C.L. 1973. *Weed Res.* **13,** 114–117.

Hamlen, R.A., Lukesic, F.L., and Bloom, J.R. 1972. *Can. J. Plant Sci.* **52,** 633–642.

Hamlen, R.A., Bloom, J.R., and Lukesic, F.L. 1973. *J. Nematol.* **5,** 142–146.

Harley, J.L. 1969. "The Biology of Mycorrhizae." Hill, London.

Heitefuss, R., and Williams, P.H. 1976. "Physiological Plant Pathology." Springer-Verlag, Berlin and New York.

Hiltner, L. 1904. *Arb. Dtsch. Landwirtsch. Ges.* **98,** 59–78.

Holden, J. 1975. *Soil Biol. Biochem.* **8,** 333–334.

Hussain, A., and Vancura, V. 1970. *Folia Microbiol.* **15,** 468–478.

Itai, C., and Vaadia, Y. 1965. *Physiol. Plant* **18,** 941–944.

Itai, C., and Benzioni, A. 1976. *In* "Water and Plant Life" (O.L. Lange, L. Kappen, and E.-O. Schultze, eds.), pp. 225–242. Springer-Verlag, Berlin and New York.

Jalali, B.L. 1976. *Soil Biol. Biochem.* **8,** 127–129.

Jalali, B.L., and Domsch, K.H. 1975. *In* "Endomycorrhizas" (F.E. Sanders, B. Mosse, and P.G. Tinker, eds.), pp. 619–626. Academic Press, New York.

Jalali, B.L., and Suryanarayana, D. 1971. *Plant Soil* **34,** 261–267.

Jalali, B.L., and Suryanarayana, D. 1972. *Indian J. Phytopathol.* **25,** 195–199.

Jalali, B.L., and Suryanarayana, D. 1974. *Plant Soil* **41,** 425–427.

Jones, D.D., and Moore, D.J. 1967. *Z. Pflanzenphysiol.* **56,** 166–169.

Jones, D.D., and Morre, D.J. 1973. *Physiol. Plant* **28** 68–75.

Joyner, B.G. 1975. Ph.D. Dissertation, Virginia Polytechnic Institute and State Univ., Blacksburg.

Kennedy, C.D., and Harvey, J.M. 1972. *Pestic. Sci.* **3,** 715–727.

Khew, K.L., and Zentmyer, G.A. 1973. *Phytopathology* **63,** 1511–1517.

Kohl, J.G., and Matthaei, U. 1971. *Biochem. Physiol. Pflanz.* **162,** 119–126.

Kraft, J.M. 1974. *Phytopathology* **64,** 190–193.

Krupa, S.V., and Dommergues, Y.R. 1978. "Ecology of Root Pathogens." Elsevier, Amsterdam.

Krupa, S.V., and Fries, N. 1971. *Can. J. Bot.* **49,** 1425–1431.

Krupa, S.V., and Nylund, J. 1972. *Eur. J. Forest. Pathol.* **2,** 88–94.

Lalove, M., and Hall, R.H. 1973. *Plant Physiol.* **51,** 559–562.

Lee, M., and Lockwood, J.L. 1977. *Phytopathology* **67,** 1360–1367.

Leppard, G.G. 1974. *Science* **185,** 1006–1067.

Leppard, G.G., and Ramamoorthy, S. 1975. *Can. J. Biol.* **53,** 1728–1735.

Lespinat, P.A., and Berlier, Y. 1975. *Soc. Bot. Fr. Colloid. Rhizosphere* **122,** 21–30.

Magyarosy, A., and Hancock, J.G. 1974. *Phytopathology* **64,** 994–1000.

Malajzuk, N., and McCoomb, A.J. 1977. *Aust. J. Bot.* **25,** 501–514.

Manning, W.J., Feder, W.A., Papin, P.M., and Perkins, I. 1971. *Environ. Pollut.* **1,** 305–312.

Martin, J.K. 1977a. *In* "Soil Organic Matter Studies," pp. 197–203. IAEA, Vienna.

Martin, J.K. 1977b. *Soil Biol. Biochem.* **9,** 1–7.

McDougall, B.M. 1970. *New Phytol.* **69,** 37–46.

McDougall, B.M., and Rovira, A.D. 1970. *New Phytol.* **69,** 99–1003.

Mitchell, J.E. 1976. *In* "Physiological Plant Pathology" (R. Heitefuss and P.H. Williams, eds.), pp. 104–128. Springer-Verlag, Berlin and New York.

Mosse, B. 1973. *Annu. Rev. Phytopathol.* **11,** 171–196.

Muller, C.H. 1966. *Bull. Torrey Bot. Club* **93,** 332–351.

Nye, P.H., and Tinker, P.B. 1977. "Solute Movement in the Soil-Root System." Univ. Calif. Press, Los Angeles.

Pass, T., and Griffin, G.J. 1972. *Can. J. Microbiol.* **18,** 1453–1461.

Paul, R.E., and Jones, R.L. 1975a. *Plant Physiol.* **56,** 307–312.

Paul, R.E., and Jones, R.L. 1975b. *Planta* **27,** 97–110.

Paul, R.E., and Jones, R.L. 1976. *Plant Physiol.* **57,** 249–256.

Paul, R.E., Johnson, C.M., and Jones, R.L. 1975. *Plant Physiol.* **56,** 300–306.

Phillips, D.A., and Torrey, J.G. 1970. *Physiol. Plant* **23,** 1057–1063.

Phillips, D.A., and Torrey, J.G. 1972. *Plant Physiol.* **49,** 11–15.

Rao, A.V. 1976. *Zweite Naturwiss. Abt.* **131**, 79-82.

Reid, C.P.P. 1974. *Plant Physiol.* **54**, 44-49.

Reid, C.P.P., and Mexal, J.G. 1977. *Soil Biol. Biochem.* **9**, 417-422.

Rovira, A.D., 1965. *Annu. Rev. Microbiol.* **19**, 214-266.

Rovira, A.D. 1969. *Bot. Rev.* **35**, 35-37.

Rovira, A.D., and Davey, C.B. 1974. *In* "The Plant Root and Its Environment" (E.W. Carson, ed.), pp. 153-204. Univ. Press of Virginia, Charlottesville.

Rovira, A.D., and Ridge, E.H. 1973. *New Phytol.* **72**, 1081-1087.

Sauerbeck, D., and Johnson, B. 1976. *Z. Planzenernaeh. Bodenkd.* **3**, 315-328.

Schmidt, E.L. 1978. *In* "Interactions Between Non-pathogenic Soil Microorganisms and Plants" (Y.R. Dommergues and S.V. Krupa, eds.), pp. 269-304. Elsevier, Amsterdam.

Shapovalov, A.A. 1972. *Soviet Plant Physiol.* **18**, 135-138.

Shepherd, A.M. 1968. *In* "Root Diseases and Soil-borne Pathogens" (T.A. Tousson, R.V. Bega, and P.E. Nelson, eds.), pp. 134-137. Univ. Calif. Press, Los Angeles.

Short, G.E., and Lacy, M.L. 1976. *Phytopathology* **66**, 188-192.

Singh, K., and Singh, C.S. 1971. *Port. Acta Biol. Ser.* A **11**, 359-364.

Smith, W.H. 1970. *Phytopathology* **60**, 701-703.

Smith, W.H. 1971. "Influence of Artificial Defoliation on Exudates of Sugar Maple." Pergamon, Oxford.

Smith, W.H. 1972. *Soil Biol. Biochem.* **4**, 111-113.

Smith, W.H. 1976. *In* "Microbiology of Aerial Plant Surfaces" (C.H. Dickinson and T.F. Preece, eds.), pp. 75-105. Academic Press, New York.

Snyder, W.C. 1968. *In* "Root Diseases and Soil-borne Pathogens" (T.A. Tousson, R.V. Bega, and P.E. Nelson, eds.), pp. 3-7. Univ. Calif. Press, Los Angeles.

Srivastava, V.B., and Mishra, R.R. 1971. *Indian Phytopathol.* **24**, 784-786.

Stotzky, G., and Schenk, S. 1976. *Am. J. Bot.* **63**, 798-805.

Sullia, S.B. 1973. *Plant Soil* **39**, 197-300.

Thompson, L.K. 1978. M.S. Thesis, Virginia Polytechnic Institute and State Univ., Blacksburg.

Tietz, A. 1975. *Biochem. Physiol. Pflanz.* **167**, 371-378.

Tingey, D.T., and Blum, U. 1973. *J. Environ. Quality* **2**, 341-343.

Tinker, P.B.H. 1975. *Symp. Soc. Exp. Biol.* **39**, 325-350.

Tinker, P.B.H., and Sanders, F.E. 1975. *Soil Sci.* **119**, 363-368.

Vagnerova, K., and Macura, J. 1974. *Folia Microbiol.* **19**, 525-535.

Vancura, V. 1964. *Plant Soil* **21**, 231-248.

Vancura, V., and Hanzlikova, A. 1972. *Plant Soil* **36**, 271-282.

Vancura, V., and Stanek, M. 1975. *Plant Soil* **43**, 547-558.

Vancura, V., and Stotzky, G. 1976. *Can. J. Bot.* **54**, 518-532.

Vancura, V., Prikryl, Z., Kalachova, L., and Wurst, M. 1977. *Ecol. Bull. (Stockholm)* **25**, 381-386.

Van Egeraat, A.W.S.M. 1975a. *Plant Soil* **42**, 37-47.

Van Egeraat, A.W.S.M. 1975b. *Plant Soil* **42**, 381-386.

Van Staden, J. 1976. *Plant Sci. Lett.* **7**, 279-284.

Wang, E.L.H., and Bergeson, G.B. 1974. *J. Nematol.* **6**, 194-202.

Warembourg, F.R., and Morrall, R.A.A. 1978. *In* "Interactions between Non-pathogenic Soil Microorganisms and Plants" (D.Y. Dommerques and S.V. Krupa, eds.), pp. 205-242. Elsevier, Amsterdam.

Werker, E., and Kislev, M. 1978. *Ann. Bot.* **42**, 809-816.

Wheeler, H. 1976. *In* "Physiological Plant Pathology" (R. Heitefuss and P.H. Williams, eds.), pp. 413-429. Springer-Verlag, Berlin and New York.

Wheeler, H., and Hanchey, P. 1968. *Annu. Rev. Phytopathol.* **6**, 331-365.

Whittaker, R.H., and Feeny, P.P. 1971. *Science* **171,** 757-770.
Wood, A., and Paleg, L.G. 1972. *Plant Physiol.* **50,** 103-108.
Wyse, D.L., Meggitt, W.F., and Penner, D. 1976. *Weed Sci.* **24,** 11-15.
Yoshida, R., and Takashi, O. 1974. *Proc. Crop Sci. Soc. Jpn* **43,** 47-51.
Young, B.R., Newhook, F.J., and Allen, R.N. 1977. *N. Z. J. Bot.* **15,** 189-191.
Zentmyer, G.A. 1968. *In* "Root Diseases and Soil-borne Pathogens" (T.A. Tousson, R.V. Bega, and P.E. Nelson, eds.), pp. 109-111. Univ. Calif. Press, Los Angeles.

ADVANCES IN AGRONOMY, VOL. 31

RED CLOVER BREEDING AND GENETICS

N. L. Taylor* and R. R. Smith†

*Department of Agronomy, University of Kentucky, Lexington, Kentucky
and
†United States Department of Agriculture, Madison, Wisconsin

I. Introduction

Red clover, *Trifolium pratense* L., for many years has been a significant forage legume among agricultural crops. In much of central and western Europe

125

it is the leading legume in forage production, and it maintains a significant position in forage production in the United States. Red clover is adapted to a wide range of soil types, pH levels, and environmental conditions. This plasticity has enabled red clover to retain its usefulness for hay, silage, pasture, and soil improvement in much of the temperate region of the world.

A. ECONOMIC IMPORTANCE

The overall economic importance of red clover to world agriculture is difficult to assess. Red clover is consumed by animals in the form of hay, silage, or pasture fodder, and is marketed to the consumer in the form of meat, fiber, and dairy products. Its introduction to central and northern Europe and the United States had a profound impact on civilization and agriculture by providing a stabilized supply of feed to livestock (Piper, 1924; Gras, 1940; Fergus and Hollowell, 1960).

Red clover is generally grown with timothy, *Phleum pratense* L. In the United States, red clover is grown for hay, silage, pasture, and soil improvement on about 5.4 million hectares. According to United States Department of Agriculture Crop Production and Agricultural Statistics Reports, the annual yield is approximately 21.5 million metric tons, or about 4.0 metric tons per hectare. Maximum yields using improved varieties should easily be twice these values. The average value received in 1976 per metric ton of hay was $66.44. Therefore, the economic worth of red clover is approximately 1.4 billion dollars annually, or $265 per hectare. This does not consider the value of the red clover as pasture. The value of the annual red clover seed crop in the United States is between 23 and 27 million dollars.

Red clover is a nitrogen-fixing legume, thus contributing to the supply of nitrogen available in the soil for subsequent crops. It is estimated that red clover provides between 125 and 200 kg of nitrogen per hectare (Rohweder *et al.*, 1977).

B. ORIGIN AND DISTRIBUTION

Red clover is thought to have originated in southeastern Europe and Asia Minor. Local ecotypes probably developed over 2000 years ago, but documentation is limited. A review of the earliest references to the utilization of red clover as forage is provided by Fergus and Hollowell (1960) and Pieters and Hollowell (1937). Red clover was probably introduced to northern Europe during the fifteenth century A.D. and into North America in the late seventeenth or early eighteenth century A.D. Today, it is an important forage legume throughout the temperate regions of Europe, the Soviet Union, Australia, New Zealand, Argentina, Chile, Canada, Japan, Mexico, Columbia, southern Canada, and eastern and central United States.

II. Taxonomy

Red clover is a species of the genus *Trifolium*, which contains 8 subgenera or sections (Hossain, 1961). The subgenus *Trifolium*, containing about 70 species, was divided into 17 subsections by Zohary (1971, 1972). Within *T. pratense*, two main types of clover are grown in the United States: medium or double-cut, and mammoth or single-cut. Mammoth usually flowers later, is taller, yields more in the first growth, and produces less aftermath growth than medium. Double-cut red clover, contrary to its name, may produce several flushes of growth per year, depending on the length of the growing season. Most red clover grown in the United States is of the medium type.

A. RELATED SPECIES AND CHROMOSOME NUMBERS

Red clover has a diploid chromosome number of 14 ($n = 7$). Classified by Zohary (1971) as closely related to red clover are species with base numbers of 8. *Trifolium diffusum* Ehrh. and *T. pallidum* Waldst. and Kit., two annual species, have a diploid number of 16. A closely related perennial is *T. noricum* Wulf., which also has 16 diploid chromosomes. Somewhat more distantly related are *T. medium* L. (zigzag clover, $2n = 64$–80), *T. alpestre* L. ($2n = 16$), *T. rubens* L. ($2n = 16$), and *T. heldreichianum* ($2n = 16$). Another species sometimes considered to be a form of *T. medium* is *T. sarosiense* Hazsl. ($2n = 48$). Two other annual species perhaps more distantly related are *T. hirtum* All. and *T. cherleri* L. ($2n = 10$).

B. EVOLUTION

Probably the best evidence on the evolution of red clover is provided by interspecific hybridization. Although many interspecific crosses have been attempted, the only verified hybrids involving red clover were obtained with the diploid annuals *T. diffusum* and *T. pallidum*. The diploid hybrid of *T. pratense* × *T. diffusum* was sterile, but the amphidiploid of $4x$ *T. pratense* × $4x$ *T. diffusum* was fertile (Taylor *et al.*, 1963). Pollination of $2n$ *T. pratense* with $2n$ *T. pallidum* yielded only shriveled seeds, but the cross of $4x$ *T. pratense* × $2n$ *T. pallidum* resulted in a sterile triploid hybrid (Armstrong and Cleveland, 1970).

Meiotic analyses of diploid and tetraploid hybrids of *T. pratense* × *T. diffusum* suggested that the chromosomes of the two species differ by a series of complex structural interchanges. This is based on the presence of chain multivalents in the diploid and amphidiploid hybrids (Schwer and Cleveland, 1972a,b). Analyses of chromosome pairing in the triploid hybrid of $4x$ *T. pratense* × $2x$ *T. pallidum* revealed mostly bivalents and univalents, but the presence of multivalents suggested some homology between the two species. These results apparently indicate that red clover is more closely related to annuals than to perennials, and to *T. diffusum* than to *T. pallidum*. Cytological analyses of interspecific hybrids suggest that speciation has resulted from a complex series of structural interchanges causing chromosome differentiation and eventual loss of one chromosome pair in *T. pratense*.

Crosses among perennial species related to red clover have been reported by Maizonnier (1972) and Quesenberry and Taylor (1976, 1977, 1978); they are summarized in Table I. Observation at metaphase-1 of diploid hybrids indicated good pairing, with most PMCs having eight bivalents. Pollen stainability was about 50%, indicating that the chromosomes of the diploid species differ by small structural changes. Although crossability barriers among these species were weak, barriers after fertilization were strong, ranging from embryo abortion to poor germination and low F_2 survival. Analysis of meiosis of polyploid hybrids also showed a low frequency of multivalents. Fertility of pollen from hybrids was 70–85%, and no major barriers to hybridization were exhibited except for partial chlorosis.

TABLE I

Interspecific Hybrids of Perennial *Trifolium* Species in Section *Trifolium* Zoh

Species and chromosome number $(2n)$[a]	Hybrid chromosome number $(2n)$	Fertility[b]	References[c]
sar (48) × alp (16)	24	Sterile	1
med (80) × sar (48) & rec.	?	?	1
med (64) × sar (48) & rec.	?	?	1
alp (16) × held (16) & rec.	16	Slightly fertile	2
alp (16) × rub (16)	16	Slightly fertile	2
med (72) × sar (48)	58–60	Slightly fertile	3
sar (48) × alp (32)	32–40	Fertile	4

[a] Species abbreviations: sar = *T. sarosiense* Hazsl.; alp = *T. alpestre* L.; med = *T. medium* L.; held = *T. heldreichianum* Hausskn.; rub = *T. rubens* L.

[b] Male and female fertility as measured by cross-pollination among hybrid plants.

[c] Refers to literature cited: (1) Maizonnier (1972); (2) Quesenberry and Taylor (1976); (3) Quesenberry and Taylor (1977); (4) Quesenberry and Taylor (1978).

Trifolium pratense produced no hybrids with the perennial species, indicating that little gene exchange has occurred. Hybridization may have been involved in the evolution of the perennial species, but the presence of F_1 sterility and the absence of natural polyploids among the species most closely related to red clover indicate that little hybridization has occurred in the more recent evolution of *T. pratense*.

III. Reproduction

A. FLORAL STRUCTURE

Red clover has the complete leguminous flower, consisting of calyx, corolla, ten stamens, and a pistil. The calyx tube terminates in five lobes, or teeth. Five petals unite at the base to form a corolla tube consisting of a standard petal, two wing petals, and two keel petals. Nine stamens and the one stigma unite to form the sexual column. The tenth anther is free. Up to 125 or more flowers are situated on peduncles to form a single head or, in rare cases, multiple heads. Flowers usually are reddish-pink but vary from white to deep reddish-purple.

B. POLLEN DEVELOPMENT AND TRANSFER

Anthers of red clover consist of four microsporangia. Prior to dehiscence the walls between the microsporangia break down, forming a common cavity on each side of the anther. Dehiscence of the anther occurs before the petals have reached their full size. Pollen fills the space between the fused keel petals and is held in place until disturbed (Hindmarsh, 1963).

Transfer of pollen generally is by bees, but the flowers of red clover are so large that honeybees (*Apis mellifera* L.) will avoid them if other nectar sources are available. Honeybees may collect pollen, however. Bumblebees (*Bombus* spp.) are efficient pollinators of red clover. Alkali bees (*Nomia melanderi* Ckll.) and leaf cutter bees (*Megachile rotundata* [F.]) also have been used for red clover pollination. When red clover is pollinated by a bee, the sexual column protrudes from the interior of the flower, with the pistil extending slightly beyond the stamens. When the weight of the bee is removed, the sexual column returns to its original position. Each flower may be pollinated several times, as long as the stigma remains receptive.

C. SEED SET AND DEVELOPMENT

The stage of bloom for optimum seed set appears to be when flowers are about half open. Stigma receptivity and pollen viability continue for about 10 days under greenhouse and field conditions. In liquid nitrogen, pollen may be stored up to 26 weeks without loss of viability (Engelke and Smith, 1974). The length of time between pollination and fertilization of the egg cell in diploid red clover is between 28 and 35 hours, and between 17 and 26 hours in tetraploids (Mackiewicz, 1965). Seeds are physiologically mature 14 days after pollination and are dry enough for harvesting at about 21 days. Two ovules occur in each ovary, but, except for some bred strains, one usually aborts (Van Bogaert, 1958; Schieblick, 1966; Bingefors and Quittenbaum, 1959). In rare cases, up to four ovules per ovary occur (Povilaitis and Boyes, 1959).

D. MECHANISMS CONTROLLING CROSS-FERTILIZATION

Red clover is a self-incompatible cross-pollinated species that under natural conditions produces very few self-seed. However, self-compatible, autogamous genotypes have been found (Diachun and Henson, 1966). The mechanism of self-incompatibility (reviewed by Fergus and Hollowell, 1960) is the one-locus, gametophytic S-allele system in which plants with the same S-alleles are prevented from selfing by slow growth of pollen tubes through styles. The system also prevents cross-fertilization of plants that have the same S-allele genotype.

Details of the S-allele systems in red clover have been investigated by inbreeding via pseudo-self-compatibility (PSC), a term used to indicate the production of self-seed by normally self-incompatible genotypes. Leffel (1963) showed that selfing heterozygous (S_1S_2) clones produced a 1 – 1 ratio of homozygotes (S_1S_1 and/or S_2S_2) to heterozygotes (S_1S_2). However, the progeny of one clone significantly deviated from this ratio, producing fewer homozygotes than expected. Johnston et al. (1968) also found a deficiency of homozygotes in I_1 (selfed) progenies of 15 heterozygous (S_1S_2) clones. Detailed examination of three families indicated that two possessed both homozygous classes (S_1S_1 and S_2S_2) as well as the heterozygous class (S_1S_2), and one possessed only one homozygous class and the heterozygous class. Consequently, it appears that the S-allele system in red clover deviates in some plants from the classical system. The reason for the deviation is unknown.

An extensive series of S-alleles exists in red clover, and much research has centered on the means by which such a large number are generated. Irradiation induces mutations to self-compatibility (S_f) rather than to new S-specificities (de Nettancourt, 1969). Denward (1963) observed what were thought to be new S-alleles among I_1 sib plants from selfed (I_0) clones, but he could not be certain that they were not contaminants. Johnston et al. (1968) found no changes in I_1's of three red clover families. Anderson et al. (1974a) reported one changed S-specificity in eight I_2 families. Pandey (1970) speculated that the principle method of generation of new S-alleles during inbreeding was intracistronic recombination in the structural cistrons of the S-gene complex, with a minor role of point mutations and deletions. Regulatory genes in the heterozygous condition during outbreeding suppress recombination. Inbreeding produces homozygous recessive regulatory genes, thus eliminating suppression of recombination. Experimental verification of this hypothesis has not been obtained, in part because of the low frequency of recombinants.

IV. Heritability and Gene Action

A. MORPHOLOGICAL CHARACTERS

Polyphylly, leaves with five leaflets, is conditioned by two recessive genes, f and n (Simon, 1962). Polyphylly results when either or both of these genes are present in the homozygous condition. Artemenko (1972) developed a population with four to nine leaflets on 89% of the plants.

One of the first reports of the inheritance of male sterility was that of Smith (1971), who described a male-sterile plant with shriveled and dark-orange anther sacs. The male-sterile condition is controlled by a single recessive gene designated ms_1ms_1. A second gene for male sterility isolated by Taylor et al. (1978c) was designated ms_2ms_2. The two loci interacted to give a duplicate recessive epistatic type of gene action. Macewicz (1976a,b) reported a male-sterile type in which the anthers were either incompletely developed or entirely absent. Male sterility was controlled by the complementary action of a recessive gene, rf_1, and a dominant one, Rf_2, in a sterile cytoplasm. Restorers of fertility were readily detected in the population. Shcheglov and Zvyagina (1975) and Zvyagina (1973) reported cytoplasmic male sterility induced by colchicine. Whittington (1958) described a male-sterile type resulting from asynapsis. Since the original source was also female-sterile, no genetic information was obtained.

Liang (1965) reported that leafmarking is inherited on a monofactorial basis with developmental and suppressor genes active. Annual habit of growth was reported to be controlled by a single recessive gene in some cases, and by two or more in others (Strzyzewska, 1974).

B. PHYSIOLOGICAL CHARACTERS

Inheritance of a gibberellin-responsive dwarf mutant in red clover was described by Smith (1974). The dwarf characteristic is controlled by one gene, dw_1, and is recessive to normal growth.

Nutman (1968) described two independent recessive host genes, n and d, each of which prevents nitrogen fixation in the nodules of red clover. Each factor is specific in its ineffectiveness of nitrogen fixation for specific strains of the bacteria *Rhizobium trifolii*. The two factors, *n* and *d*, are independent and nonallelic to factors i_1 and *ie* reported earlier by Nutman (1954, 1957).

Graham and Newton (1971) reported internal necrosis of crown pith tissue of red clover, which they called internal breakdown (IB). Selection studies by Zeiders *et al.* (1971) have shown that resistance or susceptibility to IB can be readily obtained, suggesting that few genes control this character.

C. PEST RESISTANCE

Disease resistance in red clover generally is controlled by one to a few genes. Inheritance of resistance to southern anthracnose caused by *Colletotrichum trifolii* B. & E. is conditioned by one recessive gene, according to Athow and Davis (1958). However, resistance to most diseases in red clover is conditioned by dominant genes. For northern anthracnose, caused by *Kabatiella caulivora* (Kirch.) Kavak., two (Sakuma *et al.*, 1973) or more than three (Smith and Maxwell, 1973) dominant genes are involved. Resistance to powdery mildew caused by *Erysiphe polygoni* DC is dominant and, for five races of the fungus, is monogenic. In two other races, resistance seems to be controlled by two genes, and in a third race, inheritance of resistance varies among red clover clones (Hanson, 1966; Stavely and Hanson, 1967). Inheritance of resistance to rust caused by *Uromyces trifolii* var. *fallens* Arth. is controlled by a single dominant gene (Diachun and Henson, 1974a,b). However, this source of resistance cannot be used for cultivar development because it is linked with a seedling lethality factor (Engelke *et al.*, 1977). Other types of resistance to rust are inherited in a quantitative manner (Engelke *et al.*, 1975). Although earlier investigations indicated that resistance to crown rot caused by *Sclerotinia trifoliorum* Eriks. is heritable, no detailed inheritance investigations have resulted. Autotetraploid cultivars are more resistant to crown rot than are comparable diploids (Vestad, 1960). The effect of induced tetraploidy differs by genotype, suggesting that dosage effects of genes for resistance may be important.

Resistance to virus in red clover also appears to be qualitatively inherited. Diachun and Henson (1974a,b) reported three types of resistance to bean yellow mosaic virus, each controlled by a different dominant gene: necrotic local lesion (hypersensitive) reaction; resistance to mottling and systemic necrosis; and resistance to general mottling, controlled by a gene that appears to be epistatic to the gene for hypersensitive reaction. Khan *et al.* (1978) determined that the resistance to red clover vein mosaic virus was controlled by a single dominant gene, *Rc*.

Resistance to the stem nematode (*Ditylenchus dipsaci*) was reported by Nordenskiold (1971) to be regulated by two dominant genes. One of the genes is closely linked to the S-locus (self-incompatibility).

V. Sources of Genetic Variation

A. GERMPLASM COLLECTIONS AND POOLS

The maintenance of genetic materials is important as a source of genes for important agronomic characters, and for the prevention of genetic vulnerability. Germplasm resources and vulnerability in *Trifolium* in the United States, including red clover, were reviewed by Taylor *et al.* (1977). They concluded that genetic resources were abundant because red clover is cross-pollinated, heterogeneous, and heterozygous. It is also a prolific seed producer and is hard-seeded, so that seeds remain viable in the soil for several years; it usually is not grown in monoculture. Red clover is a diploid; its plasticity is evidenced by genetic shifts in response to different environments (Akerberg, 1974; Taylor *et al.*, 1966; Wexelsen, 1965). On the other hand, the number of cultivars in the United States declined from about 36 (Fergus and Hollowell, 1960) to about 10 in 1974 (Garrison and Newton, 1974), and many of the latter may be genetically related. Recently, the number of cultivars has increased, largely through the efforts of private plant breeders.

Red clover germplasm collections in the United States are maintained primarily at the Regional Plant Introduction Station at Geneva, New York. About 300 accessions of perennial *Trifolium* species, including *T. pratense* and its perennial relatives, are being maintained (Taylor *et al.*, 1977). Seeds of a few germplasm collections and older cultivars are maintained at the National Seed Storage Laboratory in Fort Collins, Colorado.

The only germplasm pools being maintained in the United States to the authors' knowledge are one in Kentucky in which 44 worldwide introductions have been increased for 10 generations with minimal selection (Taylor, 1979), and two in Wisconsin consisting of second-generation crosses of adapted cultivars and introduced accessions.

The tremendous wealth of germplasm and genetic variation of red clover that is available worldwide is indicated by investigations in Poland (Goral, 1972), Italy (Ceccarelli, 1968), the USSR (Anikeenko, 1971), Czechoslovakia (Jahnova, 1973), Australia (Lovett and Neal-Smith, 1974), Rumania (Pamfil and Savatti, 1971), Finland (Ravantii, 1965), Japan (Kaneko, 1972), Sweden (Bingefors and Akerberg, 1961) and Wales, Great Britain (Evans, 1957). In a population of 2200 plants examined in Russia, collections from the Mediterranean region were chiefly annual; from central Europe and northern Europe, mixed biennial and perennial types; from northern Russia, mostly early-maturing perennial types; and from Siberia and Scandinavia, mainly late-maturing perennial types (Horasajlov, 1964). Other characteristics examined included winter-hardiness, seed and forage yield, maturity, disease and insect resistance, general adaptability, and chemical content. Undoubtedly, many of these collections are being maintained for distribution. Notable is the Swedish collection of 43 land races that have been collected and maintained by the Swedish Seed Growers Association (Bingefors and Akerberg, 1961).

Seed collections may be stored without significant loss of viability for several years under low temperatures and humidity. One of the most ideal storage conditions is in a sealed glass container at −5 to −15°C (Bingefors and Akerberg, 1961). By this method, red clover seeds may be maintained for at least 10 years with little loss of viability, even if allowed to thaw for one day each week (Rincker, 1974).

The primary center of diversity of red clover is thought to be the Mediterranean area, where red clover has been intensively collected. One of the more recent collecting trips by United States personnel was during June to August 1977 in Greece, Crete, and Italy (Smith et al., 1978). Genetic materials also are exchanged by botanic gardens and plant introduction agencies of various countries.

B. MUTATIONS

Red clover is very unfavorable material for mutagenesis because of cross-pollination and heterozygosity. Consequently, most recent studies have been concerned with the effect of the mutagen upon the immediate generation (see Fergus and Hollowell, 1960, for a review of earlier research). Types and amounts of mutagens applied to dry seeds include 30–150 kR of gamma radiation (Taylor et al., 1961; Manner, 1965; Jones and Plummer, 1960), 40–100 kR of X rays (Wojciechowska, 1970) and 1–8 hours of thermal neutrons (Taylor et al., 1961). X rays (5–20 kR) also were applied to wet seeds, and to pollen and ovules (Liang, 1965). Ethyl methanesulfonate (EMS) in concentrations ranging from 2×10^{-2} M to 8×10^{-2} M also was applied to red clover seeds by Liang (1965). Most mutants obtained have been deleterious, with sterility as a common mutant type, increasing in frequency with increments of the mutagen. Phenotypic correlations between pairs of characters from X-ray- and EMS-treated seeds were lower for EMS than for X rays, suggesting that mutations from EMS were predominately genic, and mutations from X rays were predominately of the gross deletion and/or duplication type (Liang, 1965). Therefore, EMS might be favored for creating genetic variability useful to most breeding programs.

The effect of X rays was greater on pollen than on ovules, perhaps because ovules are protected by maternal tissue. Seed set was reduced and shriveled seed increased with increments of dosages beginning at 2.5 kR (Liang, 1965).

A short-flowered mutant induced by X rays was examined for attractiveness to bees. The character was inherited in a recessive manner and remained unchanged in hybrid combinations. However, bees under natural conditions avoided the short-corolla type (Bugge, 1970). Other mutants that have been produced include a mutifoliolate type (Artemenko, 1972); a type showing changes in histone content (Mal'tsev and Kuzin, 1974); and a type showing earliness, greater productivity, and resistance to *Erysiphe* (Novoselova and Malasenko, 1965).

C. POLYPLOIDY

Colchicine is the primary agent for the induction of autotetraploidy in red clover (Fergus and Hollowell, 1960). Another method is to treat seeds at a concentration of 0.008% of ethyleneimine for 20 hours (Vlasyuk *et al.*, 1977). Also, treatment of fertilized flowers with nitrous oxide at 6 bars atmospheric pressure beginning 24 hours after pollination and lasting 24 hours produces up to 100% tetraploidy (Berthaut, 1965; Matsu-ura *et al.*, 1974; Taylor *et al.*, 1976).

Methods of identification of tetraploids include pollen pore number (Dijkstra and Speckmann, 1965); chloroplast number, shape, and structure (Najcenska and Speckmann, 1968; Nuesch, 1966; Marek and Riman, 1969); stomata size (Vavilov *et al.*, 1976a); pollen shape (Riman and Polak, 1968; Taylor *et al.*, 1976); turgor of leaves (Riman and Mistinova, 1971); and number of heteropycnotic chromocenters in root cap cells (Reitberger, 1964). Although these methods are useful for preliminary screening, none is as reliable as chromosome counts for establishing ploidy (Vavilov *et al.*, 1976a; Taylor *et al.*, 1976).

Generally, most investigators are convinced that tetraploid cultivars frequently outyield diploids and often are more disease-resistant (Navalikhina, 1965; Thomas, 1969; Riman and Mistinova, 1970; Bellmann, 1966; Makarov, 1973). It is not always clear whether tetraploids are superior per se or whether the higher yields and greater disease resistance are the result of more tetraploid than diploid breeding research. Vestad (1960) obtained chimera plants—plants with both tetraploid and diploid shoots—and produced a series of corresponding diploid and tetraploid F_2 families. The tetraploids had 67.0% and the diploids 53.9% plants surviving after infection with *Sclerotinia trifoliorum*, the causal agent of crown rot.

Bred tetraploid cultivars in many countries are superior to diploids. In the Irish Republic, the tetraploid cultivars HUNGAROPOLY, TEROBA, TETRI, and RED HEAD yielded up to 20% more than did older diploid cultivars and were more persistent (Crowley, 1975). In Scotland, HUNGAROPOLY, NORSE-MAN, and REA 4N gave higher digestible dry matter yields than did diploid cultivars (Hunt *et al.*, 1974). In the Ukrainian SSR, a new cultivar AN-TETRA-1 outyielded diploid cultivars by 10–14% (Zosimovich *et al.*, 1973). In Switzerland, tetraploid materials exceeded diploids by 8% in green matter yield (Nuesch, 1974). In the USSR, the yields of TETRAPLOIDNYI VIK were 18.9% and 30.3% higher than that of the diploid cultivar MOSKOVSKYI-1 when used in rotation and pasture, respectively (Novoselova and Matskiv, 1974). In Sweden, tetraploid late SVO34 and ULVA were superior to corresponding diploid cultivars (Julen, 1969). In New Zealand, the tetraploid PAWERA was clearly superior to diploid cultivars at two cutting heights (Anderson, 1973). No tetraploid cultivars have been bred in the United States.

Although tetraploid red clover is superior to diploid red clover in many respects, the change from a well-adjusted diploid to the raw autotetraploid results in lower seed yields. In families from chimera plants, tetraploids produced 22.5–23.8% of the seed yield of the diploids (Picard and Berthaut, 1966; Wexelsen and Aastveit, 1967).

Seed yield of tetraploids has been increased to a limited extent by long-term breeding programs, and seed of some cultivars has become available (Eskilsson and Bingefors, 1972). Frame (1976) concluded that tetraploid red clover has value in the United Kingdom particularly for hay as contrasted to grazing management, but its use is likely to be restricted by low seed yields.

Low seed yields in tetraploid red clover seem to be associated with two primary factors: low fertility, and external or morphological problems. Low fertility may be revealed by poor pollen quality and abnormal megasporogenesis (Evans, 1961; Mackiewicz, 1965). No correlation has been found between frequency of abnormal disjunction and pollen fertility (Eskilsson, 1971). Pollen fertility and embryo survival are under genetic control and are responsive to selection (Papenhagen, 1967; Povilaitis and Boyes, 1959, 1960). However, direct selection for high seed yield may be more effective (Shcheglov et al., 1975; Bragdo-Aas, 1970).

Aneuploidy also may be a factor in low seed production of autotetraploids. A 1–1 euploid/aneuploid equilibrium, which appeared in the first generation after doubling, was not altered by generation number or genetic structure (Maizonnier and Picard, 1970). Maizonnier et al. (1970) concluded that aneuploid plants were less fertile and less competitive than euploids, and that measurements of aneuploidy furnish a basis for the estimation of fertility. Zosimovich et al. (1974) eliminated small seeds from the euploid–aneuploid mixture prior to sowing and reduced the aneuploid content from 50% to 20%. They also showed that the aneuploid gametes were viable and that selection for fertility was not possible by evaluation of pollen fertility alone.

Correlated selection for cross-fertility and self-fertility may result in inbreeding depression and consequently lowered forage and seed yields (Laczynska-Hulewicz, 1963; Bragdo-Aas and Wexelson, 1973). Since the level of outcrossing is high under field conditions, it is possible that a significant degree of self-fertility will not reduce forage and seed yields.

One portion of the external differences between diploids and tetraploids (dubbed reduced sexuality by Bragdo-Aas [1970]) involves the number of ovules per flower, per head, per plant, and per unit of area. Mackiewicz (1965) obtained the following figures for number of florets per head: diploids, 112.4, and tetraploids, 97.2. By far the largest difference, however, was in the number of heads per unit area. Eskilsson and Bingefors (1972) found that the tetraploid cultivar ULVA produced only two-thirds as many heads as the diploid DISA, which was of similar origin. Similar results were found by Gikic (1972) and by Nuesch (1969), who observed that tetraploids were more susceptible to unfavorable weather than diploids.

Another group of external or morphological differences between tetraploids and diploids includes nectar production, corolla tube length and diameter, and preference of bees. Tetraploids have longer and wider corolla tubes than diploids. Greater nectar volume in tetraploids is not enough to compensate for the greater length of tetraploid corollas (Goral, 1968; Skirde, 1963). The longer corolla tubes make insect pollination more difficult, particularly for honeybees, which prefer the diploid form (Goral, 1965). Bumblebees are effective tetraploid pollinators but may be in short supply (Vestad, 1962). Breeding for shorter corolla tubes in tetraploid red clover results in lower seed yields, so this avenue of overcoming lower seed yields is not promising (Julen, 1970). However, some variability among plants in volume of nectar was independent of corolla tube length, suggesting means of overcoming the pollination problem (Bond, 1968).

A nongenetic factor that influences the seed yield of tetraploids is the presence of diploids. Schweiger (1959) showed that the isolation distance should be at least 8 m. If the frequency of diploids is below 3%, the population is self-purifying, but if it is above 3%, the frequency of diploids rises at an increasing rate (Hagberg, 1962).

Finally, properties of the initial materials that are to be doubled have an effect on both forage and seed yield. Odintsova (1970) showed that, in populations derived from chimera plants, intensity of characters of the $2x$ and $4x$ forms were highly correlated and suggested the possibility of selection before doubling. Picard et al. (1970) also found that diploid material selected for high fertility before chromosome doubling resulted in a higher degree of fertility than did unselected materials. Hybridization of genetically diverse material prior or subsequent to doubling seems important for maximizing both forage and seed yields (Polak, 1974; Novoselova and Malasenko, 1970).

Triploids, hexaploids, and octoploids have been found from matings of tetraploids (Maizonnier and Picard, 1970). Triploids and hexaploids are sterile and rather nonvigorous, and octoploids, although possibly possessing low female fertility, are male-sterile (Maizonnier and Picard, 1970; Taylor et al., 1976). Similar euploid and aneuploid numbers have been found from sib-matings of diploid plants (Strzyzewska, 1974). Most aberrant plants have unusual morphological features and are of little agronomic value.

D. INTERSPECIFIC HYBRIDIZATION

Red clover has been crossed with only two species, both annuals (see Section II,B). The hybrid $4x$ T. pratense × $2x$ T. pallidum is a sterile triploid and has been of no value as a source of genetic variation. The other hybrid, $4x$ $T.$ *pratense* $\times$ $4x$ $T.$ *diffusum*, is less perennial than $T.$ *pratense* itself and is of little agronomic value. Backcrosses of the amphidiploid to $4x$ $T.$ *pratense* are possible, but little use has been made of these populations. The backcross of the diploid hybrid to diploid red clover also has been made (Taylor *et al.*, 1978b), but not evaluated. It is possible that $T.$ *diffusum* does not possess sufficiently different genes to be useful as a source of genetic material for improvement of $T.$ *pratense*.

VI. Alteration of Populations through Selection and Hybridization

Red clover, like many other forage crop species, is naturally cross-pollinated. The species is basically self-sterile, owing to the gametophytic incompatibility system. Therefore, natural selection and breeding procedures, such as mass selection, maternal line selection, recurrent selection, and hybridization, have contributed to the development of cultivars or strains in commercial use.

A. NATURAL SELECTION

Fergus and Hollowell (1960) suggested that the genetic plasticity of red clover has contributed to its wide distribution. Through natural selection, new ecotypes developed in response to selective factors of new environments. Specific "land races" or ecotypes are still in existence in almost all red clover growing areas. Natural selection alone, however, cannot be expected to produce types of plants needed by the breeder to develop cultivars with wide adaptation and for diverse agricultural use.

B. MASS SELECTION

In red clover many of the cultivars have been developed by some form of mass selection. Older, improved cultivars are the result of identifying superior ecotypes that have evolved through natural selection. In recent years the cultivars have been developed by breeders primarily using controlled mass selection schemes.

Large populations of space-planted or solid-planted individuals are established and evaluated for the desired characteristics. In some instances, as for disease and insect resistance, large populations are evaluated in greenhouses or controlled environment chambers. Selected plants may then be handled by using several schemes.

Selected plants may be intercrossed among themselves to develop a new population for testing and/or selection. Continued repetition of this process is referred to as phenotypic recurrent selection.

Progeny tests may be conducted by using open-pollinated seed from the selected individuals in the original source nursery. Further selection can then be applied on a maternal line basis (best maternal line continued into next generation or cycle, or incorporated into a synthetic) or on a single-plant basis within a family.

A third scheme is to bulk open-pollinated seed from selected plants in the source nursery and repeat the selection process in successive generations.

Fergus and Hollowell (1960) reviewed the use of these breeding schemes in red clover prior to the late 1950s. Plants resulting from three cycles of phenotypic recurrent selection were superior for persistence, growth habit, and vigor (Malm and Hittle, 1963). Each cycle consisted of three growing seasons, and selection against root-rotting organisms was effective. The phenotypic recurrent selection scheme also has been effective in breeding for resistance to northern anthracnose (Maxwell and Smith, 1971; Smith and Maxwell, 1973; Smith *et al.*, 1973), powdery mildew (Owen, 1977), clover rot (Ludin and Jonsson, 1974), yellow clover aphid [*Theroaphis trifolii* (Monell)], and pea aphid [*Acyrthosiphum pisum* (Harris)] (Gorz *et al.*, 1979), and for reduced formononetin content (Francis and Quinlivan, 1974).

Mokhtarzadek *et al.* (1967) reported that maternal line selection and phenotypic selection were effective in improving persistence in red clover. Improved persistence was accompanied by decreased forage yield and flowering in first-year stands but increased forage yield and delayed flowering in the second harvest year. Selection for resistance to the stem nematode was effective when maternal line selection was used (Bingefors, 1956; Dijkstra, 1956). Dijkstra (1969) used this same method to increase the percentage of two-seeded pods in diploid red clover. It was not effective for tetraploid germplasm. However, Anderson (1973) effectively used maternal line selection to develop a late-flowering tetraploid red clover, GRASSLANDS PAWERA.

Mass selection has been employed effectively to increase height of nectar in the corolla (Hawkins, 1971), dry matter yield (Novoselova and Matskiv, 1974; Julen, 1971), and protein content (Novoselova and Matskiv, 1974). Mass selection was effective in developing germplasm adapted to the Altai and Kalinin Regions of the USSR (Burdina, 1971; Sorokin, 1972). Mass selection followed by progeny testing was used to identify the superior clones of the cultivars NORLAC (Folkins *et al.*, 1976) and KENSTAR (Taylor and Anderson, 1973a).

Combinations of mass and individual selection with and without progeny testing have improved dry matter yield, protein percentage, digestibility, disease and insect resistance, and seed yield in red clover (Julen and Lager, 1966; Schieblich, 1966; Novoselova and Piskovatskaya, 1972; Pamfil *et al.*, 1972;

Cerny and Vasak, 1973; Rogash *et al.*, 1973; Litvinenko, 1974; Vestad, 1974; Julen, 1974, 1975; Novoselova and Cheprasova, 1975; Goral, 1976).

C. PROGENY TESTING, COMBINING ABILITY, AND HERITABILITY

Knowledge of the combining ability of clones or lines becomes important when the breeder considers the basic material to be used in developing a hybrid or synthetic cultivar. Testing of progenies to evaluate breeding value of parents has not been conducted extensively in red clover because of the difficulty of maintaining parents. Red clover propagules are short-lived and susceptible to virus diseases. Taylor *et al.* (1962) found that clones tolerant of viruses and other diseases could be maintained up to 5 years. Of 1500 third-year plants dug from fields of KENLAND red clover, all except 20 were eventually eliminated through clonal evaluation for response to viruses and other diseases (Taylor *et al.*, 1968).

Progeny testing by use of selfs has not been conducted extensively, because most plants are self-incompatible. Limited amounts of seed may be produced via PSC after heat treatment (Kendall and Taylor, 1969; Kendall, 1973). Taylor *et al.* (1970) found that one clone of four tested was nonpersistent both as an I_0 and an I_1 clone. Other I_1 clones showed segregational effects, however.

Top crosses and open-pollinated crosses, although useful in maternal line selection, have not been used extensively for progeny testing. Top-cross progenies were evaluated by Dijkstra (1970) to assess the potential of using reciprocal recurrent selection for the improvement of red clover. He concluded that the performance of the progenies did not warrant a program of reciprocal recurrent selection. Dry matter yield, growth habit, and corolla tube length as measured on open-pollinated progeny of nine clonal genotypes had high heritability values (Lawson, 1971). The cultivar KENSTAR was developed on the basis of the polycross progeny test to evaluate parental clones (Taylor and Anderson, 1973a). The polycross nursery, consisting of 10 plants of 20 clones in 20 replications of a randomized block design, was maintained for 3 years, during which seed was harvested for subsequent polycross progeny testing. Subsequent progeny tests showed significant parent–progeny correlations for persistence, vigor or yield, blooming or seed yield, and mildew resistance. Ten clones were selected for the synthetic cultivar KENSTAR (Taylor *et al.*, 1968). Competitive effects relative to the polycross progeny test have been evaluated by Taylor and Kendall (1965). They reported that seed of outstanding progenies do not always produce similar performance in polycross tests as when composited in synthetics. The polycross progeny test also has been used successfully to evaluate sib lines used to maintain parental genotypes (Torrie *et al.*, 1952).

Diallel crosses of parental clones probably have been used more extensively in red clover than has any other type of progeny testing. Anderson (1960) made diallel crosses of seven noninbred, selected parents and tested the progeny in field spacings of a 7 × 7 balanced lattice design with six replications. He observed significant general combining ability (GCA) and specific combining ability (SCA) between the seven parents for yield, growth habit, persistence, and flowering. GCA variances were greater than SCA variances, and both GCA and SCA interacted with the season. Relatively high heritabilities were obtained for yield, habit of growth, persistence, and flowering time. Anderson et al. (1974b) progeny-tested 10 noninbred clones of red clover by diallel crosses. Seeds of the 45 single crosses were produced by hand-crossing in a greenhouse. GCA was the most important source of genetic variation in this study, and additive genetic variance constituted over 81% of genetic variance. Heritability estimates for persistence, yield, vigor, and date of first bloom ranged from 17 to 42%. In another study, additive genetic variance was a significant portion of the genetic variance among diallel progeny from plants selected from wild and cultivated populations (Ceccarelli, 1971). Genetic variance was greater among populations than within, and there were reciprocal differences for forage yield. Significant GCA effects were observed among male-sterile types of red clover from the variety MOSCOW 1 (Zvyagina, 1973).

Smith and Puskulcu (1976) evaluated the combining ability of six I_1sib_1 lines of red clover by testing diallel progeny in both space-planted and broadcast-planted experiments. Both GCA and SCA were significant for dry matter yield, maturity, visual yield, and regrowth after harvest. Specific effects were large for some combinations, suggesting that nonadditive genetic variance was of importance in this material. However, the SCA × location interaction was significant. Little association was observed between space-planted and broadcast-planted material. Reciprocal effects were of little importance.

Evaluating diallel progeny of 10 I_1 red clover clones, Cornelius et al. (1977) also stressed the importance of nonadditive genetic variance. Reciprocal effects were important in this case. Heritabilities were estimated on individual plants and on hybrid means, and performances of all possible double crosses were predicted. They concluded that progeny testing of I_0 clones was not as useful as testing of I_1 clones because of segregational effects during the inbreeding process.

Julen (1974) reported high heritability values for crude protein content of red clover progeny from a diallel among high and low parents. Selection should be effective in improving protein content, but this could lead to decreased digestibility, particularly of the leaves. High heritability values were observed for qualitative characters and low values for quantitative characters in tetraploid red clover (Jaranowski and Broda, 1977).

D. BACKCROSS BREEDING

Backcross breeding has not been utilized much in red clover, perhaps because of the lack of suitable recurrent parents and the possibility of selection for desirable characters during breeding by other procedures. The only program to the authors' knowledge is the transfer of mildew resistance to the cultivar KENSTAR (Taylor and Anderson, 1974). LAKELAND as the donor parent was crossed with each of 10 clones of Kenstar in 10 separate cages. Seed was harvested only from the clonal parent in each cage, thus ensuring that cross seed was produced. The F_1 and five subsequent backcross generation plants were screened as seedlings for mildew resistance in a greenhouse according to the method of Hanson (1966). Since mildew resistance is inherited as a dominant gene (Stavely and Hanson, 1967), it was necessary to sib-mate the fifth backcross generation, reselect for mildew resistance, and then to progeny-test the selected plants. Heterosis was exhibited after intermating the 10 improved clones and reconstituting the synthetic. Subsequent cultivar evaluation indicated complete transferral of mildew resistance and restoration of forage yields to the level of the original synthetic. Forage yields of advanced synthetic generations were not determined (Taylor and Anderson, 1974).

E. INBREEDING AND HETEROSIS

Perhaps the most important advance in the inbreeding of red clover is the increase in self-seed set by high-temperature treatments. Pseudo-self-compatibility (PSC) increases when red clover clones are grown at 32–38°C (Leffel, 1963; Leffel and Muntjan, 1970). In an excised stem technique, self-seed production ranged from 2 to 20% for 10 clones maintained at 40°C (Kendall and Taylor, 1969; Kendall, 1973). The technique was as follows: Stems were cut at a length of 14 cm when petal color first became visible in the flower bud, or when most florets were suitable for pollination. Usually, about 5 to 10 stems were placed in glass bottles containing an aqueous solution of 2.5% sucrose and 22 ppm of boric acid. During the period of anthesis, the culture bottles were partially submerged in a water bath at 25°C, which was held in an incubator at 40°C. After anthesis, the flower heads were selfed by hand manipulation (rolling) or by tripping the individual flowers with a dissecting needle. During seed development the media and plants were held at 20°C. Another satisfactory technique is to leave the flower heads intact and insert young heads in an incubator maintained at 40°C. Holes suitable for insertion of heads are drilled in the sides of the incubator.

Although PSC is increased by high temperature, it apparently is an inherited characteristic and is transmitted from clones to progenies (Brandon and Leffel,

1968). PSC also decreases with inbreeding, so that it is difficult to maintain inbred lines beyond two generations of selfing (Duncan et al., 1973). It is now apparent that PSC under high temperatures can supply sufficient self-seed for most breeding purposes.

Inbreeding of red clover leads to a decrease in vigor similar to that in other cross-pollinated diploid plants (Fergus and Hollowell, 1960). Considerable variation exists, however, and some inbred families contain plants that equal or exceed their noninbred parents (Taylor et al., 1970). In general, inbred lines are difficult to maintain because of lack of vigor and virus susceptibility.

Red clover also may be inbred by sib-mating for identification and elimination of aberrant types. Among genotypes isolated in this manner are chlorophyll deficiencies, annual growth habit, aneuploid lines with two supernumerary chromosomes ($2n = 16$), and spontaneous polyploids ($2n = 21, 35$, and 42) (Strzyzewska, 1974).

Inbreeding also has been conducted with tetraploid red clover, as discussed in Section V,C. Self-fertility of tetraploid cultivars ranges up to 13% and declines in subsequent generations. Inbreeding depression may be up to 30–40% of the fresh forage weight and 23–26% for seed yield, but, as in diploids, it varies among inbred lines (Laczynska-Hulewicz, 1963).

The primary reason for inbreeding red clover is to obtain inbred lines, which, when hybridized, will give a maximum expression of heterosis. Heterosis also may be obtained by crossing noninbred cultivars. Manner (1963, 1966) reported that bulk crosses gave higher green matter yields than both the mother and the parental means. Control of pollination in this case is not complete, since a part of the seed is derived from intracultivar crosses. Novoselova and Malasenko (1967) and Bekuzarova and Mamsurov (1974) reported that hybrids obtained from crosses of geographically distant forms resulted in heterosis expressed by higher seed and forage yields, leaf size, and winter-hardiness. Shcheglov and Zvyagina (1975) also reported heterosis in six of nine F_1 hybrids of late-maturing cultivars and cytoplasmic male-sterile genotypes. Generally, crosses of inbred parents will result in heterosis, and the vigor of the F_1 will on the average equal that of the noninbred parent. Some combinations may be significantly superior to the noninbred parents (Taylor et al., 1970; Krstic, 1972). Utilization of heterosis, then, involves a search for superior combining ability (discussed in previous section).

A secondary reason to inbreed red clover is to obtain homozygous S-allele genotypes necessary for control of crossing (see Section VII,B,2).

F. TISSUE CULTURE

The use of tissue culture for breeding and improvement of red clover has been limited. However, recent research on growth media and redifferentiation suggests that procedures are now available for variant selection, meristem culture, and embryo culture (Bingham *et al.*, 1977; Beach and Smith, 1978; Phillips and Collins, 1977, 1978; Ranga Rao, 1976). Zakrzewski and Zakrzewska (1976) reported the use of red clover callus tissue as a medium for propagating the stem nematode. Freeing breeding clones of viruses, and long-term cold storage of meristematic tissue of valuable genotypes, appear to be two of the most immediate practical uses of tissue culture.

VII. Use of Selected Materials

The ultimate goal of any plant breeding program is to develop and make available to the grower germplasm that is superior in performance to that being used at the current time.

A. GERMPLASM RELEASES

Often material is developed by the breeder that possesses unique characteristics, such as insect or disease resistance, or persistence, but that has not been developed into a cultivar. Breeders may officially release and provide limited amounts of seed of these populations to other breeders for further selection and use. Formal release procedures are similar to those for a cultivar, except that extensive testing and maintenance of a continual supply of initial seed are not required. A specific example is the release of the hybrid between *Trifolium pratense* and *T. diffusum* Ehrh. (Taylor and Anderson, 1973b).

Germplasm of other seed lots is interchanged among breeders on an informal basis. It is commonly understood that any user or developer of cultivars from either formally or informally shared germplasm will recognize the organization and breeder responsible for developing the initial germplasm.

B. CULTIVAR DEVELOPMENT

Older cultivars were developed by direct use of specific ecotypes or selection among open-pollinated progenies of parent material that had been subject to natural selection. In recent years the cultivars are the result of breeding programs that apply conscious selection for the choice of parental material. Cultivars originating from these programs are synthesized from the selected parent material. In recent years emphasis has been placed on the development of hybrids in red clover.

1. Synthetic Cultivars

The recombination of selected clones or lines into synthetic cultivars is an effective method of utilizing stocks with superior characteristics. The initial clones or lines (Syn 0) are used to produce the F_1 (Syn 1). Because of limited seed quantities in Syn 1, the material may be advanced by random mating to the Syn 3 or Syn 4 generation before it is available to the farmer. Factors such as yeild or performance of parent clones or lines, number of parent clones or lines combined, yield of F_1 crosses, and extent of natural cross-pollination may affect performance of the synthetic cultivar. The maximum performance should be realized in the Syn 1 generation, with a decline in subsequent generations depending upon the above factors.

In the absence of natural selection, the yield of a synthetic cultivar should not decline beyond the Syn 2 generation. However, if natural selection is important, the cultivar may be maintained in a limited generation program. ARLINGTON (Smith *et al.*, 1973), KENSTAR (Taylor and Anderson, 1973b), and NORLAC (Folkins *et al.*, 1976) are recent examples of cultivars that have been developed as synthetics. ARLINGTON is the result of intercrossing six lines, and KENSTAR and NORLAC are advanced generations of intercrossing 10 and 11 initial clones, respectively.

2. Hybrids

Combining highly selected materials into single- or double-cross hybrids maximizes genetic gain, since both additive and dominance genetic variance are utilized. Genetic male sterility is one method of controlling crossing in red clover, and several genetic male steriles have been isolated (Smith, 1971; Macewicz, 1976a, b; Taylor *et al.*, 1978). Smith concluded that the male-sterile gene could be employed with the self-fertility gene (S_f) in a hybrid breeding program. It is doubtful, however, that genetic male sterility could compete with other systems for controlling crossing in red clover.

In common with other crops, cytoplasmic male sterility (CMS) may be more useful for hybridization of red clover than genetic male sterility. Shcheglov and Zvyagina (1975) and Zvyagina (1973) have reported CMS induced by colchicine. In crosses with late-maturing cultivars, most of the forms used as male parents were total or partial fertility restorers.

Considerable research has been conducted on the possibility of using the S-allele system for the control of pollination for single- and/or double-cross hybrid red clover. One method for control of crossing and theoretical expectations of S-alleles in I_1 single and double crosses are shown in Fig. 1 (Anderson *et al.*, 1972). I_0 clones are inbred one generation by PSC, producing homozygous

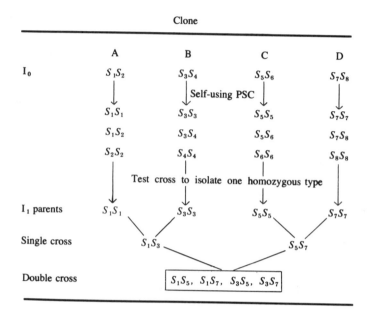

FIG. 1. Genotypic expectations of S-alleles in the production of double-cross hybrid red clover.

S-allele genotypes maintained vegetatively for use in crossing (Leffel, 1963; Johnston *et al.*, 1968). Results of control of crossing by the S-allele system conform to theoretical expectations according to a gene marker analysis conducted by Anderson *et al.* (1972). One of the most difficult problems is maintenance of parental lines, which are low in vigor and easily infected with viruses.

Leffel (1963) suggested that seed maintenance of parental lines might be feasible if mutation to the self-fertility allele (S_f) does not occur (Fig. 2). Leffel and Muntjan (1970) elaborated the seed maintenance scheme as follows: I_0 plants are vegetatively increased and are selfed at high temperature. Each I_1 line is maintained and increased by the composite sib method. All seed produced by this sib-mating would be heterozygous—that is, S_1S_2. The I_1sib_1 population would be sib-mated with bees at high temperatures, perhaps in large greenhouses or in the southwestern United States. The expected segregation of S-alleles in the I_1sib_2 generation would be the same as that in the I_1 generation. The F_1 single-cross $S_1S_3 \times S_5S_7$ would result from the cross of two lines in the I_1sib_1 or subsequent alternate generations. Each of the two lines would be self- and sib-incompatible in the normal environment. A heritable degree of PSC of about 8–10% probably would be necessary.

Because of the difficulties of isolating lines for seed maintenance with 8–10% PSC, Duncan *et al.* (1973) proposed vegetatively increasing superior I_1 plants with high PSC. These I_1 clones then would be isolated under high-temperature field conditions, and self-seed would produce the I_2. Selfed seed from two I_1 lines (I_2's) would be mixed and sown to produce single-cross seed. Single-cross seed from different clonal sources would be mixed and sown for the production of double-cross hybrid red clover.

Implicit in the seed maintenance scheme is the stability of S-alleles. Anderson *et al.* (1974) observed a change to a new S-allele during inbreeding and cautioned that continued seed maintenance by PSC and sib-mating in alternate generations may cause an increase in frequency of new S-alleles; thus two isolated inbreds would sib-mate as well as cross.

Evidence exists that hybridization of inbred lines without prior selection for combining ability does not result in superior hybrids (Anderson *et al.*, 1972; Taylor and Anderson, 1974). A test of combining ability in 10 I_1 single crosses of red clover showed that the best and poorest performing single crosses for persistence were, respectively, 197% and 33% of the check cultivar, KENSTAR (Cornelius *et al.*, 1977). Predicted performance of all possible double crosses of the 10 I_1 clones indicated that the best and poorest could be expected to be 169% and 61%, respectively, of the check. Cornelius *et al.* also found that genetic variance among modified single-cross hybrids (both I_1 parents from a single I_0 clone) is considerably greater than it is among double-cross hybrids. They concluded that high-seed-yielding and vigorous I_1 clones should be isolated for comparison of their performance both in modified single crosses and in double crosses.

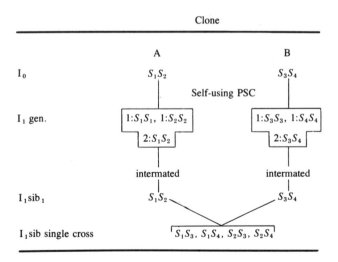

FIG. 2. Genotypic expectation of S-alleles in the production of I_1-sib$_1$ single-cross hybrids.

Theoretically, advanced generations of single- or double-cross hybrids would be expected to decline in forage yield. Smith (1978) examined advanced generations of 15 single crosses derived from a diallel cross of I_1sib_1 lines of six red clover clones and observed an average of 4% decline in the Syn 2 and 9% in the Syn 3. However, some families did not decline in yield in advanced generations. He concluded that, with enough breeding effort, it should be possible to develop high performing combinations that will retain their performance in advanced generations of seed increase.

C. CULTIVAR OR HYBRID RELEASE

In many countries, performance testing of cultivars prior to release is required. In the United States, testing is required by the state experiment stations and the federal government prior to release of publicly developed cultivars (Fergus and Hollowell, 1960). Privately developed cultivars are tested by the commercial companies themselves, but no uniform standards have been established. A cultivar review board has been suggested for red clover cultivars prior to certification by the Association of Official Seed Certifying Agencies (AOSCA, Seed Certification Handbook, 1971). Cultivars may be registered by the Crop Science Society, provided certain criteria, primarily uniqueness, are met (Garrison, 1968). Breeders' rights to released cultivars also may be protected under the Plant Variety Protection Act (PVPA, Anonymous, 1973) administered by Federal Seed Regulatory officials, often acting in cooperation with state regulatory officials. A cultivar may be protected under PVPA either by private litigation or under the seed certification option. The certification option requires that the protected cultivar be sold by cultivar name only as a class of certified seed. In this option, the Federal Seed Regulatory Office assumes the responsibility for enforcement. The primary requirement for plant variety protection is distinctiveness from other red clover cultivars in existence. KENSTAR and REDLAND are examples of protected cultivars.

VIII. Maintenance of Genetic Stability during Seed Multiplication

Frequently, to increase rapidly and to maintain seed supplies of new cultivars, the seed must be produced outside the area of forage adaptation. High-seed-yielding areas include the western United States, Canada, and Israel. Because most red clover cultivars are heterogeneous and heterozygous, and the environments of the seed- and forage-producing areas are likely to be different, opportunity exists for shifts away from the bred characteristics. Trueness-to-type tests must be conducted, comparing the increased seed with the original.

Causal factors involved in genetic shifts are not completely understood. Apparently, a shift toward earliness and winter susceptibility occurs with increases at southern locations (Bula et al., 1965, 1969). According to Taylor et al. (1966), early-flowering genotypes, particularly at southern locations, produce the most seed, shifting the cultivar toward an earlier type. Earliness and lack of persistence have been shown to be related (Taylor et al., 1966). Other factors, such as lack of attention to land history, volunteer seedlings, lack of isolation, and seed mixtures in harvesting and seed-cleaning equipment, cannot be ruled out. Differential survival of genotypes due to diseases in diverse environments also is a factor, particularly in old stands.

Prevention of genetic shifts is important because of the necessity of continuing seed production in high producing areas. Steps that have been taken to prevent or minimize shifts include:

(1) Limiting the number of generations of seed increase (example, KENSTAR).

(2) Restricting the area of seed increase to more northern locations (example, KENSTAR).

(3) Clipping the first growth to allow more equal flowering of genotypes. This practice reduces seed yield in the western United States (Dade, 1966; Rincker et al., 1977) but is a prevalent practice in the forage-producing area of the eastern United States (Taylor et al., 1966).

(4) Preventing seed production on first-year stands (Example, KENSTAR). Some evidence indicates that differences in seed production exist between stands that have or have not been subjected to freezing (Bula et al., 1965, 1969).

(5) Restriction to less heterogeneous cultivars (Example, TEPA vs. ALAS-KALAND, Dovrat and Waldman, 1966). Apparently, the less diverse cultivars are less subject to changes in population structure.

(6) Strict adherence to AOSCA certification requirements, consisting in land history, isolation, control of volunteering, and seed mixtures.

In summary, most investigations, while indicating the possibility of genetic shifts in red clover cultivars, also suggest means of limiting shifts. It appears likely that seed increases can be continued outside the area of forage adaptation without serious loss of desirable agronomic characteristics.

REFERENCES

Akerberg, E. 1974. Hereditas 77, 177–182.
Anderson, L.B. 1960. N. Z. J. Agric. Res. 3(4), 680–692.
Anderson, L.B. 1973. N. Z. J. Agric. Res. 16(3), 395–398.
Anderson, M.K., Taylor, N.L., and Kirthavip, R. 1972. Crop Sci. 12, 240–242.

Anderson, M.K., Taylor, N.L., and Duncan, J.F. 1974a. *Euphytica* **23**, 140-148.

Anderson, M.K., Taylor, N.L., and Hill, R.R., Jr. 1974b. *Crop Sci.* **14**, 417-419.

Anikeenko, A.P. 1971. *Sel' skokhoz Biol.* **6**, 771-772.

Anonymous. 1973. U.S. Plant Variety Protection Act of Dec. 24, 1970. U.S.D.A. Agric. Mkt. Service.

Armstrong, K.C., and Cleveland R.W. 1970. *Crop Sci.* **10**, 354-357.

Artemenko, M.I. 1972. *Ukr. Kii Bot. Zh.* **29**(2), 232-234.

Athow, K.L., and Davis, R.L. 1958. *Phytopathology* **48**, 437-438.

Beach, K., and Smith, R.R. 1978. *Agron. Abstr.* **70**.

Bekuzarova, S.A., and Mamsurov, B.K. 1974. *Tr. Sev. Kavk. NII Gorn. Predgorn. S. Kh.* **1**, 78-81.

Bellmann, K. 1966. *Zuchter* **36**, 126-135.

Berthaut, J. 1965. *Ann. Amelior. Plant* **15**, 37-51.

Bingefors, S. 1956. *Nematologica* **1**, 102-108.

Bingefors, S., and Akerberg, E. 1961. *Euphytica* **10**, 147-151.

Bingefors, S., and Quittenbaum, G. 1959. *Z. Pflanzenzuecht.* **42**, 214-222.

Bingham, E.T., Smith, R.R., Rupert, E.A., and Tomes, D.T. 1977. *Agron. Abstr.* **49**.

Bond, D.A. 1968. *J. Agric. Sci. (Cambridge)* **71**, 113-116.

Bragdo-Aas, M. 1970. *Meld. Norg. Landbr. Hogsk.* **49**, 1-34.

Bragdo-Aas, M., and Wexelsen, H. 1973. *Meld. Norg. Landbr. Hogsk.* **52**, 21.

Brandon, R.A., and Leffel, R.C. 1968. *Crop Sci.* **8**, 185-186.

Bugge, G. 1970. *Z. Pflanzenzuecht.* **63**, 196-208.

Bula, R.J., May, R.G., Garrison, C.S., Rincker, C.M., and Dean, J.G. 1965. *Crop Sci.* **5**, 425-428.

Bula, R.J., May, R.G., Garrison, C.S., Rincker, C.M., and Dean, J.G. 1969. *Crop Sci.* **9**, 181-184.

Burdina, V.M. 1971. *Tr. VNII Lna.* **9**, 109-113.

Ceccarelli, S. 1968. *Genet. Agr. (Pavia)* **22**, 81-88.

Ceccarelli, S. 1971. *Rev. Agron.* **5**, 2-3 and 89-97.

Cerny, J., and Vasak, J. 1973. *Sb. Vys. Sk. Zemed. Praze A* **1**, 149-159.

Cornelius, P.L., Taylor, N.L., and Anderson, M.K. 1977. *Crop Sci.* **17**, 709-713.

Crowley, J.G. 1975. *Farm Food Res.* **6**, 38-40.

Dade, E. 1966. *Crop Sci.* **6**, 348-50.

de Nettancourt, D. 1969. *Theor. Appl. Genet.* **39**, 187-196.

Denward, T. 1963. *Hereditas* 49, 189-202; 203-236; 285-329; 330-334.

Diachun, S., and Henson, L. 1966. *Phytopathology* **56**, 345.

Diachun, S., and Henson, L. 1974a. *Proc. Int. Grassl. Congr., 12th* **3**, 750-753.

Diachun, S., and Henson, L. 1974b. *Phytopathology* **64**(5), 758-759.

Dijkstra, J. 1956. *Euphytica* **5**, 298-307.

Dijkstra, J. 1969. *Euphytica* **18**, 340-351.

Dijkstra, J. 1970. *Euphytica* **19**, 40-46.

Dijkstra, J., and Speckmann, G.J. 1965. *Euphytica* **14**, 244-248.

Dovrat, A., and Waldman, M. 1966. *Proc. Int. Grassl. Congr., 10th* pp. 787-792.

Duncan, J.F., Anderson, M.K., and Taylor, N.L. 1973. *Euphytica* **22**, 535-542.

Engelke, M.C., and Smith, R.R. 1974. *Agron. Abstr.* **52**.

Engelke, M.C., Smith, R.R., and Maxwell, D.P. 1975. *Plant. Dis. Rep.* **59**, 959-963.

Engelke, M.C., Smith, R.R., and Maxwell, D.P. 1977. *Crop Sci.* **17**, 465-468.

Eskilsson, L. 1971. *Z. Pflanzenzuecht.* **66**, 221-234.

Eskilsson, L., and Bingefors, S. 1972. *Z. Pflanzenzuecht.* **67**, 103-119.

Evans, A.M. 1961. *Eucarpia. Proc. Symp. Fertil. Tetraploid Clover* pp. 20-26.

Evans, G. 1957. *J. Agric. Soc. Univ. Coll. Wales* **38**, 30-35.

Fergus, E.N., and Hollowell, E.A. 1960. *Adv. Agron.* **12**, 365–436.
Folkins, L.P., Berkenkamp, B.B., and Baenziger, H. 1976. *Can. J. Plant. Sci.* **56**, 757–758.
Frame, J. 1976. *J. Br. Grassl. Soc.* **31**, 139–152.
Francis, C.M., and Quinlivan, B.J. 1974. *Proc. Int. Grassl. Congr., 12th* **3**, 754–758.
Garrison, C.S. 1968. *Crop Sci.* **8**, 261–262.
Garrison, R.H., and Newton, S.B. 1974. Prod. Publ. No. 28. Assoc. Off. Seed Cert. Agencies.
Gikic, M. 1972. *Poljop. Znan. Smotra* **28**, 173–188.
Goral, S. 1965. *Postep. Nauk Roln.* **12**, 21–30.
Goral, S. 1968. *Roczn. Nauk* **94**, 457–473.
Goral, S. 1972. *Biul. Inst. Hodowli Aklimat. Roslin* No. 5/6, 159–161.
Goral, S. 1976. *Hod. Ros. Aklim. Nasienn.* **20**, 145–155.
Gorz, H.J., Manglitz, G.R., and Haskins, F.A. 1979. *Črop Sci.* **19**, 257–260.
Graham, J.H., and Newton, R.C. 1971. *Plant. Dis. Rep.* **43**, 1114–1116.
Gras, N.S.B. 1940. "A History of Agriculture in Europe and America." Crafts, New York.
Hagberg, A. 1962. *Z. Pflanzenzuecht.* **47**, 277–285.
Hanson, E.W. 1966. *Proc. Int. Grassl. Congr., 10th* pp. 734–737.
Hawkins, R.P. 1971. *J. Agric. Sci.* **77**, 347–350.
Hindmarsh, G.J. 1963. *Aust. J. Bot. 1964* **12**, 1–14.
Horasajlov, N.C. 1964. *Bull. Appl. Bot. Genet. Plant. Breed.* **36**, 14–58.
Hossain, M. 1961. *Notes R. Bot. Garden* **23**, 387–481.
Hunt, I.V., Frame, J., and Harkess, R.D. 1974. Experimental Record, Advisory and Development Ser., West of Scotland Agric. College, No. 38.
Jahnova, V. 1973. *Sb. Ved. Pr. Vyzk. Stanice Picn. Troubsk. Brna* No. 3, 57–71.
Jaranowski, J.K., and Broda, Z. 1977. *Genet. Pol.* **18**, 331–336.
Johnston, K., Taylor, N.L., and Kendall, W.A. 1968. *Crop Sci.* **8**, 611–614.
Jones, I., and Plummer, G.L. 1960. *Agron. J.* **52**, 462–464.
Julen, G. 1969. *Tidsk.* **79**, 80–111.
Julen, G. 1970. *Sart. Sverges Utsad. Sonderdr. Ausg.* **101**, 217–224.
Julen, G. 1971. *Eucarpia. Rep. Meet. Stat. Amelior. Plant. Fourag*ères, Lusignan, France.
Julen, G. 1974. *Vaxtodling* No. 29, 121–125.
Julen, G. 1975. *Nord. Jordbrugsforsk.* No. 4, 931–932.
Julen, G., and Lager, A. 1966. *Sart. Sverges Utsad. Tidsk.* (5–6), 324–339.
Kaneko, K. 1972. *Res. Bull. Hokkaido Natl. Agric. Exp. Sta.* No. 101, 1–50.
Kendall, W.A. 1973. *Crop Sci.* **13**, 559–561.
Kendall, W.A., and Taylor, N.L. 1969. *Theor. Appl. Genet.* **39**, 123–126.
Khan, M.A., Maxwell, D.P., and Smith, R.R. 1978. *Phytopathology* **68**, 1084–1086.
Krstic, O. 1972. *Arlllv Za Poljoprivr. Nauk.* **25**, 107–113.
Laczynska-Hulewicz, T. 1963. *Genet. Pol.* **4**, 97–119.
Lawson, N.C. 1971. *Res. Agron. Quebec* No. 16, 69.
Leffel, R.C. 1963. *Crop Sci.* **3**, 377–380.
Leffel, R.C., and Muntjan, A.I. 1970. *Crop Sci.* **10**, 655–658.
Liang, G.H.L. 1965. Ph.D. Thesis, Univ. Wisconsin. *Diss. Abstr.* **25**, 6173.
Litvinenko, F.P. 1974. *Byul. Mironov NII Sel. Semenovo. Pshenit.* **5**, 12–13.
Lovett, J.V., and Neal-Smith, C.A. 1974. *Plant. Introduct. Rev.* **10**, 20–32.
Ludin, P., and Jonsson, H.A. 1974. *Agr. Hort. Genet.* **32**(1/4), 44–54.
Macewicz, J. 1976a. *Hodowla Rosl. Aklim. Nasienn.* **20** 175–194.
Macewicz, J. 1976b. *Hodowla Rosl. Aklim. Nasienn.* **20** 195–209.
Mackiewicz, T. 1965. *Genet. Pol.* **6**, 5–40.
Maizonnier, D. 1972. *Ann. Amelior. Plant.* **22**, 375–387.
Maizonnier, D., and Picard, J. 1970. *Ann. Amelior Plant.* **20**, 407–420.

Maizonnier, D., Picard, J., and Berthaut, J. 1970. *Ann. Amelior. Plant.* **20**, 421-431.

Makarov, N.M. 1973. *Sib. Vestn. Skh. Nauk* No. 3, 23-26 and 114.

Malm, N.R., and Hittle, C.N. 1963. *Crop Sci.* **3**, 285-288.

Mal'tsev, A.V., and Kuzin, A.M. 1974. *Radiobiologiya* **14**, 480-485.

Manner, R. 1963. *J. Sci. Agric. Soc. Finland* **35**, 47-55.

Manner, R. 1965. *Suom. Maataloust Seur. Julk.* **107**, 171-174.

Manner, R. 1966. *Proc. Int. Grassl. Congr., 10th* pp. 726-729.

Marek, J., and Riman, L. 1969. *Pol. Hospodarst.* **15**, 615-620.

Matsu-ura, M., Maki, Y., and Hayakawa, R. 1974. *Res. Bull. Hokkaido Natl. Agric. Exp. Sta.* No. 108, 99-105.

Maxwell, D.P., and Smith, R.R. 1971. *Plant Dis. Rep.* **55**, 920-922.

Mokhtarzadek, A., Leffel, R.C., and Beyer, E.H. 1967. *Crop Sci.* **7**, 264-266.

Najcevska, C.M., and Speckmann, G.J. 1968. *Euphytica* **17**, 357-362.

Navalikhina, N.K. 1965. *Ukrain. J. Bot.* **22**, 8-14.

Nordenskiold, H. 1971. *Hereditas* **69**, 301-302.

Novoselova, A.S., and Cheprasova, S.N. 1975. *Ref. Zh.* **9**, 258.

Novoselova, A.S., and Malasenko, V.S. 1965. Mutagenesis in higher plants. *Abstr. Rep. Symp. Moscow*, pp. 115-117.

Novoselova, A.S., and Malasenko, V.S. 1967. *Proc. All Un. Lenin Acad. Agric. Sci.* **4**, 16-18.

Novoselova, A.S., and Malasenko, V.S. 1970. *Breed. Seed Grow.* No. 1, 39-41.

Novoselova, A.S., and Matskiv, O.I. 1974. *Proc. Int. Grassl. Congr., 12th* **3**, 925-928.

Novoselova, A.S., and Piskovatskaya, R.G. 1972. *Sb. Nauch. Rabot VNII Kormov.* **40**, 87-97.

Nuesch, B. 1966. *Proc. Int. Grassl. Congr., 10th* pp. 661-663.

Nuesch, B. 1969. *Schweiz. Landw. Forsch.* **8**, 284-298.

Nuesch, B. 1974. *Schweiz. Landw. Forsch.* **13**, 59-75.

Nutman, P.S. 1954. *Heredity* **8**, 47-60.

Nutman, P.S. 1957. *Heredity* **11**, 157-174.

Nutman, P.S. 1968. *Heredity* **23**, 537-551.

Odintsova, I.G. 1970. *Genet. Moskva* **6**, 44-49.

Owen, C.R. 1977. *La. Agric. Exp. Sta. Bull.* 702.

Pamfil, C., and Savatti, M. 1971. *Agricultura* **27**, 227-235.

Pamfil, C., Savatti, M., Sebok, C., and Goia, E. 1972. *Bull. Acad. Sci. Agr. For.* **1**, 31-38.

Pandey, K.K. 1970. *Nature (London)* **227**, 689-690.

Papenhagen, F. 1967. *Tagungsb. Dtsch. Akad. Landw. Wiss. Berlin* No. 95, 273-279.

Phillips, G.C., and Collins, G.B. 1977. *Agron. Abstr.* 67.

Phillips, G.C., and Collins, G.B. 1978. *Agron. Abstr.* 83.

Picard, J., and Berthaut, J. 1966. *Proc. Int. Grassl. Congr., 10th* pp. 664-666.

Picard, J., Maizonnier, D., and Berthaut, J. 1970. *Ann. Amelior. Plant.* **20**, 389-406.

Pieters, A.J., and Hollowell, E.A. 1937. U.S. Dept. Agric. Yearbook, pp. 1190-1201.

Piper, C.V. 1924. "Forage Plants and Their Culture," 2nd ed. Macmillan, New York.

Polak, J. 1974. *Poda Uroda* **22**, *Suppl.* 3-4.

Povilaitis, B., and Boyes, V.W. 1959. *Can. J. Plant Sci.* **39**, 364-374.

Povilaitis, B., and Boyes, J.W. 1960. *Can. J. Bot.* **38**, 507-539.

Quesenberry, K.H., and Taylor, N.L. 1976. *Crop Sci.* **16**, 382-386.

Quesenberry, K.H., and Taylor, N.L. 1977. *Crop Sci.* **17**, 141-145.

Quesenberry, K.H., and Taylor, N.L. 1978. *Crop Sci.* **18**, 536-540.

Ranga Rao, V. 1976. *Plant Sci. Lett.* **6**, 77-83.

Ravantii, S. 1965. *Suom. Maataloust. Seur. Julk.* **107**, 272-299.

Reitberger, A. 1964. *Zuechter* **34**, 129-135.

Riman, L., and Mistinova, A. 1970. *Ved. Prac. Vysk. Ustav. Rast. Vyrob. Piest.* **8**, 355-388.

Riman, L., and Mistinova, A. 1971. *Ved. Prac. Vysk. Ustav. Rast. Vyrob. Piest.* **9**, 57–71.

Riman, L., and Polak, J. 1968. *Pol. Nahospod.* **14**, 836–851.

Rincker, C.M. 1974. *Crop Sci.* **14**, 749–750.

Rincker, C.M., Dean, J.G., Garrison, C.S., and May, R.G. 1977. *Crop Sci.* **17**, 58–60.

Rogash, A.R., Razin, N.S., and Burdina, V.M. 1973. *Tr. VNII Lna.* **11**, 89–93.

Rohweder, D.A., Shrader, W.D., and Templeton, W.C., Jr. 1977. *Crops Soils* **29**, 11–15.

Sakuma, T., Shimanuki, T., and Suginobu, K. 1973. *Jn. Soc. Grassl. Sci. J.* **7**, 242–244.

Schieblich, J. 1966. *Zuchter* **36**, 236–39.

Schwer, J.F., and Cleveland, R.W. 1972a. *Crop Sci.* **12**, 321–324.

Schwer, J.F., and Cleveland, R.W. 1972b. *Crop Sci.* **12**, 419–422.

Schweiger, W. 1959. *Zuechter* **29**, 225–229.

Shcheglov, Yu. S., and Zvyagina, G.N. 1975. *Vestn. Selsk. Vennoi Nauk.* No. 6, 31–36.

Shcheglov, Yu. S., Aleksashova, V.S., and Strekalova, V.A. 1975. *Sel. Semen. (MOSC)* **6**, 26–28.

Simon, U. 1962. *Crop Sci.* **2**, 258.

Skirde, W. 1963. *Ann. Agric. Fenn.* **20**, 73–90.

Smith, R.R. 1971. *Crop Sci.* **11**, 326–327.

Smith, R.R. 1974. *Euphytica* **23**, 597–600.

Smith, R.R. 1978. *Agron. Abstr.* 62.

Smith, R.R., and Maxwell, D.P. 1973. *Crop Sci.* **13**, 271–273.

Smith, R.R., and Puskulcu, H. 1976. *Agron. Abstr.* 62.

Smith, R.R., Maxwell, D.P., Hanson, E.W., and Smith, W.K. 1973. *Crop Sci.* **13**, 771.

Smith, R.R., Taylor, N.L., and Langford, W.R. 1978. *Proc. South. Past. For. Crop Improv. Conf.,* *35th Sarasota, Fla.* pp. 146–155.

Sorokin, V.D. 1972. *Tr. Blisk. Opyt Selekts. St.* **3**, 145–160.

Stavely, J.R., and Hanson, E.W. 1967. *Phytopathology* **57**, 193–197.

Strzyzewska, C. 1974. *Genet. Pol.* **15**, 255–293.

Taylor, N.L. 1979. *Crop Sci.* **19**, 564.

Taylor, N.L., and Anderson, M.K. 1973a. *Crop Sci.* **13**, 772.

Taylor, N.L., and Anderson, M.K. 1973b. *Crop Sci.* **13**, 777.

Taylor, N.L., and Anderson, M.K. 1974. *Proc. Int. Grassl. Congr., 12th* **3**, 985–990.

Taylor, N.L., and Kendall, W.A. 1965. *Crop Sci.* **5**, 50–52.

Taylor, N.L., Gray, E., Stroube, W.H., and Kendall, W.A. 1961. *Crop Sci.* **1**, 458–460.

Taylor, N.L., Stroube, W.H., Kendall, W.A., and Fergus, E.N. 1962. *Crop Sci.* **2**, 303–305.

Taylor, N.L., Stroube, W.H., Collins, G.B., and Kendall, W.A. 1963. *Crop Sci.* **3**, 549–552.

Taylor, N.L., Dade, E., and Garrison, C.S. 1966. *Crop Sci.* **6**, 535–538.

Taylor, N.L., Kendall, W.A., and Stroube, W.H. 1968. *Crop Sci.* **8**, 451–454.

Taylor, N.L., Johnston, K., Anderson, M.K., and Williams, J.C. 1970. *Crop Sci.* **10**, 522–525.

Taylor, N.L., Anderson, M.K., Quesenberry, K.H., and Watson, L. 1976. *Crop Sci.* **16**, 516–518.

Taylor, N.L., Gibson, P.B., and Knight, W.E. 1977. *Crop Sci.* **17**, 632–634.

Taylor, N.L., Kitbamroong, C., and Anderson, M.K. 1978a. *Crop Sci.* **18**, 1033–1036.

Taylor, N.L., Quarles, R.F., and Anderson, M.K. 1978b. *Agron. Abstr.* pp. 64–65.

Taylor, N.L., May, R.G., Decker, A.M., Rincker, C.M., and Garrison, C.S. 1979. *Crop Sci.* **19**, 429–434.

Thomas, H.L. 1969. *Crop Sci.* **9**, 365–366.

Torrie, J.H., Hanson, E.W., and Allison, J.L. 1952. *Agron. J.* **44**, 569–573.

Van Bogaert, G. 1958. *Rev. Agric. (Brussels)* **11**, 947–955.

Vavilov, P.P., Kabysh, V.A., and Putnikov, L.I. 1976a. *Izv. Timiryaz. Selsk. Akad.* **1**, 145–153.

Vavilov, P.P., Kabysh, V.A., and Putnikov, L.I. 1976b. *Selsk. Nauk Imeni V.1. Lenina* **12**, 8–9.

Vestad, R. 1960. *Euphytica* **9**, 35–38.

Vestad, R. 1962. *Sart. Medd. Sveriges. Frood.* pp. 1–7.

Vestad, R. 1974. *Nord. Jordbrugsfors.* **56**, 366–369.
Vlasyuk, I.I., Pomogaibo, V.M., and Pestova, T.M. 1977. *Tsitol. Genet.* **11**, 169–171.
Wexelsen, H. 1965. *Suom. Maataloust. Seur. Julk.* **107**, 30–43.
Wexelsen, H., and Aastveit, K. 1967. *Meld. Norg. Landbr. Hogsk.* **46**(3), 1–20.
Whittington, W.J. 1958. *J. Hered.* **49**, 202.
Wojciechowska, W. 1970. *Genet. Pol.* **11**, 385–386.
Zakrzewski, J., and Zakrzewska, E. 1976. *Hod. Ros. Aklimat. Nasien.* **20**, 97–104.
Zeiders, K.E., Graham, J.H., Sprague, V.G., and Wilkinson, S.R. 1971. *U.S.D.A., Plant. Sci. Res. ARS* No. 34–126.
Zohary, M. 1971. *Candollea* **26**, 297–308.
Zohary, M. 1972. *Candollea* **27**, 99–158.
Zosimovich, V.P., Navalikhina, N.K., and Sidak, V.A. 1973. *Vestn. Sel'sk. Nauki, Moscow* **8**, 50–54.
Zosimovich, V.P., Navalikhina, N.K., Marekha, L.N., and Pavlenko, G.S. 1974. *Moscow Nauka.* pp. 172–184.
Zvyagina, G.N. 1973. *Nauch. Tr. NII Kh. Tsentr. Rnov Nechern.* No. 29, 98–100.

ADVANCES IN AGRONOMY, VOL. 31

THE AVAILABILITY OF NUTRIENTS IN THE SOIL AS DETERMINED BY ELECTRO-ULTRAFILTRATION (EUF)

K. Németh

Büntehof Agricultural Research Station, Hannover, Federal Republic of Germany

I. Introduction

In a fertile soil, it is understood that the most important plant nutrients (including water and air) are effectively available in the course of the vegetation period. The practical farmer needs to know the amount of effective available nutrients in his soil before the crop is grown so that he can adapt his fertilizer measures accordingly.

Methods of soil analysis should be simple to apply, rapid, and cheap. For this reason, attempts have been made to determine the soil nutrients chemically by single extractions (for example, NH_4 acetate for potassium) affording speed and simple operation. With these rapid methods (Bray, 1937; Egner et al., 1960; Olsen et al., 1954; etc.) it was possible even a few decades ago to single out nutrient-deficient soils for corrective application of fertilizer to obtain higher yields. The nutrient contents of most agricultural soils are higher today than they were a decade ago. Also, high yields are necessary for farming to be economic and to raise world food production. These yields are possible because of the genetic potential of modern cultivated plants. A more detailed evaluation of the status of nutrients in the soil by a more precise method is required, therefore, to ascertain the reliability of fertilizer recommendations. After numerous tests, the electro-ultrafiltration (EUF) method with varying voltage and temperature presented here appears especially suitable to characterize the nutrient status of a soil comprehensively.

155

II. Problems of Conventional Soil Testing Practice

In soil analysis, only methods that can well define the amount of effectively available nutrients are suitable for the determination of the fertilizer requirements of a plant. Which nutrients are *effectively available?* Certainly those nutrients are effectively available that are present in the *soil solution* in a dissolved ionic form, as illustrated in Fig. 1. This figure shows a root hair, the soil solution with dissolved ions, and a clay mineral with adsorbed ions. The ions of the soil solution are moving permanently (in a thermal motion). The principle is, therefore, easy to understand: The more ions that are present in the soil solution—that is, the higher the concentration—the more ions can reach the plant roots per unit of time by diffusion and mass flow, and consequently the more ions can be taken up by the plant (Barber, 1962; Barber *et al.*, 1963; Rowell *et al.*, 1967; Mengel *et al.*, 1969; Fox and Kamprath, 1970; Grimme *et al.*, 1971; Németh, 1975).

The mineral lattice ions are also available, but somewhat less readily. They must first go into the soil solution by diffusion from the interlattice sites of the clay minerals. The mobility of ions in soil—for instance, of K—may be measured by their diffusion coefficient, which ranges from less than 10^{-20} cm^2/sec for interlayer K in unexpanded illite, to approximately 10^{-8} cm^2/sec for K in a dry soil and 10^{-6} cm^2/sec in a moist one (Nye, 1972). The mobility of nutrients as mineral lattice ions is thus so low that the life span of the root hairs is often not long enough for these ions to arrive in time. Nutrients that cannot go into the soil solution in the course of the vegetation period by desorption or solution processes are, therefore, only *potentially available*. This is what makes soil analysis so

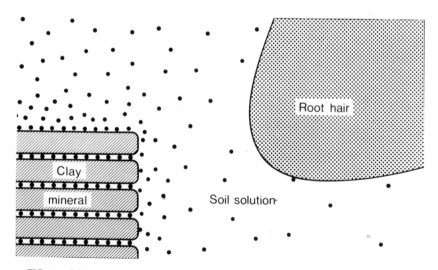

FIG. 1.   Schematic representation of root hair, soil solution, and soluble and sorbed ions.

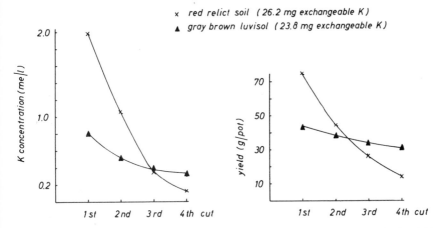

FIG. 2.   Differences in the decrease in K conconcentration in the soil solution and in the yield depressions of ryegrass in two different soils.

problematic. If strong solutions are used for extraction, both the effectively as well as the potentially available fractions will be obtained together (for example, 20 mg of K per 100 g of soil). It is possible that up to 80% of this amount is effectively available. But even if the effectively available portion is only 20%, this level is not without importance for the K uptake rate of the plant.

Therefore, it is reasonable to expect from soil analysis that the *ionic concentration in the soil solution* should be measured, as the soil solution is the *nutrient solution* of the plant under natural conditions. There are several methods of obtaining the soil solution, as described by Briggs and McCall (1904), Van Zyl (1916), Lipmann (1918), Northrup (1918), Parker (1921), von Wrangell (1930), Magistad *et al.* (1945), and Adam and Winsor (1973). Being very time-consuming, these methods are not suitable for routine investigations. Moreover, the ionic concentration in the soil solution is only a "momentary value." It remains unknown how this value changes in the course of the vegetation period. Yet this is a highly important factor for the nutrient uptake of the plant.

An example is given in Fig. 2. This figure shows the differences in the decrease in the K concentration in the soil solution and in the depressions in yield of ryegrass in two different soils during the vegetation period. Although the contents of exchangeable K are almost equal, the K concentration in the soil solution decreases more rapidly in the red relict soil than in the gray-brown luvisol. This can be ascribed to the different mineral composition of the clay fractions. The clay minerals of the red relict soil belong to the kaolinitic group and have a low capacity to release K into the soil solution, whereas the illitic clay minerals of the gray-brown luvisol are capable of keeping the K concentration in the soil solution relatively stable up to a certain level. For this soil the yield depression rate is consequently very slow. The contents of exchangeable K alone, therefore, cannot indicate the expected change in the K concentration in the soil solution during the vegetation period.

It is even more difficult to determine the amount of K fertilizer that has to be applied in order to raise the K concentration in the soil solution by a determined value (for example, from 0.2 to 1.0 me/l). Figure 3 shows clearly the considerable differences in the increase in the K concentration in the soil solution of the various soils at equal K fertilizer levels, although the contents of exchangeable K were practically the same (9–10.5 mg per 100 g). The contents of exchangeable K are consequently not sufficient information for calculating the K fertilizer requirements.

Figure 3 furthermore shows that the K fertilizer requirements can be very different even at comparable clay contents (19–21%) and almost equal K concentrations (0.2–0.3 me/l) in the soil solution, if the mineral composition of the clay fraction is different.

In agricultural practice, soil analysis should therefore provide the following information:

(1) What is the nutrient concentration in the soil solution, or what is the effectively available amount of nutrients in the soil?

(2) Which of the changes of these effectively available amounts in the course of the vegetation period are due to nutrient removal, leaching, weathering, etc.?

(3) What is the amount of nutrients to be applied to the soil in order to raise the effectively available amount to the required value (fertilizer requirement)?

It will be the objective of this paper to explain how the requirements specified above can be achieved.

III. Electro-ultrafiltration

A. COMPONENT PROCESSES OF ELECTRO-ULTRAFILTRATION

1. Dialysis, Ultrafiltration, and Electrodialysis

To provide information on the processes taking place in electro-ultrafiltration, the component processes, which have been actually known for a long time, will first be described.

A century ago, *dialysis* had already been used for the separation of ions from colloids (for example, clay minerals). This method is based on the principle of diffusion along a concentration gradient through a semipermeable membrane.

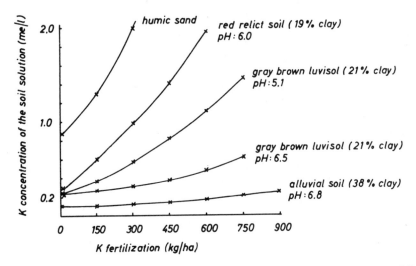

FIG. 3. Relationship between the K concentration in the soil solution and the amounts of K fertilizer of soils with equal contents of exchangeable K (9–10 mg per 100 g of soil)

The process is very slow, but it can be accelerated by increasing the surface of the membrane, the concentration gradient, and the temperature.

A second method of separating plant nutrients from soil is based on the principle of *ultrafiltration*. This is a filtration process by means of which the soil colloids are collected on a filter and the sorbed ions are removed by leaching. This method is also time-consuming, because the filtration rate decreases considerably with increasing dispersity of the soil.

After the development of *electrodialysis* by Morse and Pierce (cited by Bechold, 1925) in 1903, ion diffusion through membranes could be accelerated considerably. This method could also be used for soil tests (König *et al.*, 1913; Mattson, 1926; Norman *et al.*, 1927; Rost, 1928; Bradfield, 1928). It has been found that the various soils release very different amounts of ions. Investigations with synthetic permutite and feldspar powder showed that mainly exchangeable ions were detected by electrodialysis. The feldspar powder released very small amounts of cations (Bradfield, 1928). Mattson (1926) as well as Bradfield (1928) demonstrated that the exchangeable cations can be determined quantitatively. The time of dialysis, however, amounted to 20–50 hours.

2. Electro-ultrafiltration at Constant Voltage and Temperature during the Extraction Process

In addition to requiring a considerable expenditure of time, there were undesirable effects involved in electrodialysis; of these, the decrease in pH during the extraction proved to be most important. To overcome this undesirable decrease in pH, Bechold (1925) suggested combining ultrafiltration with electrodialysis. He called the new method *electro-ultrafiltration*. It differs basically from electrodialysis in that the secondary products of the electrodialysis (hydroxides and acids) are removed by suction and cannot enter the middle cell, so that the change in pH is reduced to the minimum.

The electro-ultrafiltration equipment used by Bechold and König was improved by Köttgen in the thirties and forties and used for soil tests (Köttgen and Diehl, 1929; Köttgen, 1937, 1940; Köttgen and Jung, 1941; Jung and Németh, 1966, 1969). More recently, Grimme (1978, 1979) has worked with constant voltage with the objective of evaluating the kinetics of nutrient desorption.

B. ELECTRO-ULTRAFILTRATION WITH VARYING VOLTAGE AND TEMPERATURE DURING THE EXTRACTION PROCESS

Once the significance of the soil solution in nutrient transport (diffusion and mass flow) to the plant root had been recognized, means of measurement were sought that would allow the nutrient concentration in the soil solution and its buffering to be measured in routine analysis. A further development of the EUF procedure appeared to be suitable for this purpose.

1. Principle of the Method

When an electrical potential is applied to a soil suspension, the following reactions can occur at the cathode:

$$2Na^+ + 2e \rightarrow 2Na^0 \quad \text{(metal reduction)}$$
$$2Na^0 + 2H_2O \rightarrow 2NaOH + H_2 + 67.4 \text{ kcal}$$

Similar reactions are possible for K, Mg, Ca, etc.

At the anode, where oxidative processes take place, the following reactions can occur:

$$NO_3^- - e = [NO_3]$$
$$[NO_3]\cdot + H_2O \rightarrow NO_3^- \text{ a } [OH]$$
$$2[OH]\cdot = H_2O_2 \rightarrow H_2O - \frac{1}{2}O_2$$

However, the processes taking place at the electrodes at high voltages (200 and 400 V) and different temperatures need further investigation. More details are given by Németh (1972, 1976).

2. EUF Apparatus and Extraction

A three-cell apparatus is used in electro-ultrafiltration. The middle cell containing the soil suspension (soil:water = 1:10) has a stirrer and a water inflow

FIG. 4. Part of the automatic EUF equipment.

(Fig. 4). Each side of this middle cell is provided with a micropore filter attached to the platinum electrodes that separate the middle cell from the two outside compartments (Fig. 5). These two other cells have vacuum connections. Therefore, the hydroxides [Na(OH), K(OH), Ca(OH)$_2$, NH$_4$(OH), etc.] accumulating at the cathode and the acids (HNO$_3$, H$_2$SO$_4$ etc.) accumulating at the anode are washed away by the continuous stream of water to the collecting tanks. The

FIG. 5.   Application of the micropore filters to the platinum electrodes.

procedure is regulated in such a way as to obtain fractions in 5 min or other intervals. In contrast to the situation that prevails with electrodialysis (König *et al.*, 1913; Mattson, 1926; Norman *et al.*, 1927; Bradfield, 1928), the pH of the soil suspension remains constant during EUF extraction. This is very important, because the pH exerts a decisive influence on the desorption and solubility rates.

The negatively charged *clay minerals* of the soil migrate to the anode and are deposited on the anode filter (Fig. 6). The water permeability of the anode filter is reduced by the accumulation of clay. The amount of water flowing through this filter is therefore inversely proportional to the clay content, so that the latter can be determined indirectly, as indicated in Fig. 7 (Németh, 1976). The clay content determined by the conventional method is plotted on the ordinate. The abscissa indicates the flow of water through the anode during the period from 10 to 35 min. As can be seen, the correlation is very close, and thus a rapid determination of the clay content for practical purposes is made possible, which is necessary for the assessment of fertilizer requirements.

In contrast to Ca, K, Na, etc., the heavy metals (Cu, Cd, Fe, Mn, Ni, Zn, etc.) are collected by the cathode filter as hydroxides and hydrated oxides (Fig. 6). Thus, a seperation can be made between the hydroxides of Ca, K, Na, etc., and the hydrated oxides of heavy metals. (Mg has an intermediate position; see Section III,C,3.) Heavy metals determined by EUF represent their effectively

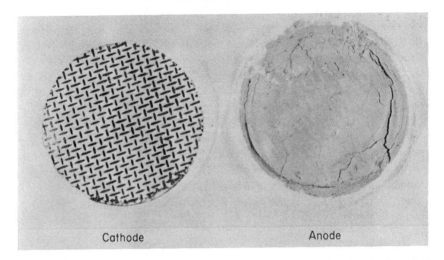

FIG. 6. Clay minerals accumulated at the anode filter and heavy metals collected at the cathode filter.

plant-available fractions, since only soluble or desorbed ions can move in an electric field.

For practical purposes, an extraction duration of 35 min is adequate (seven fractions of 5 min each). As a result of intensive research and comparison with conventional soil testing methods, the voltage is changed as follows: 0–5 min, 50 V (11.1 V/cm); 5–30 min, 200 V (44.4 V/cm); 30–35 min, 400 V (88.8 V/cm). In routine investigations three fractions are sufficient: EUF_I collected within

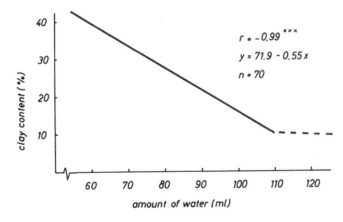

FIG. 7. Correlation between the clay content and the amount of water that flows through the anode filter during 10–35 min.

0-10 min, EUF_{II} collected within 10-30 min, EUF_{III} collected within 30-35 min. The voltage is varied, as stated above. On the other hand, for research purposes it is more appropriate to obtain the fractions at 5-min intervals in order to get more information on the rate of supply of nutrients from the soil.

3. Determination of Nutrients in the EUF Fractions

The determination of nutrients in the EUF fractions can be carried out by conventional methods. For serial tests, a combination with the AutoAnalyzer has proved useful. The determination of K, Na, and Ca can be carried out with an emission spectrophotometer, and that of Mg, Mn, Zn, Fe, etc., by means of an atomic absorption spectrophotometer, whereas B, PO_4, SO_4, NO_3, NH_4 can be determined colorimetrically. NH_4 and NO_3 can also be determined by the micro-Kjeldahl method.

4. Presentation of Results

The quantities determined by EUF in 5-min intervals can be plotted against time (duration of extraction), so that the amount desorbed (milligrams of nutrient per 100 g of soil per 5 min) appears on the ordinate, and the duration of desorption and the voltages applied can be read off from the abscissa.

Figure 8 shows two possibile ways of presenting the results. Example (a) shows the values in the form of a curve, and example (b) shows these values in the form of columns. In routine tests (three or two fractions), the values are given in the form of columns or are listed in a table, which is even simpler. From

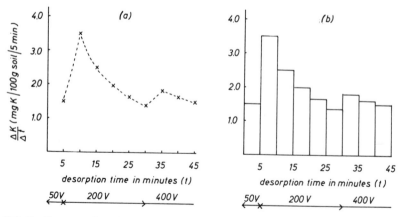

FIG. 8.   Representation of the EUF values at varied voltages (using the nutrient K as an example).

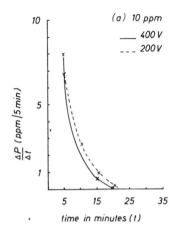 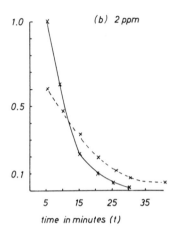

FIG. 9.   Quantities of phosphorus transported by means of EUF from a solution at different voltages and different P concentrations.

the shape of the curves, however, information on the desorption and solubility rates can also be obtained.

5. Significance of Altering the Voltage

The speed of ion migration in electro-ultrafiltration (in the middle cell) is proportional to the field strength and inversely proportional to the frictional forces. In turn, the field strength depends on the voltage and the distance between the electrodes. Since the latter is kept constant, the field strength is raised with an increase in voltage, and the ions are transported more rapidly to the electrodes. This is illustrated by Fig. 9, which shows the quantities of phosphorus removed from a solution by means of EUF at different voltages (200 and 400 V) and various concentrations. The transported amounts, in parts per million per 5 min, are plotted on the ordinate, and the duration of extraction can be read off from the abscissa. It is shown clearly that the phosphate ions are removed more rapidly from the middle cell of the EUF apparatus with increasing voltage. The difference in the transport rate due to raised voltage is greater, the lower the ion concentration in the middle cell.

It can further be seen from Fig. 9 that the transport of ions at a given voltage is more rapid, the higher their concentration. At a P concentration of 10 ppm, larger quantities of P, compared with the total amount, are transported to the electrodes in the time unit (for example, 5 min) than are transported at a P concentration of 2 ppm.

The rate of ion migration in electro-ultrafiltration is therefore not constant at a given voltage and temperature. In the course of electro-ultrafiltration it decreases to such an extent that the concentration in the middle cell is lowered. At low concentrations, the rate of transfer attains an almost constant state in which the field strength and the frictional forces maintain an equilibrium (Fig. 9b, 200 V). In order to accelerate the transport of ions, the voltage must therefore be raised. Relatively high voltages must be applied in order to raise the field strength (volts per centimeter) significantly, since the distance between the electrodes must be 4.5 cm in order to have enough space for the soil sample in the middle cell.

The attainment of an equilibrium between field strength and frictional forces is more complicated in the soil than in a solution. The attainment of constant rates of extraction is counteracted by the release of ions from the soil reserves. The faster the release from the reserve and the higher the voltage, the higher are the rates of extraction (Fig. 10).

Figure 10 shows the rates of P extraction in milligrams per 100 g per 5 min for two soils—(a) eroded gray-brown luvisol and (b) pararendzina—with equal P concentrations in the soil solution (in the saturation extract). At 50 V, these EUF-P values are low, but comparable, for the two soils. The P availability is accordingly well-characterized at 50 V, since the P concentration in the soil solution of the two soils is likewise equal. It could be concluded from this result that the soils should be investigated at 50 V, since the concentration is characterized well and their buffering could also be registered after a long time of extraction. Unfortunately, the rates of extraction of most nutrients are very low (hardly measurable) at 50 V, so that a long extraction would be required to record the buffering as well.

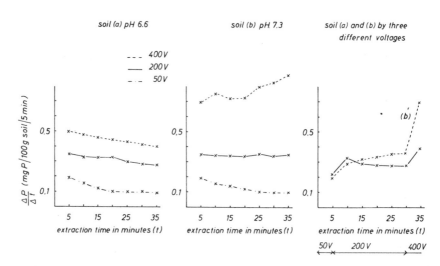

FIG. 10.   Amounts of P extracted by means of EUF from two soils with equal P concentrations (1.5 ppm) in the soil solution at constant and varied voltages.

At 200 V the rates of P extraction are considerably higher, but the P curves of the two soils are still almost the same. However, a more constant supply becomes noticeable in soil (b), whereas a slight falling off is noticed in soil (a).

The amounts of P extracted from the two soils are entirely different at 400 V. The rates of desorption from soil (b) are almost twice as high as those of soil (a), although the P concentrations in the soil solution of the two soils were comparable. Hence, the P concentration in the soil solution is not correctly assessed at 400 V. On the other hand, the P reserves are not registered at 50 V within 35 min. The difference between the two soils at 50 V is evident only after an extraction for several hours.

By raising the voltage stepwise (50 V, 200 V, and 400 V), both the P concentration in the soil solution and its buffering can be characterized in one operation in which the duration of extraction required for practical purposes is only 35 min. This is the most important advantage of varying the voltage.

6. Significance of Altering the Temperature

The speed of ion migration in the electrical field is raised with an increase in temperature, so that more nutrients can be extracted per unit of time. Care has to be taken not to confound this increase in the speed of ion migration through the increase in temperature with high amounts of effectively available nutrients in the analyzed soil. It is, therefore, advisable to keep the temperature below 25°C during extraction by EUF. Higher temperatures during extraction can, on the other hand, be suitable for the determination of larger quantities from the nutrient reserves (from the potentially available reserves), thus giving a more precise insight into the nutrient dynamics of the soil.

Therefore, the EUF procedure was modified by varying the temperature during the desorption processes. The modified procedure is as follows:

A 20- to 30-min extraction at 200 V and 20°C is suggested for the determination of the effectively available nutrient fractions. Subsequently the temperature is raised from 20°C to 80°C, the voltage is increased from 200 V to 400 V, and the extraction is maintained for another 10–15 min. The effectively available amount of nutrients and the nutrient reserves can be characterized more precisely when the fractions at 20°C and 80°C are collected at 5-min intervals. Reliable information will thus be obtained on the rate of supply from the reserves.

Table I shows the K and P desorption of five soils at different K fertilizer levels (K_1 and K_4) as influenced by voltage and temperature. From this table, it can be gathered that considerable amounts of K and P are extracted when the temperature is raised from 20°C to 80°C and at the same time the voltage is increased from 200 V to 400 V. The K amounts in the second fraction (80°C) are for many soils higher than the effectively available K contents in the first fraction (20°C). These amounts are higher, the higher the contents of K-selective clay minerals in the soil.

TABLE I

K and P Desorption of Soils Fertilized with Different Amounts of K (K_1 and K_4) as Influenced by Temperature and Voltage

Soil	pH (CaCl$_2$)	Clay (%)	Exch. K (mg/100 g)	P-H$_2$O (mg/100 g)	P$_2$O$_5$-AL[a] (mg/100 g)	P$_2$O$_5$-CAL[b] (mg/100 g)	EUF-K 20°C (200 V)	EUF-K 80°C (400 V)	EUF-P 20°C (200 V)	EUF-P 80°C (400 V)
Chernozem										
K_1	6.4	17.5	6.0	1.4	22.5	13.5	1.7	2.9	1.4	2.5
K_4	6.4	17.5	25.2	1.4	22.5	13.5	12.7	15.5		
Pararendzina										
K_1	7.3	17.4	4.6	0.9	32.1	12.1	1.6	3.9	0.7	3.0
K_4	7.3	17.4	18.4	0.9	32.1	12.1	10.1	18.2		
Gray-brown luvisol										
K_1	6.4	14.4	5.2	2.3	23.5	16.5	2.2	3.3	2.3	2.9
K_4	6.4	14.4	16.0	2.3	23.5	16.5	10.1	11.7		
Alluvial soil										
K_1	6.6	36.6	9.2	0.8	20.4	9.5	1.2	4.4	0.4	1.9
K_4	6.6	36.6	23.0	0.8	20.4	9.5	7.2	31.0		
Humic sand										
K_1	5.5	3.0	2.0	2.3	19.2	18.5	2.5	1.8	1.9	2.0
K_4	5.5	3.0	8.0	2.3	19.2	18.5	8.2	6.0		

[a] P$_2$O$_5$-AL = NH$_4$-lactate extraction.
[b] P$_2$O$_5$-CAL = Ca-lactate extraction.

The increase in voltage alone only increases the rate of K desorption, which means that the half-time period of the desorption decreases, whereas the desorbable total amounts remain almost the same (Németh, 1976; Grimme, 1978). If, however, the temperature is also raised, the desorbable K amounts increase considerably. An increase in voltage and temperature is, therefore, highly suitable to define the K reserves.

The extraction of P is also considerably affected by varying the voltage and the temperature. The acceleration is higher, the higher the pH values, the clay contents, and the P_2O_5-AL contents in comparison with the P_2O_5-CAL values. Consequently, in this case mainly the accelerated extraction of Ca phosphates and sorbed phosphates is concerned.

A simultaneous increase in voltage and temperature can, therefore, very well indicate the K and P reserves (as well as the reserves of B, Ca, NO_3, NH_4, etc.). More details will be given later.

C. INTERPRETATION OF EUF RESULTS FOR PLANT NUTRITION

1. Potassium

In a solution the ions K^+ and NH_4^+ are more rapidly transported to the electrodes than are Ca^{2+}, Mg^{2+}, and Na^+. The faster transfer of K^+ and NH_4^+ ions is probably due to their lower hydration shell. In an equimolar solution the K^+ and NH_4^+ ions are almost quantitatively transferred after 5 min, whereas only approximately 50% of the CA^{2+} ions is extracted after 5 min. The CA^{2+} and Mg^{2+} ions occupy an intermediate position (Németh, 1976). The differences in the rate of transfer are even greater in the soil suspension, where the velocity of ion transport decreases through friction with the surrounding soil particles. Care has to be taken not to confound the decrease in the speed of ion migration through friction with a lasting release of ions from the reserves into the solution. This mistake will not happen when the voltage is increased to 400 V and the temperature to 80°C, so that the speed of ion transport will be accelerated and the low reserves quickly exhausted.

a. EUF-K Curves at Low K Selectivity of the Adsorption Complex. The K^+ ions of a solution are almost quantitatively transferred after 5 min. Therefore, a continuous supply of K^+ after a longer extraction period (for example, 30 min) can give information on the rate of K desorption from clay minerals.

For clay minerals that have no sites for the selective binding of K as kaolinite, the K desorption is practically terminated after 25 min at 200 V and 20°C, and it is independent of the K saturation of these clay minerals (Németh, 1972). Similar K desorption curves will be obtained with many tropical soils in which the clay fraction is predominantly kaolinitic. When the voltage is raised to 400 V and the temperature to 80°C, in these soils, which are unlike the illitic soils in this respect (Table II), no increase in the release of K is observed after 20 min.

The EUF-K curves for humic sandy soils are similar to those for kaolinitic soils, as indicated in Fig. 11. It shows that the K amounts transferred after 35 min correspond to the contents of exchangeable K and that there is practically no further release after the voltage has been raised to 400 V. The accumulated clay at the anode filter allows the determination of soil texture (Fig. 6).

For soils with a long acid phase in their development, the K desorption proceeds in a similar way to that described above (Németh, 1976). For a given K status of these soils, K desorption is faster, the higher the content of exchangeable Al (Németh, 1972). With increasing Al content there is a competition for the sites that selectively bind K (Rich and Black, 1964; Rich, 1972). As these soils contain soluble Al, Fe, Mn, and other heavy metals in the soil solution, which will be collected at the cathode filter (Fig. 6), they can easily be distinguished from the other soils with low K selectivity.

These types of soil do not manifest any K fixation, even when low in K. It must be admitted, however, that fertilizer K is being leached more readily (Németh, 1972; Thiagalingam and Grimme, 1976; Wolpers and Tobing, 1977). Liming, on the other hand, gives rise to K fixation, and the EUF-K curves take a different course.

b. EUF-K Curves at High K Selectivity of the Adsorption Complex. The desorption of K from clay minerals with sites for selective sorption of K in the interlattice (Schachtschabel, 1940; Schouwenburg and Schuffelen, 1963) is not terminated after 30 min, as ions released from the interlattice sites still go into the solution. This is particularly obvious when voltage and temperature are raised simultaneously (Table I). These clay minerals can, therefore, keep the K concen-

TABLE II

K Desorption of Soils from India with Low K Selectivity of the Adsorption Complex as Influenced by Temperature and Voltage

Soil	Clay content (%)	Exchangeable K (mg/100 g)	EUF-K		EUF-K (mg/100 g)
			20°C (200 V)	80°C (400 V)	
Haplustalfs (Trichy)	17.4	13.2	9.8	2.2	12.0
Haplustalfs (Valparai)	18.5	16.8	12.5	4.6	17.1
Rhodulstalfs (Coimbatore)	17.8	12.1	9.5	3.4	12.9

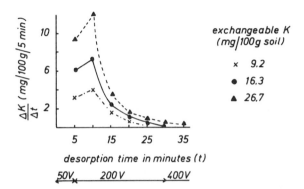

FIG. 11.   EUF-K desorption curves of a humic sand with different content of exchangeable K.

tration in the soil solution stable at a certain level. In such soils K is also subject to fixation, when the level of K saturation has considerably decreased.

Figure 12 shows the EUF-K curves of a chernozem at different K fertilizer levels, the corresponding K saturation of the inorganic CED (cation exchange capacity), and the amounts of fixed K obtained by conventional methods. At a low K saturation of the CEC (1.4%), the course of the curve is very flat. The first peak after 10 min at 200 V is hardly higher than the second one at 400|V within 30–35 min.

With a higher K saturation, the first peak increases; that is, the K concentration in the soil solution increases and the K fixation decreases. At a K saturation of the inorganic CEC of 6%, practically no K is fixed any more. In this case the first peak of the EUF-K curve will be about three times as high as the second one, and the absolute value for the first peak is > 3.0 mg of K per 100 g per 10 min at 20°C (see Section IV,A,1 for further details).

c. Correlations between K Uptake, Yield, and EUF Fractions. The more K ions that are desorbed within 10 min, the higher is the K concentration in the soil solution (Németh, 1971, 1972), and the more K^+ ions can be taken up by the plant (Németh, 1972, 1975, 1977; Németh and Harrach, 1974; Jankoviç and Németh, 1974; Mengel, 1975; Németh and Forster, 1976).

The best correlation between EUF-K desorbed within the first 10 min and K uptake was found for plants that remove high amounts of K within a relatively short period of time—for example, cereals. In the case of other plants, which take up K continuously during the vegetation period—for example, sugar beet—the shape of the EUF-K curves is an important indicator.

Figure 13 shows the EUF-K curves of three soils with different K dynamics and the decrease in yield of ryegrass during four cuts on these soils. The course of the EUF-K curves as well as the decrease in yield shows differences between

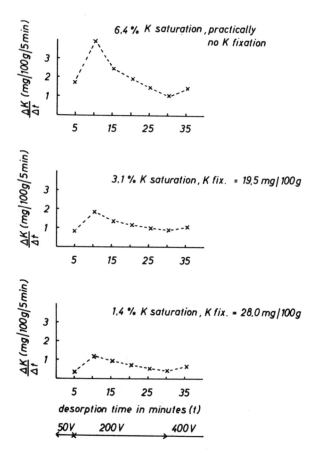

FIG. 12. EUF-K desorption curves of a chernozem with different K saturation of the inorganic CEC.

the various soils, although the contents of exchangeable K were similar (23.8–26.2 mg per 100 g). The faster the K desorption takes place, the higher is the yield of the first cut. This is the case for the oxisol. If, however, K desorption rates fall off rapidly, the yields obtained from the subsequent cuts also diminish considerably. The slower the decrease in K desorption rates (gray-brown luvisol, pH 6.8), the lower is the yield depression of the subsequent cuts. The EUF-K curves can, therefore, give valuable information on the *course of yield* to be expected. The contents of exchangeable K, on the other hand, do not furnish reliable data for assessing the yield of the first and the subsequent cuts (see Fig. 13). These values were similar for the different soils, whereas the yields obtained from the various cuts showed wide variations (see Section IV for further details).

2. Sodium

In contrast to curves for K, the EUF-Na curves of soils have only one peak after 10 min, as the desorption rates are comparatively low, when the voltage is raised from 200 V to 400 V. Na desorption from sandy soils as well as from clay soils is terminated after 35 min. Most of the sodium is already set free after 10 min. The course of the EUF-Na curves is uniform also for soils with different types of clay minerals, as the Na^+ ion is not selectively bound by clay minerals or by humus.

The evaluation of the Na desorption curves in the daily practice of soil testing is rather simple. The content of exchangeable Na is equivalent to the amount of Na desorbed within 30 min at 200 V and 20°C. There is a statistically significant correlation between the exchangeable Na and the Na concentration in the soil solution, which is independent of the clay content and its mineral composition (Mengel and Németh, 1971).

3. Magnesium

In contrast to KOH, NaOH, and $Ca(OH)_2$ the $Mg(OH)_2$ is less readily soluble in water, so that most of the Mg released by EUF will be found in the cathode

▲   gray brown luvisol (pH 6.8 ; 23.8 mg exchangeable K)

●   gray brown luvisol (pH 5.0 ; 24.1 mg exchangeable K)

×   oxisol (pH 6.3 ; 26.2 mg exchangeable K)

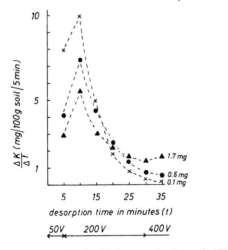

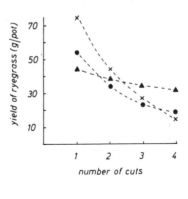

FIG. 13.   Relationship between the shape of EUF-K curves and the decrease of ryegrass yields in the course of four cuts.

filter as precipitate. It is, therefore, necessary to determine the EUF-Mg amounts in the cathode filtrate as well as in the cathode filter and to calculate the sum of both values (Németh, 1976; Nair and Grimme, 1978).

In accordance with previous investigations, a soil can be considered as well fertilized with Mg when the sum of the Mg values obtained by EUF is above 5 mg of Mg per 100 g per 35 min.

There are statistically significant correlations between the amounts of Mg obtained at 200 V and 20°C during 30 min and the Mg content of soil extracted with 0.025 N CaCl$_2$ solution. There are, however, two regressions lines—one for the soil samples from field experiments, and the other for soil samples from pot experiments. According to this finding, the bonding of Mg—that is, the ratio of sorbed to soluble Mg—at equal amounts of exchangeable Mg in the soil is lower in soil samples from pot experiments than in those from field experiments. This is due to the different anion concentration in the soil solution of pot experiments on the one hand, and under field conditions on the other. Different anion concentrations give rise to different cation concentrations in the soil solution, which can be measured by EUF. The conventional exchange methods cannot differentiate between soluble and sorbed ions.

4. Calcium

The course of the EUF-Ca curves depends on the Ca reserves and on the content and type of humus, the CaCO$_3$ content, and the anion concentration of the soil solution, all of which soil properties determine the binding power of Ca in the soil. If these soil properties vary considerably, the EUF-Ca curves can be very different for comparable amounts of exchangeable Ca (Németh, 1972). In most cases the Ca concentration in the soil solution is sufficient for an adequate Ca supply of the plant. The Ca status of a soil can, however, influence the effective availability of other plant nutrients. The EUF-Ca reserves should, therefore, also be consulted for the assessment of nutrient dynamics in the soil, particularly those relating to P and to Mn and other trace elements (see Sections III,C,6 and IV,B for further details).

The Ca reserves can easily be determined by EUF. The amounts of Ca extracted at 400 V and 80°C clearly indicate the presence of these reserves, since CaCO$_3$ is intensively transferred into solution at 400 V and 80°C, and the desorption of the Ca ions from the soil is accelerated as well. If the Ca amounts extracted at 400 V and 80°C are twice the amounts released at 200 V and 20°C, the soil in question contains easily soluble CaCO$_3$. The higher the absolute values are, the greater is the CaCO$_3$ solubility. If the amounts of Ca ions collected at 400 V and 80°C during 15 min are below 10 mg per 100 g of soil, the examined soil contains very low Ca reserves. In such cases there is the possibility of a heavy metal toxicity if the soil contains heavy metals by nature. Sandy soils, for example, have low metal contents despite their low Ca reserves. The heavy metal contents of most soils increase with increasing clay contents.

5. Nitrogen

From a pure NH_4NO_3 solution the NO_3^- ions are quantitatively transferred to the anode, whereas a quantitative determination of the NH_4^+ ions is problematic, as a loss of 20–25% may occur in the form of NH_3 (Németh, 1976). The transfer of NH_4^+ and NO_3^- ions from a soil suspension is not the same as that from a solution. A clean separation between NH_4^+ and NO_3^- ions from a soil suspension according to electrical charge is not possible, owing to the accumulation of clay particles obstructing the anode filter, so that the transfer of NO_3^- ions is slowed down. As the cathode filter is not obstructed, water can pass freely, and it may transport NO_3^- ions cathodically during the extraction run. In order to determine quantitatively the NH_4^+ and NO_3^- ions, the cathode filtrates as well as the anode filtrates have to be analyzed for NH_4^+ and NO_3^-. Mixing the filtrates for subsequent analysis has proved highly suitable.

a. *EUF-NH$_4$ and EUF-NO$_3$ in Relation to Soil Type and Fertilizer Application.* The EUF-NH$_4$ contents of well-drained soils are often lower than the EUF-NO$_3$ fractions. Some EUF-NH$_4$ values of various soils are given in Table III. Equal amounts of N fertilizer had been applied to these soils for 10 years. It can be seen that the EUF-NH$_4$ contents increase with increasing clay content. These samples belong to soils poor in K, so that a selective binding of the NH_4^+ ions has to be expected in the interlayer of the clay minerals. The desorption rate of NH_4^+ ions is similar to that of K ions, increasing with increasing voltage and temperature; thus, an increase in voltage and temperature seems to be advisable

TABLE III
EUF-NH$_4$ Values of Different Soils with Different Clay Content

Soil	Clay content (%)	Exchangeable K (mg/100 g)	EUF-NH$_4$ (mg N/100 g) 25°C	EUF-NH$_4$ (mg N/100 g) 0°C
1. Gray-brown luvisol	14.4	6.8	0.30	0.81
2. Pararendzina	17.4	5.4	0.36	0.95
3. Chernozem	17.7	8.0	0.78	1.64
4. Eroded gray-brown luvisol	20.0	9.2	0.74	1.80
5. Alluvial soil	38.6	11.8	1.27	2.47

[a] Results of long-term fertilizer trials with N applied at equal amounts.

to obtain information on possible N release from the NH_4 reserves (Németh *et al.*, 1979).

An accelerated NH_4 desorption was also observed in humic sandy soils ($< 3\%$ clay), when voltage and temperature were raised. An example is given in Table IV.

It is unlikely that the desorbed NH_4 amounts come from clay minerals, as the K saturation in these sandy soils was above 6% of the inorganic CEC. In this case the K-selective sites are occupied by K^+ ions, so that selective NH_4 sorption can be excluded.

It is known that amino acids can be extracted from soils by means of EUF. These amino acids (serine, glycine, alanine, asparagine, glutamine, etc.) are found mainly in the anode filtrates. In the soil suspension (middle cell of the EUF apparatus), less of the amino acids could be found after the extraction process than before. At increasing voltage and temperature, a marked increase in amino acids was noted. This explains the observation that the EUF total N values, which include NH_4, NO_3, and other compounds, are higher than the sum of the EUF-NH_4 and EUF-NO_3, and that these higher values were founded at 20°C as well as at 80°C (Németh *et al.*, 1980).

In order to study the behavior of amino acids under EUF conditions, experiments were carried out with synthetic amino acids. After EUF extraction, NH_4 amounts that could stem only from amino acids were found at the cathode filtrate. These results allow the conclusion that part of EUF-NH_4 is derived from amino acids. A release of NO_3 from amino acids was not observed.

It is, however, not known from which N compounds the considerable EUF-NO_3 amounts derive that are obtained at 80°C and 400 V. It seems unlikely that it is dissolved NO_3, as the dissolved NO_3^- ions have already been extracted at 20°C and 200 V. An anodic oxidation of low-molecular N compounds to NO_3 might be possible, but this has not yet been proved.

TABLE IV
EUF-NH$_4$ Values of Humic Sandy Soils (3.0% Clay)

	EUF-NH$_4$ (mg N/100 g) on:			
	6/3/78		1/8/78	
Sample	20°C (200 V)	80°C (400 V)	20°C (200 V)	80°C (400 V)
1	1.3	1.5	1.0	1.3
2	1.1	1.5	1.0	1.2
3	0.5	1.0	0.6	1.0
4	1.1	1.3	0.8	1.3

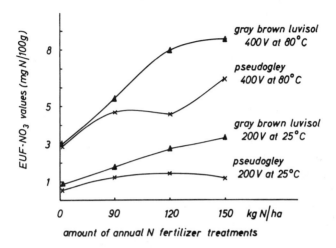

FIG. 14   Increase of EUF-NO₃ fractions of two soils due to the 15-year application of N fertilizer.

For plant nutrition it is not the question of the origin of EUF-N fractions that is of primary importance, but rather the relationship between these values and the yield as well as the N fertilizer levels.

Soil analysis of 15-year N fertilizer trials showed that the EUF-NO₃ fractions were changed, owing to the 15-year application of N fertilizer (Fig. 14). With increasing N fertilization, the EUF-NO₃ values increased clearly at 20°C. For the brown luvisol these values increased more than for a pseudogley. Probable N losses due to denitrification may have occurred under the occasionally anaerobic conditions of the psudogley. For both soils N fertilization substantially increased the EUF-NO₃ values obtained at 400 V and 80°C. Consequently, this fraction can also give information on possible N release from the N reserves of the soil (Németh *et al.*, 1979).

b. Time for Taking of Soil Samples To Be Used for EUF-N Tests.   It has proved useful to take soil samples for EUF-N tests in June and July in temperate regions. At this time an equilibrium between N incorporation into the biomass, N mineralization, and N translocation is to be expected, so that the NH₄ and NO₃ values determined by EUF are characteristic for a specific site. The N supplied by the soil can be calculated on the basis of trials carried out for several years as follows: N uptake minus N supplied by fertilizer = N supplied by the soil. In this way the EUF-NO₃ values are "calibrated," and the N supplied by the soil can also be expressed in kilograms per hectare. These trials, carried out on 4000 farms in Austria (Tulln Sugar Factory), show that 1 mg of EUF-N (extracted at 20°C and 200 V) = 90 kg of N per hectare.

A close correlation can be found between these EUF-N values determined in June and July on the one hand, and the sugar beet yield and quality in the following year on the other hand; this correlation was observed for a period of 5 years. These EUF-N values also correlated well with the yields of maize and wheat.

6. Phosphorus

Compared with lactate extraction as practiced in German soil testing laboratories, little P can be extracted by EUF. Depending on the level of P concentration in the soil solution, about 5–20% of the lactate P is obtained by EUF extraction. There is consequently only a general correlation between the EUF-P and the lactate P values. In contrast, the correlation between the P values obtained by the water method according to van de Paauw (1969) and the EUF-P values at 200 V and 20°C is very close. About the same P amounts are extracted by EUF after 35 min as by the Dutch water method. The best correlations are those between the P amounts extracted by EUF in 30 min at 200 V and 20°C and the P amounts obtained by sixfold extraction with 0.5 N NH_4Cl (Németh, 1976).

When the voltage is increased from 200 V to 400 V and the temperature is raised to 80°C, the amount of P extracted by EUF can be five times as high as the amounts extracted at 200 V and 20°C.

To get more information about the influence of temperature on P extraction, the maximal desorbable P amounts and their half-time values were calculated from some soil samples by means of an equation analogous to the Langmuir equation (Grimme, 1978). The maximal desorbable P amounts (P_{dm} in milligrams per 100 g) and their half-time values ($t_{0.5}$ in min), as obtained by EUF, are given in Table V for a chernozem and a gray-brown luvisol. It can be seen that the P_{dm} values at constant temperature increase with increasing voltage. Raising the temperature at constant voltage causes the amounts of extracted P to increase again. Finally, an increase in the P_{dm} values and a decrease in the half-time

TABLE V

Maximal Desorbable P Amounts (P_{dm}) and Their Half-Time Values ($t_{0.5}$) at Varying Temperatures

	200 V		400V	
Soil	10°C	25°C	10°C	25°C
Chernozem				
P_{dm} (mg/100 g)	3.4	5.4	4.0	6.8
$t_{0.5}$ (min)	19.9	37.2	22.1	26.4
Gray-brown luvisol				
P_{dm} (mg/100 g)	6.7	8.4	10.8	11.6
$t_{0.5}$ (min)	44.3	38.6	53.7	34.9

TABLE VI
Amounts of EUF-P Extracted at 200 V and 20°C and at 400 V and 80°C,
As Well As Some Other Soil Properties

Soil	pH	Clay (%)	P_2O_5-AL (mg/100 g)	P-H_2O (mg/100 g)	EUF-P (mg/100 g) 20°C	EUF-P (mg/100 g) 80°C
1. Chernozem	6.4	17.5	22.5	1.4	1.4	2.5
2. Gray-brown luvisol	6.4	14.4	23.5	2.3	2.3	3.0
3. Pararendzina	7.3	17.9	32.1	0.9	0.8	3.0
4. Alluvial soil	6.6	36.6	20.4	0.8	0.5	2.0
5. Humic sandy soil	5.5	3.0	19.2	2.3	1.9	2.0
6. Chromusterts (India)	7.2	29.0	26.5	0.7	0.4	1.8
7. Ustortherts (India)	6.3	33.2	3.8	0.15	0.15	0.5
8. Haplustalfs (India)	4.9	18.5	2.0	0.05	0.03	0.15

values are observed when voltage and temperature are raised simultaneously. Therefore, simultaneous increases in voltage and temperature seem to be a suitable way to characterize the P reserves more precisely.

For plant nutrition, however, it is not the amounts of extracted nutrients but their effective availability that is of primary importance. The findings are similar to those for K desorption. The effectively available P amounts can be obtained at 200 V and 20°C within 20 min. The subsequent second fraction, which is extracted at 400 V and 80°C within 15 min, can give information on the P reserves.

Table VI shows that an increase in voltage and temperature accelerates P extraction considerably. The differences between EUF-P at 20°C and EUF-P at 80°C are higher, the higher the pH values and the clay contents are. Probably the solubility of Ca phosphates and the desorption of sorbed phosphates will be accelerated with increasing voltage and temperature.

IV. The EUF Values Required for Optimal Plant Nutrition and Their
Calculation

A. THE EUF-K AND EUF-P VALUES

The relationship between the EUF-K and EUF-P values at 20°C and the K and P uptake of various plants has already been described in several publications

(Jankoviç and Németh, 1974; Németh and Harrach, 1974; Németh, 1972, 1975, 1977; Németh and Forster, 1976). In this connection, reference should be made to the several thousand results of 5-year field experiments conducted by the Tulln Sugar Factory in Austria. These studies show, for example, a sugar beet yield of 55 tons/ha or a sugar yield of 8–9 tons/ha can be obtained from a soil with the following EUF-K and EUF-P values at 200 V and 20°C : 15–18 mg of K per 100 g and 1.4–1.6 mg of P per 100 g (Németh, 1977). This sugar yield was obtained at exchangeable K contents between 15 and 30 mg per 100 g of soil, depending on the clay content and the type of clay. The lactate P_2O_5 values with maximum yields were between 12 and 24 mg per 100 g of soil. In contrast to these values, the optimal values for EUF-K and EUF-P can be more precisely determined, because these nutrients extracted at 200 V and 20°C are effectively available. As the availability of nutrients extracted by lactate solutions or NH_4 acetate solution, depending on soil properties, can be very different, the optimal lactate values can vary within the above-mentioned wide range.

Investigations carried out over the last few years, moreover, revealed that high yields (wheat, maize, sugar beet, etc.) can also be obtained at lower EUF-K and EUF-P values, if the K and P amounts desorbed at 80°C and 400 V are high (Jankoviç and Németh, 1979). It is, therefore, advisable to consider both the EUF-K and EUF-P values obtained at 20°C and those obtained at 80°C. If the K and P reserves desorbed at 80°C are low in comparison with the K and P values at 20°C, these reserves can be neglected. In such cases, the already mentioned values of 15–18 mg of K per 100 g and 1.4–1.6 mg of P per 100 g should be considered (for wheat and rice 10–12 mg of K per 100 g per 20 min will be sufficient). If, however, the EUF-K reserves at 80°C exceed 5 mg per 100 g per 15 min and the P reserves are above 1.0 mg per 100 g per 15 min, then 8–10 mg of K and 1.0 mg of P per 100 g at 20°C are sufficiently high. The higher the reserves are at 80°C, therefore, the lower the values at 20°C can be to ensure optimal plant nutrition. The effectively available amounts at 20°C should not be too low, however, as in that case an adequate supply of nutrients to the plant roots will not be guaranteed despite high reserves.

1. Calculation of the Amount of K Fertilizer

The K requirements depend on the quantity of clay minerals that bind K more selectively in the soil than other cations as, for example, Ca, Mg, and Na. Soils with low K selectivity—light soils, kaolinitic soils in tropical regions, and soils with low pH values—release K very quickly by means of EUF (within 10–15 min at 20°C). Therefore, the EUF_I fraction (0–10 min) is at least three times the EUF_{III} fraction (30–35 min). Often this third fraction is altogether missing (Fig. 13, oxisol). In this case there is no K fixation. If the soil adsorbs K selectively, the K release by means of EUF at 20°C is very low. The course of the K-EUF curve is correspondingly flat. Considerable amounts of K are still released within

30–35 min, as compared with the EUF_I fraction (Fig. 13, gray-brown luvisol, pH 6.8).

The relationships between the amounts of EUF_I and EUF_{III} and between the amounts of EUF-K at 20°C and EUF-K at 80°C for these soils are as follows:

$$EUF_I:EUF_{III} = 1:1 \text{ or } 2:1$$
$$\text{EUF-K } (20°C):\text{EUF-K } (80°C) = 1:1 \text{ or } 1:1$$
$$(200 \text{ V}) \qquad (400 \text{ V})$$

Most soils developed from sediments (for example, of loess) belong to this group. Consequently, the soils can be classified by means of EUF into two categories according to their K adsorbing power: (*a*) no K fixation, and (*b*) possibility of K fixation (selective binding). This is, however, only a qualitative classification. In the case of K fixation, the amount of K fertilizer in kilograms per hectare has to be determined. It is, therefore, necessary to ascertain the relationship between the amounts of fertilizers and the amounts of nutrients extracted by EUF for given soil properties.

The influence of soil properties on fertilizer nutrients has been studied in numerous field, pot, and laboratory experiments. As a result, tables have been drawn up that show the amounts of fertilizer in kilograms per hectare that should be applied, so that the given amounts of nutrients extracted by EUF will increase by a determined value (for example, by 1 mg per 100 g of soil) (Table VII).

TABLE VII
Quantity of K Required To Raise the EUF-K Value to
15 mg per 100 g of Soil
K requirements (kg/ha)
↓

Value found in analysis	Clay content (%)			
	10	10–20	20–30	30–40
1	600	1200	1600	3000
2	560	1050	1300	1800
3	480	900	1100	1400
4	420	800	900	1100
5	370	700	800	900
6	330	600	700	800
7	270	450	550	650
8	240	300	400	500
9	210	250	300	350
10	190	200	250	300
11	150	150	150	200
12	120	120	120	150
13	90	90	90	90
14	60	60	60	60

EUF-K (mg/100 g/35 min)

↑
K supplied by the soil (buffering)

Example. Soil test results from samples taken in July revealed an EUF-K rate of 10 mg of K per 100 g per 20 min at 20°C. The rate for the following year was fixed at 15 mg of K per 100 g per 20 min at 20°C. Question: How many kilograms of K fertilizer have to be applied per hectare to obtain 15 mg of K per 100 g of soil? The amount of K fertilizer to be applied depends on the amount of clay minerals that bind K selectively. If less than 10% of these clay minerals are present in the soil, 190 kg/ha will have to be applied. Soils with a clay content of 30–40% require 300 kg/ha. When only 13 mg instead of 15 mg is envisaged for a soil with a clay content of 30% (high amount of K at 80°C), the K requirements according to Table VII will be $300 - 90 = 210$ kg/ha. The amount of 90 kg/ha, which is necessary to increase the rate of 13 mg of K to 15 mg per 100 g, has to be deducted from the 300-kg K value. Table VII can be used to determine not only the quantities of fertilizer K needed but also the K supplied by the soil (buffering). If a sugar beet yield of 55 tons/ha (8–9 tons of sugar per hectare) removes from the soil 450 kg of K per hectare, the EUF-K value will decrease from 15 to 8–9 mg of K per 100 g in a soil with 30–40% clay of high K selectivity, and to 3–4 mg of K per 100 g in a soil with less than 10% clay.

2. Calculation of the Amount of P Fertilizer

The P requirements of a soil depend on the clay content (Table VIII) also.

Example. Soil test results from samples taken in July revealed an EUF-P rate of 1 mg per 100 g per 20 min at 20°C. For the following year the expected value

TABLE VIII
Quantity of P Required to Raise the EUF-P Value to
1.6 mg per 100 g of Soil

P requirement (kg/ha)
↓

		Clay content (%)			
		10	10–20	20–30	30–40
EUF-P (mg/100 g/35 min)	0.2	200	300	350	500
	0.4	150	180	200	350
	0.6	70	85	100	200
	0.8	65	80	80	120
	1.0	40	45	45	65
	1.2	30	30	30	40
	1.4	30	30	30	30
	1.6				

↑
P supplied by the soil

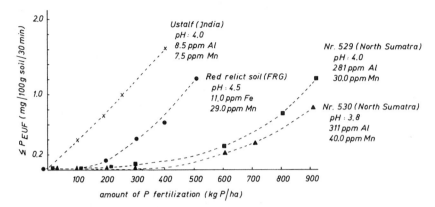

FIG. 15. Relationship between the EUF-P values at 200 V and 20°C and the amounts of P fertilizers of soils with nearly equal clay content and pH.

for this soil was fixed at 1.4 mg of P per 100 g. Table VIII shows that, with a clay content of 10%, P fertilizer has to be applied at a rate of 40 kg/ha in order to raise the EUF-P value from 1.0 mg to 1.4 mg of P per 100 g. Soils with a clay content of 40% require 60 kg of P per hectare.

The higher the amounts of fertilizer required to increase the amounts of EUF-P, the less will be the change of these EUF values in the course of the vegetation period. The expected decline of the EUF values can, therefore, also be read off from Table VIII—namely, from the bottom to the top.

Table VIII, however, is suitable only for the calculation of P requirements in soils with pH values above 5.5 or with low Al, Fe, and Mn concentration in the soil solution. For, the higher the Al, Fe, and Mn concentration in the soil solution, the more P is fixed at nearly equal amounts of clay and pH values (Fig. 15). The EUF heavy metal fraction (Fig. 6) should, therefore, also be taken into account for the determination of P requirements in acid soils.

B. CALCULATION OF PHYSIOLOGICAL LIME REQUIREMENT

The limited growth of plants on acid soils is caused not so much by the high H^+ ion concentration of the soil solution as by the toxic effect of Al, Fe, Mn, Zn, etc. ions. The physiological lime requirement of a soil is, therefore, determined by the need to reduce toxic concentrations of heavy metals. But in soil testing practice liming recommendations are often based on the titration of both H^+ and Al^{3+}, as well as other heavy metal ions. There is, however, a loose correlation

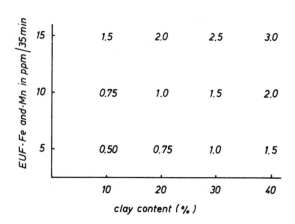

FIG. 16.   Quantity of CaO required to eliminate heavy metal toxicity.

between the soluble and easily desorbable heavy metal fraction and the lime requirement (titration with base). The physiological lime requirement differs, therefore, from the conventional lime requirement.

Because of the large number of liming experiments, a table has been worked out showing the amounts of CaO (in kilograms per hectare) needed to eliminate heavy metal toxicity. Here the clay content plays a decisive role. With increasing clay content, increasing amounts of CaO or $CaCO_3$ must be applied to eliminate the toxicity of a given quantity of EUF-Fe and EUF-Mn. Thus, only a small part of the total desorbable heavy metal will be extracted by EUF. The higher the clay content, the less heavy metal can be desorbed by EUF at a given time. Therefore, with increasing clay content higher amounts of CaO must be given in order to inactivate the heavy metals not extractable by EUF in 35 min. Since the clay content also can be determined by means of EUF, Fig. 16 can be used to calculate the physiological lime requirement.

According to investigations by Németh (1976), toxic symptoms were found in red clover when the EUF-Mn and EUF-Zn contents had reached 15 ppm in different soil types. The yield of red clover in pot experiments had been decreased about 70% by 40 ppm of Mn and Zn. For EUF-Cu this toxic limit was 4 ppm. These critical toxic values should be further investigated for plant species and varieties in order to obtain detailed information on the necessity of liming (Brown and Jones, 1977). In any case, optimum plant nutrition with macronutrients—Ca, K, and Mg—plays a decisive role (Vlamis, 1953; Németh and Grimme, 1974; Németh, 1976).

V. Conclusions for Practical Soil Analysis

Modern plant nutrition requires from soil analysis information on the concentration and composition of the soil solution and on the amount of effectively available nutrients. It must further be ascertained to what extent the concentration of nutrients in the solution is changed by removal of nutrients by the plants or by the addition of definite amounts of fertilizer (fertilizer requirements).

Conventional soil analysis (single chemical extraction) can only partly fulfill these requirements. If weak extraction agents are used (water, 0.025 M $CaCl_2$, etc.), approximate statements on the nutrient concentration in the soil solution are possible, but it remains unknown how these concentrations are changed by removal of the plants during the vegetation period. For instance, a certain amount of K extracted with 0.025 M $CaCl_2$ may be associated with a sandy soil without buffering, and the same amount with a heavy clay soil with high reserves. For the same reason, it also cannot be stated to what extent these values determined by weak extraction are altered by specific amounts of fertilizer. In a clay soil, such changes would be much less than in a sandy soil. Nevertheless, these methods are preferable to those that work with stronger extraction solutions (for example, with acid or salt solutions). Larger amounts are extracted with the latter, but their effective availability remains unknown. The correlations between the uptake of the plants and the values of these soil analyses remain, therefore, unsatisfactory. However, the great advantage of the rapid chemical methods is that they can be performed easily and quickly. Their ease of performance in routine investigations, however, should not cause their importance to be overestimated. All one-sidedness in soil analysis should be avoided, since it does not constitute progress, but a step backward. Speed may be attained at the expense of gaining information.

If rapid chemical methods are to be retained in practical soil testing, the analysis should be extended to other soil properties. The relationships between the values obtained by chemical analysis (lactate values, exchangeable cations, and the nutrient concentrations in the soil solution) and other soil properties (clay content, type of clay, etc.) have been explored thoroughly. These relationships enable a better interpretation of the values of conventional soil analysis. An extended soil analysis covering clay content, kind of clay, nutrient fixation, etc., entails a considerable expenditure of time, work, and materials, however.

A further possibility is to develop methods in which these soil properties are incorporated into the results of the analysis. This possibility is afforded by the EUF technique. The quantity of effectively available nutrients is reproduced very well at low voltages (50 and 200 V) and at 20°C. On the other hand, the rates of desorption at 400 V and 80°C characterize the nutrient reserve.

The relationship between the effectively available fractions (0–20 min at 20°C) and the reserves (at 80°C) can give information on important soil properties such as clay content, type of clay, type of soil phosphate, and $CaCO_3$ content. Therefore, the fertilizer requirements (K, P, and CaO) of a soil can be determined by means of EUF (Tables VII and VIII and Fig. 16).

The EUF values required for optimal plant nutrition can be determined more precisely when the temperature and voltage are varied. If the reserves at 80°C are high, the effectively available fraction can be lower. High reserves are, however, of little use at low contents of effectively available nutrients, as an adequate supply of nutrients to the roots will not be guaranteed.

Since soil analysis (whatever the method) determines only a few of the components of nutrient availability, there are limits to its prognostic value. Apart from the amount of available nutrients that are measurable with EUF, the transport of nutrients to the plant roots, for instance, depends on the actual water content of the soil and the average diffusion paths. The more intensively the soil is penetrated by roots, the shorter is the average diffusion path.

In determining the nutrient requirements, the expected yield or yield potential of the location—that is, higher possible yield—should be taken into account as well as the nutrient supply. The yield potential depending on the physical properties of the soil can be characterized according to Harrach (1970) by the utilizable water capacity of the soil that can be penetrated by the roots. Wittman and Grottenthaler (1975) demonstrated correlations between the "degree of ecological moisture" and the yield potential. The nutrient requirement for this yield potential can be calculated by the EUF method.

In conclusion, it can be observed that nutrient extraction by means of the EUF technique is very similar to the processes of nutrient supply to the roots that actually take place in the soil. The technique is thus more universal in application than the conventional methods are. It is suitable for very different types of soil, including those in unknown locations, where the nutrient status and fertilizer requirements can be only partly determined by the usual rapid methods (single extraction).

In principle the EUF technique can also be used for nutrient fractioning in plant material and fertilizers. Furthermore, it can be applied successfully under ecological aspects, since the solution rates of most of the toxic and useful substances in sewage sludge, compost, etc., can also be measured by EUF.

ACKNOWLEDGMENTS

I am grateful to Prof. Dr. H. Beringer and Dr. H. Grimme for their helpful suggestions, and to Mrs. M. Labrenz and Mrs. I. Mühlhausen for assisting in the preparation of this manuscript.

REFERENCES

Adams, P., and Winsor, G.W. 1973. *Plant Soil* **39,** 649–659.
Barber, S.A. 1962. *Soil Sci.* **93,** 39–42.

Barber, S.A., Walker, J.M., and Vasey, E.H. 1963. *J. Agric. Food Chem.* **11**, 204-207.
Bechold, H. 1925. *Z. Elektrochem. Angew. Physik. Chem.* **31**, 496-498.
Bradfield, R. 1928. *Proc. Int. Congr. Soil Sci., 1st, Washington* II, 264-278.
Bray, R.H. 1937. *Soil Sci. Soc. Am. Proc.* **2**, 175-179.
Brown, J.C., and Jones, W.E. 1977. *Agron. J.* **69**, 396-414.
Briggs, L., and McCall, A.G. 1904. *Science* **20**, 566-569.
Egner, H., Riehm, H., and Domingo, W.R. 1960. *Kungl. Lantbrukshögsk. Ann.* **26**, 199-215.
Fox, R.L., and Kamprath, E.J. 1970. *Soil Sci. Soc. Am. Proc.* **34**, 902-907.
Grimme, H. 1978. *Bodenkd. Pflanzenernähr.* (in press)
Grimme, H. 1979. *Bodenkd. Pflanzenernähr.* (in press)
Grimme, H., Németh, K., and v. Braunschweig, L.C. 1971. *Landw. Forsch. Sonderh.* **26**(I), 165-176.
Harrach, T. 1970. *Ergebn. Landw. Forsch. Justus Liebig-Univ. Gießen* No. XI.
Jankoviç, M., and Németh, K. 1974. *Proc. Int. Congr. Potash Inst., 10th* pp. 63-71.
Jankoviç, M., and Németh, K. 1979. *Landw. Forsch.* **32**, 283-291.
Jung, L., and Németh, K. 1966. *Bodenkd. Pflanzenernähr.* **113**, 132-140.
Jung, L., and Németh, K. 1969. *Bodenkd. Pflanzenernähr.* **122**, 33-42.
König, J., Hasenbäumer, J., and Glenk, K. 1913. *Landw. Vers. Sta.* 79-80, 491-534.
Köttgen, P. 1937. *Z. Pflanzenernähr.* **3**, 56-89.
Köttgen, P. 1940. *Bodenkd. Pflanzenernähr.* **18**, 57-64.
Köttgen, P., and Diehl, R. 1929. *Bodenkd. Pflanzenernähr.* A(4), 65-105.
Köttgen, P., and Jung, L. 1941. *Bodenk. Pflanzenernähr.* **25**, 57-64.
Lipmann, C.B. 1918. *Univ. Calif. Publ. Agric. Sci.* **3**, 131-134.
Magistad, O.C., Reitemeier, R.F., and Wilcox, L.V. 1945. *Soil Sci.* **59**, 65-75.
Mattson, S. 1926. *J. Agric. Res.* **33**, 552-568.
Mengel, K. 1975. *Neth. J. Agric. Sci.* **22**, 283-294.
Mengel, K., and Németh, K. 1971. *Landw. Forsch.* **24**, 151-158.
Mengel, K., Grimme, H., and Németh, K. 1969. *Landw. Forsch. Sonderh.* **23**(I), 79-81.
Nair, P.K.R., and Grimme, H. 1978. *Commun. Soil Sci. Plant Anal.* **9**, 755-769.
Németh, K. 1971a. *Geoderma* **5**, 99-109.
Németh, K. 1971b. *Landw. Forsch. Sonderh.* **26**(I), 192-198.
Németh, K. 1972. *Proc. Colloq. Int. Potash Inst., 9th* pp. 3-12.
Németh, K. 1975. *Plant Soil* **42**, 97-107.
Németh, K. 1976. Habilitationsschrift, Universität Gießen.
Németh, K. 1977. *Proc. Int. Seminar Soil Environ. Fertil. Man. Intensive Agric. (ISSS)* pp. 803-810.
Németh, K., and Forster, H. 1976. *Bodenkultur* **27**, 111-119.
Németh, K., and Harrach, T. 1974. *Landw. Forsch. Sonderh.* **30**(I), 131-137.
Németh, K., and Grimme, H. 1974. *Trans. 10th Int. Congr. Soil Sci., Moscow*, IV, pp. 376-382.
Németh, K., Makhdum, I.Q., Koch, K., and Beringer, H. 1980. *Plant and Soil* (in press).
Norman, A.C., Humerfield, H., and Alben, A.O. 1927. *Soil Sci.* **24**, 291-306.
Northrup, Z. 1918. *Science* **47**, 638-639.
Nye, P.H. 1972. *Proc. Colloq. Int. Potash Inst., 9th* pp. 147-155.
Olsen, S.R., Cole, C.V., Watanabe, F.S., and Dean, C.A. 1954. *U.S. Dep. Agric. Cir.* No. 939, p. 19.
Parker, F.W. 1921. *Soil Sci.* **12**, 209-232.
Rich, C.J. 1972. *Proc. Colloq. Int. Potash Inst., 9th* pp. 3-12.
Rich, C.J., and Black, W.R. 1964. *Soil Sci.* **97**, 384-390.
Rost, C.O. 1928. *Proc. 1st Int. Congr. Soil Sci., Washington*, II, pp. 334-341.
Rowell, D.L., Martin, U.W., and Nye, P.H. 1967. *J. Soil Sci.* **18**, 204-222.
Schachtschabel, P. 1940. *Kolloid-Beih.* **51**, 199-276.

Schouwenburg, J. Ch., van, and Schuffelen, A.C. 1963. *Neth. J. Agric. Sci.* **11**, 13–22.

Thiagalingam, K., and Grimme, H. 1976. *Planter Kuala Lumpur* **52**, 83–89.

Vaidynathan, L.V., Drew, M.C., and Nye, P.H. 1968. *J. Soil Sci.* **19**, 94–107.

van der Paauw, F. 1969. *Landw. Forsch. Sonderh.* **23**(I), 102–109.

van Zyl, P.P. 1916. *J. Landwirtsch.* **64**, 201–275.

Vlamis, J. 1953. *Soil Sci.* **75**, 383–394.

von Wrangell, M. 1930. *Landw. Jahrbücher* pp. 149–169.

Wittmann, O., and Grottenthaler, W. 1975. *Mitt. Dtsch. Bodenkl. Ges.* **21**, 222–234.

Wolpers, K., and Tobing, E.L. 1977. *Publikasi Marihat Res. Sta Indonesia.*

ADVANCES IN AGRONOMY, VOL. 31

VOLATILIZATION LOSSES OF NITROGEN AS AMMONIA FROM SURFACE-APPLIED FERTILIZERS, ORGANIC AMENDMENTS, AND CROP RESIDUES

G. L. Terman

Soils and Fertilizer Research Branch, National Fertilizer Development Center, Tennessee Valley Authority, Muscle Shoals, Alabama

I. Introduction

Loss of nitrogen (N) as NH_3 from applied fertilizers, organic amendments, and crop residues is only one of several possible fates of soil and applied N. Crop utilization usually ranges from about 30 to 70% of the applied N, and may average about 50% (Allison, 1955, 1964) the first season. More may be used in succeeding seasons. Leaching losses of N, largely as nitrate (NO_3-N), may be severe on sandy, gravelly, or lateritic soils under conditions of heavy rainfall or excessive irrigation. Most leaching losses with natural rainfall occur during winter or early spring, when crops are not growing and precipitation exceeds

189

TABLE I
Fertilizers Mentioned in This Chapter, and Abbreviations Used

N fertilizer	Chemical formula	Approximate composition (%)	Abbreviation
Ammonium nitrate	NH_4NO_3	33.5 N	AN
Ammonium sulfate	$(NH_4)_2SO_4$	21 N	AS
Calcium nitrate	$Ca(NO_3)_2$	16 N	CN
Diammonium phosphate	$(NH_4)_2HPO_4$	18 N; 20 P	DAP
Monoammonium phosphate	$NH_4H_2PO_4$	11 N; 21 P	MAP
Ammonium polyphosphate	$NH_4H_2PO_4$; $(NH_3)_3HP_2O_7$	12 N; 24 P	APP
Urea	$CO(NH_2)_2$	46 N	Urea
Urea ammonium phosphate	$CO(NH_2)_2$; $NH_4H_2PO_4$	35 N; 8 P	UAP

evapotranspiration. Denitrification losses of soil and applied N to the atmosphere may be substantial under reducing conditions caused by poor drainage, especially from waterlogged soils. Applied N may also be incorporated into soil organic matter, or NH_4-N may be fixed by clay minerals.

Numerous investigators have reported equal effectiveness of urea and other soluble N fertilizers incorporated into well-drained soils for various crops. The mobility of ammonium (NH_4-N) is much less than it is for NO_3-N from fertilizers when first applied to soil. However, NH_4-N rapidly nitrifies to NO_3-N under conditions optimum for growth of upland crops. As a result of denitrification of NO_3-N under flooded rice (*Oryza sativa* L.), urea and NH_4-N fertilizers incorporated rather deeply into the soil are much more effective than are NO_3-N fertilizers. This chapter will be concerned largely with upland crops grown on well-drained soils, unless flooded rice is specified.

Earlier information on the effectiveness of fertilizer N in relation to volatilization losses of N as NH_3 from surface-applied fertilizers has been summarized by Gasser (1964a,b), Terman (1965), and others. Substantial potential losses from both urea and NH_4-N sources were indicated.

Total annual use of fertilizer N has increased to about 11,000,000 tons in the United States alone. Urea is comprising an increasingly large portion of total N use, to about 10% in 1977 in the United States. In tropical Asia, urea production may constitute 85% of the total N fertilizer production capacity by 1985 (Stangel, 1977). The purpose of this review is to summarize research results on NH_3 volatile losses from applied N fertilizers and if possible to assess the impact of greater use of urea N. Losses of NH_3 from crop residues and applied manures and other organic amendments will also be discussed, as well as techniques for measuring N volatilization losses directly as NH_3 and indirectly by measuring crop uptake of applied N, and NH_3 absorption by soils and vegetation. Urea hydrolysis is also discussed.

Various N sources, nutrient composition, and abbreviations used in this chapter are given in Table I.

II. Measurement of NH₃ Volatilization Losses

A. DIRECT LABORATORY METHODS

The common method for measuring NH_3 loss in laboratory studies is to pass air over treated soil in a closed vessel and collect the NH_3 in 2% H_3BO_3, standard H_2SO_4, or other acid (Fig. 1). The air stream is first bubbled through water or dilute H_2SO_4 to remove possible NH_3 contamination and to regulate humidity. The H_3BO_3 or standard acid is titrated with acid or base to determine NH_3 loss (Overrein and Moe, 1967; Watkins *et al.*, 1972; Fenn and Kissel, 1973; and others). Terman *et al.* (1968) reported higher NH_3 loss from urea with dry than with moist aeration. Soil moisture was lost only with dry aeration. Hargrove *et al.* (1977) found that diurnal fluctuations in NH_3 loss corresponded closely with daily fluctuations in atmospheric humidity. At low air renewal volumes loss of NH_3 is commonly proportional to airflow, largely a result of readsorption of NH_3 by soil. For example, Watkins *et al.* (1972) found that NH_3 losses were proportional to the log of airflow rate over the range of 100 ml/min to 3 liters/min (complete air replacement varying from 26 min to <1 min). An increase to 7 liters/min had little additional effect. Overrein and Moe (1967) also found NH_3

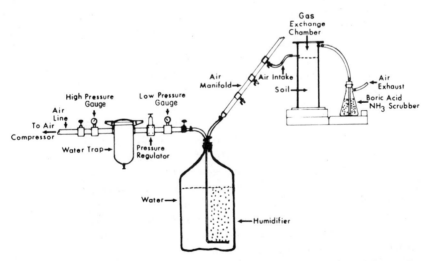

FIG. 1.   Diagrammatic sketch of a typical soil air chamber and associated air-scrubbing systems. From Fenn and Kissel (1973).

losses to increase with gas exchange rate. Fenn and Kissel (1973) recommended air renewal volumes of 14–16 per minute to avoid such proportionality and to reduce loss variability to 10% or less.

B. DIRECT FIELD METHODS

Use of closed systems in the field involves essentially an extension of laboratory methods. Open systems involve additional measurements of atmospheric conditions from which NH_3 losses are calculated. Indirect measurements involve crop yields and/or N recovery by crops grown on small experimental plots.

Robertson and Hansen (1959) used a method for measuring volatilization losses from low-pressure N solutions involving an inverted trough connected to an air pump, which forced the escaping NH_3 into H_3BO_3 solution. Baker *et al.* (1959) used a similar inverted pan to measure NH_3 losses from injected anhydrous NH_3.

Volk (1959, 1961) measured NH_3 volatilized from turf and bare soil in the field by inverting a glass dish over the treated area. Glass wool soaked with dilute acid was attached to the dish bottom. No air movement was involved. Harding *et al.* (1963) used plastic and aluminum foil-covered $1.2 \times 1.2 \times 0.6$-m frames as the field collecting chamber. Air was drawn through paired frames at a rate of 0.07 m³/min, and NH_3 was collected in standard acid. Hutchinson and Viets (1969) used an absorbing surface of 240 cm² of dilute acid to measure atmospheric NH_3 concentrations.

Equipment designed by Kissel *et al.* (1977) and used by Hargrove *et al.* (1977) included 5×10-cm or 15×22-cm steel cylinders driven into the soil flush with the soil surface. Each cylinder was fitted with five air inlets, one outlet, and a removable lid, which was closed only during NH_3 loss measurements. Ammonia in the air stream absorbed in 2% H_3BO_3 was unaffected by airflow rate if air movement through the chamber exceeded 15 exchange volumes per minute.

Denmead *et al.* (1974) described a micrometeorological method of estimating NH_3 loss in the field based on conservation of energy at the ground surface. Denmead *et al.* (1977) made calculations of the aerial transport of NH_3 across the downwind edge of the treated field from measurements of wind speed, wind direction, and atmospheric NH_3 concentrations. They detected NH_3 losses as small as 1 kg of N per hectare in 2 hours, with maximum errors of 20% or less. Soil sampling errors precluded detection of NH_3 loss by analysis.

Forster and Lippold (1975) used a difference method to measure NH_3-N losses involving urea-fertilized microplots, with urea placed in the soil 4 cm deep as the standard. Soluble N in the soil was measured after cessation of NH_3 loss. Nommik (1973a) estimated NH_3 loss as the difference between N applied as ^{15}N-labeled urea (200 kg/ha) and that recovered from the soil profile after 13, 31, and 39 days.

C. INDIRECT CROPPING METHODS

Indirect methods involve growing an N-responsive crop on an N-deficient soil following N application to measure response to the remaining available N. Either greenhouse pot or field plot experiments may be used, in which other nutrients are applied uniformly in amounts adequate for good growth. Dry matter yields may be used as the criterion of response, but N uptake by the crop is usually more appropriate. A NO_3-N source subject to little or no NH_3 volatilization loss should be used as the standard, since complete recovery from any source is not attained in crop growth experiments. Multiple N rates are also recommended, since volatilization losses tend to increase with application rate (Terman and Hunt, 1964). The N sources may be labeled with ^{14}N or ^{15}N, but labeling is usually not a great advantage unless retention of labeled N in the soil is also measured.

Terman *et al.* (1968) used a combination of measurement of NH_3 volatilized from soil in pots, followed by determination of N uptake by corn (*Zea mays* L.). Hargrove *et al.* (1977) compared direct field measurements of NH_3 loss with indirect estimates based on crop N uptake. Cropping experiments in both pots and field plots (Meyer *et al.*, 1961) indicated substantial NH_3 losses from surface-applied urea.

III. NH_3 Losses from NH_4-N and Urea Fertilizers, Including Moisture and Temperature Effects

Results from NH_3-N volatilization studies are greatly influenced by method and type of study. Consequently, results in subsequent sections are divided primarily into laboratory, pot, and field studies.

A. LABORATORY RESULTS

The earliest report of work on NH_3 loss from surface-applied urea was apparently from the Limburgerhof (Germany) experimental station in 1939 (Gasser, 1964a).

Jewitt (1942) verified results of Crowther (1941) showing higher cotton (*Gossypium hirustum* L.) yields with CN than with AS. Jewitt found that 13–87% of the N applied as AS to thin layers of alkaline Sudanese soils in bottles was lost as

NH$_3$. Water and NH$_3$ losses from moist soil varied together and over long periods. As shown in Fig. 2, NH$_3$-N losses were greatly affected by high soil pH, presumably as affected by Na and CaCO$_3$. Moisture content was not important except when it approached air-dry levels. Steenbjerg (1944) reported that losses of N as NH$_3$ from AS surface-applied to Danish soils ranged from 5% or less at a soil pH of 6.0 or below, to 50–60% at pH 8.0. Losses also increased with content of CaCO$_3$ in the soils. Mixing AS with or watering it into the soil reduced losses, even on calcareous soils.

Losses from various N sources applied in solution to the soil surface were measured by Martin and Chapman (1951). Little NH$_3$ was lost from AN or AS applied to soils with pH below 7.2 (Table II). More N was lost from AS than from urea applied to calcareous soils (pH 7.5–8.0).

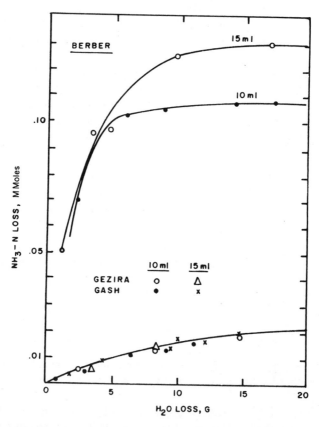

FIG. 2.   NH$_3$-N losses by aeration from three alkaline Sudanese soils (Berber, pH 10.5; Gezira, pH 9.3; and Gash, pH 8.6) in relation to H$_2$O losses. Adapted from Jewitt (1942).

TABLE II
NH$_3$ Losses from California Soils in 70 Days

N source	N applied, (mg)[b]	NH$_3$-N lost (% of applied N)			
		Meloland cl, pH 8.0	Yolo sl, pH 7.7	Hanford sl, pH 7.5	Ramona sl, pH 6.7
None	0	0	—	—	—
(NH$_4$)$_2$SO$_4$	170	23	19	26	5
(NH$_4$)$_2$SO$_4$	283	25	14	19	4
NH$_4$OH	175	27	45	36	51
NH$_4$OH	292	28	40	34	41
Urea	300	16	14	16	36
NH$_4$NO$_3$	298	11	7	7	1
NaNO$_3$	300	0	0	0	0
Dried blood (mixed)	300	3	3	3	18
Orange leaves	186	—	5	6	7

[a] Adapted from Martin and Chapman (1951).
[b] N surface-applied to 600 g of soil. The soils were wetted to 75% moisture and aerated to dryness four times.

Van Schreven (1950) found that loss of NH$_3$ from AS was most rapid from drying calcareous soil. Loss was reduced or eliminated by mixing fertilizer with or watering it into the soil, or by placing it in a furrow 5 cm deep and covering it at once. Wahhab *et al.* (1957) measured N losses of 15–25% in 6 days from AS applied to a sandy soil, and losses of 7–13% from this source applied to a sandy loam. Losses decreased with depth of placement and lower soil pH, but increased with temperature. There was no loss at pH 5.4. Loss of NH$_3$ was associated with loss of water from the soil during drying.

Ernst and Massey (1960) found that increasing temperature and pH markedly increased losses of NH$_3$ from urea applied to a Kentucky soil (Fig. 3). Losses from soil at pH 6.5 were essentially the same for urea surface-applied to the soil, mixed with the surface 0.5 cm, or surface-applied and watered (0.5 cm of water), but they decreased with depth in the soil. Greater losses occurred during drying of the soil and ceased as the surface soil became dry. Loss was directly related to the initial soil moisture content, presumably because of the effect of this variable on the duration of the drying process.

Kresge and Satchell (1960) reported large losses of N from surface-applied urea or calcium cyanamid on slightly acid Hagerstown silt loam, but practically none from AN or AS. Loss was more rapid from soils drying out from field capacity than at higher moisture contents. Watering urea into the soil or preventing moisture evaporation reduced losses. Less N was lost from urea–AN solutions than from urea alone.

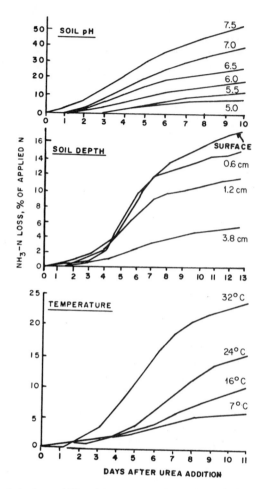

FIG. 3.   Cumulative losses of N as NH₃ from 50 ppm of urea, N applied to Dixon silt loam, as affected by soil pH, depth, and temperature. From Ernst and Massey (1960).

Little N was lost (Carter and Allison, 1961) as NH_3 during incubation periods of 6 to 10 weeks from AS mixed with soils of pH varying from 4.5 to 6.0. Losses during drying following incubation were also negligible on unlimed soil, but were as high as 15% on soils limed to pH 6.7.

Meyer *et al.* (1961) measured losses of N as NH_3 during a 1-month period from surface and mixed applications of pelleted urea, pelleted AN, crystalline AS, and a urea–AN solution to acid to alkaline Nebraska soils. Losses were most severe (up to 20%) with urea fertilizers applied to the soil surface (Fig. 4). Losses decreased with soil pH and were minor from fertilizers mixed with soil or surface-applied and watered into the soil. Losses on alkaline Crofton silty clay loam (pH 7.8) were as follows: urea–AN solution > urea > AS > AN > no N.

Complete recovery was found by Low and Piper (1961) after 4 days of N added as AS and nitro-chalk surface-applied and mixed with soil at pH 4.8, and as fine urea mixed with soil. Recovery was about 80% from surface-applied fine and pelleted urea and from urea pellets mixed with the soil. Recovery from the four sources ranged from 60 to 96% with surface application to soils at pH 6.8 and 8.0. Although it was not pointed out in their paper, the low recovery of N from urea pellets mixed with the soil may have resulted from losses to the atmosphere in forms other than NH_3 as a result of temporary NO_2-N accumulation around the pellets (Court et al., 1962; Hauck and Stephenson, 1965).

Lehr and van Wesemael (1961) reported that volatilization of NH_3 from soil was more closely correlated with $CaCO_3$ content than with pH. Loss at 40°C averaged 3.5 times the loss at 20°C. Volatilization was greater with AS than with AN, but the difference was less marked in soils containing high contents of $CaCO_3$.

Similar losses occurred from urea and AS applied to two calcareous soils at a rate of 112 kg of N per hectare (Gasser, 1964a). Nitrite tended to accumulate in sandy and calcareous soils, accumulation being favored by lower temperature and wetter soil. Ammonia was lost from calcareous soils until all the NH_4-N had been nitrified, from slightly alkaline soils until 90% had nitrified, and from neutral sandy soils until 60% had nitrified. Volatilization losses of NH_3 varied from 2% on clay loam to 13% on sandy loam soils. Khan and Haque (1965) reported usual results of highest NH_3-N losses from urea surface applied to calcareous Bangladesh soil, less from neutral soil, and least from acid soil.

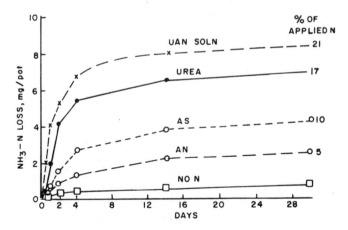

FIG. 4. NH_3-N losses from 36 mg of N in 30 days from AN, AS, urea, or UAN solution surface-applied to Crofton silty clay loam (pH 7.8) to which straw residue had been applied.

Covering the urea with 5 cm of soil reduced losses, as did application of super-phosphate with the urea. Gandhi and Paliwal (1976) found NH_3 loss to increase with soil salinity; losses were higher from urea than from AS.

Overrein and Moe (1967) found NH_3 loss rates from urea to increase exponentially with rate of application, but they decreased with depth of placement in the soil. Fenn and Escarzaga (1977) reported that NH_4-N was moved deeper by irrigation into an initially wet soil than it was if the soil was initially dry. A twofold increase in NH_3 loss was found from application of AN and AS in irrigation water over broadcast application followed by irrigation.

It may be concluded from the numerous cited results that losses of NH_3 increase with increase in intensity of drying conditions (higher temperatures, more air movement, and lower humidity), with higher soil pH, with coarse-textured soils of low exchange capacity, and with lower initial soil moisture content. Losses are very low if various N sources are incorporated into acid soils or are moved at once into the soil by rain or irrigation. Severe losses of NH_3 may occur with urea surface-applied to both acid and alkaline (or calcareous) soils and from NH_2-N and NH_4-N sources on alkaline soils. These may be either naturally occurring alkaline soils or heavily limed acid soils. Losses of N as NH_3 should not be confused with denitrification or leaching losses.

B. CHEMISTRY OF NH_3 LOSSES

Van Schreven (1956) concluded that loss of NH_3 from ammonium fertilizers applied to calcareous soils was dependent on the anion present in the salt. Loss was found to be more rapid from AS than from AN. Loss of NH_3 was postulated to be due to the equilibrium

$$NH_4^+ \rightleftharpoons NH_3 + H^+$$

which influenced the solubilities of $CaSO_4$ and $Ca(NO_3)_2$ formed by the reaction of the ammonium salts with $CaCO_3$.

Duplessis and Kroontje (1964) found that NH_3 lost from AS applied to Virginia soils increased with soil pH from 4.5 to 7.1. Since OH^- ions increase with pH, it was postulated that NH_3 losses occurred according to the equation (Jewitt, 1942)

$$NH_4^+ + OH^- \rightleftharpoons NH_3 + H_2O$$

Results of Larsen and Gunary (1962) showed that NH_3 loss from a thin layer of air-dry calcareous soil (pH 7.7) was greatest with AS, about half as great with MAP, DAP, and AN, and negligible with $MgNH_4PO_4$. Losses of NH_3 in 8 days from an acid soil heavily limed with $CaCO_3$ were AS > DAP = AN. For the same soil similarly limed with $MgCO_3$, losses were AS > AN > DAP. It was concluded from this study that factors in addition to pH affected the loss of N as NH_3.

Terman and Hunt (1964) concluded from petrographic studies and crop recoveries of applied N that volatilization losses of N as NH_3 from alkaline soils increased greatly with a decrease in solubility of the reaction products of NH_4-N sources with Ca compounds. Typical reactions may be illustrated as follows:

$$\text{DAP: } (NH_4)_2HPO_4 + CaCO_3 \rightarrow CaHPO_4 + (NH_4)_2CO_3$$
$$\text{AS: } (NH_4)_2SO_4 + CaCO_3 \rightarrow CaSO_4 + (NH_4)_2CO_3$$
$$\text{AN: } 2NH_4NO_3 + CaCO_3 \rightarrow Ca(NO_3)_2 + (NH_4)_2CO_3$$

The $(NH_4)_2CO_3$ decomposes readily as follows:

$$(NH_4)_2CO_3 \overset{H_2O}{\rightarrow} 2NH_3\uparrow + CO_2\uparrow$$

Also,

$$NH_3 + H_2O \rightleftarrows NH_4OH$$

Losses from MAP tend to be lower than losses from DAP or AS because MAP forms intermediate reaction products, such as $Ca(NH_4)_2(HPO_4)_2 \cdot H_2O$.

Both DAP and AS form Ca salts of low solubility in alkaline soils, allowing more complete formation of $(NH_4)_2CO_3$, which is unstable and decomposes to NH_3 and CO_2. Ammonium nitrate forms soluble Ca nitrate, which is stable, and which limits the reaction of the NH_4-N component to form $(NH_4)_2CO_3$. Thus, losses of NH_3 from AN are the least from these three sources and probably similar to losses from NH_4Cl. Hydrolysis of urea is not dependent on reaction with Ca compounds, but requires only H_2O and urease as follows:

$$CO(NH_2)_2 + H_2O \overset{urease}{\rightarrow} (NH_4)_2CO_3$$

Thus, NH_3 losses from surface-applied urea occur on both acid and alkaline soils and may be less on alkaline soils than from NH_4-N compounds.

Fenn and associates have verified the conclusions of Terman and Hunt (1964) and have greatly expanded research on NH_3 losses from calcareous soils. Fenn and Kissel (1973) found that NH_3 losses were in the order $NH_4F > AS > DAP > AN = NH_4Cl > NH_4I$. Solubility of the apparent reaction products was the major factor regulating NH_3 losses from a Houston black clay (pH 7.6) and from a neutral Wilson clay loam. Fenn and Kissel (1974) showed that total NH_3-N losses from AS applied to the Houston soil increased with rate of application (Fig. 5), but losses from AN did not. Losses from both sources increased with temperature, more from AN than from AS. In other studies NH_3-N losses from reactions of AS with $CaCO_3$ in Houston soil ranged from 0 to 70 or 80%.

Losses of NH_3 from surface-applied NH_4F or AS were reduced by adding either AN or MAP to Houston black clay (Fenn, 1975). However, adding MAP to $(NH_4)_2CO_3$ did not reduce NH_3 losses much. Reduced formation of unstable $(NH_4)_2CO_3$ is thus important. Fenn and Kissel (1975) found an increased NH_3 loss from AN and AS with an increase in soil $CaCO_3$ content. Nitrification of the NH_4-N in AN and AS reduced the pH of soils low in $CaCO_3$ and thus reduced NH_3 loss.

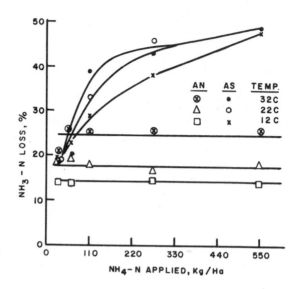

FIG. 5. NH$_3$-N losses in 100 hours from AN and AS surface-applied to Houston black clay, as affected by application rate and temperature. From Fenn and Kissel (1974).

Feagley and Hossner (1978) concluded that losses of NH$_3$ from AS on reaction with carbonates proceeded through an NH$_4$HCO$_3$ intermediate rather than directly to (NH$_4$)$_2$CO$_3$.

As was reported earlier, Fenn and Kissel (1976) found that NH$_3$ loss from calcareous soils decreased with an increase in soil CEC. Losses also decreased with increasing depth of soil incorporation. Avnimelech and Laher (1977) concluded from a theoretical study that the final concentration of NH$_4$-N held by a soil increases with an increase in H$^+$ ion activity, initial NH$_4$-N concentration, and partial NH$_3$ pressure in the air, but decreases with an increase in soil buffer capacity. Soil pH is the dominant factor controlling extent of NH$_3$ loss if the soil buffer capacity is high or if NH$_4$-N concentration is low. The dominant factor at high pH and high initial NH$_4$-N concentration is the soil buffer capacity.

Fenn *et al.* (1979) surface-applied solutions of urea alone (9% NH$_4$-N) and with Ca and Mg nitrates or chlorides. Loss of NH$_3$ was drastically reduced by these salts. It was postulated that, as soil pH rose as a result of hydrolysis of urea to (NH$_4$)$_2$CO$_3$, CaCO$_3$ and MgCO$_3$ were precipitated, as follows:

$$(NH_4)_2CO_3 + Ca(NO_3)_2 \text{ [or } Mg(NO_3)_2] \rightarrow CaCO_3 \text{ (or } MgCO_3) + 2NH_4NO_3$$

Precipitation of Ca and Mg carbonates reduced (NH$_4$)$_2$CO$_3$ concentrations and consequently NH$_3$ losses. Precipitated CaCO$_3$ + MgCO$_3$ was estimated by weight loss of CO$_2$ after addition of 3 N HCl.

Khasawneh (unpublished TVA data) showed that exchangeable Ca and Mg are precipitated as the carbonates, and equivalent amounts of NH_4 are adsorbed on the soil CEC, thereby reducing losses of NH_3-N. In this work, Decatur loam (pH 5, 16% moisture) was packed in the columns to a bulk density of 1.20, moisture was increased to 24%, granular urea (0.89 g per column) was surface-applied, and the columns were incubated for 8 days at 25°C. Each column was then sliced at 1-mm intervals to measure nutrient diffusion. Exchangeable Ca and Mg were extracted twice from the slices for 1 hour with 2 N KCl containing 5 ppm of phenylmercuric acetate (PMA) to inhibit urease activity. The residue was then extracted with 25 ml of 0.1 N HCl for 15 min. Figure 6 shows that exchangeable Ca extracted by 2 N KCl increased, whereas Ca extracted by 0.1 N HCl ($CaCO_3$) decreased with distance from the point of urea application. The sum of Ca extracted by KCl and HCl was constant (the soil exchangeable Ca level).

The duration of NH_3-N loss measurements may greatly affect NH_3 loss results and their interpretation.

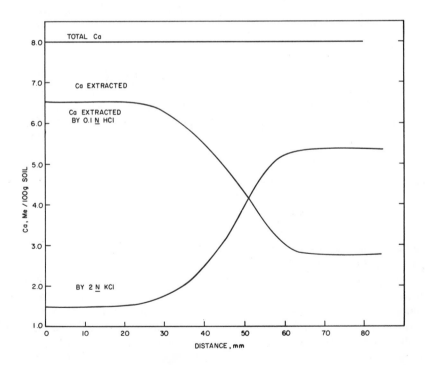

FIG. 6.   Exchangeable Ca extracted from Decatur loam by 2 N KCl and Ca extracted by 0.1 N HCl with distance from surface-applied urea.

C. POT TEST RESULTS

Wagner and Smith (1958) reported greater N recovery by sudangrass (*Sorghum* sp.) from several N fertilizers mixed with the soil than when the fertilizers were surface-applied. Greatest losses of N (total loss, including NH_3) generally occurred from urea, aqua ammonia, and nitrate fertilizers. Recovery of N by ryegrass (*Lolium multiflorum* L.) from urea applied on dry surface soil (moist soil below) was much lower than it was with application with the seed at a depth of 8 mm (Low and Piper, 1961). In a second experiment with N topdressed on the surface, recovery of N by ryegrass was about 40% from urea and 50% from nitro-chalk. Meyer *et al.* (1961) reported less recovery of N by oats (*Avena sativa* L.) from surface application than from mixed application of urea. Less N was recovered after incubation for a month than after no incubation.

It is likely that several sets of published results from pot experiments, such as those of Court *et al.* (1962), have been affected both by NH_3-N losses and by seedling toxicity in the soil resulting from the released NH_3 and formation of nitrite.

Jackson and Burton (1962) applied urea and AN to bermudagrass (*Cyanadon dactylon* L.) sod in cans, sealed the tops of the cans, and passed air over the sod through openings in opposite sides. About 13% of the N was lost as NH_3 from urea applied to both dry and moist sods. No loss from AN was found.

Terman and Hunt (1964) determined yields of corn forage and uptake of N from N fertilizers mixed with and surface-applied to moist soils a week prior to planting. Losses of N were low with mixed placement. High losses of N, presumably as NH_3, occurred from granular urea applied to Hartsells fine sandy loam limed to pH 6.2 and 7.5 and to Webster silty clay loam (pH 8.2). Losses were reduced by coating urea with sulfur (SCU) or by including phosphate in the granules (UAP). Severe losses of N also occurred with AS and DAP surface-applied to calcareous soil. Maximum recoveries of N from soils of pH 6.2 and 8.2 were obtained from AN, MAP, and APP.

Engelstad *et al.* (1964) found that corn forage yields and N recovery from surface-applied urea and fine oxamide were much lower than they were from mixed application with the soil. Best recovery of N was obtained with either surface or mixed application of AN and AS.

Four greenhouse pot experiments were conducted by Terman *et al.* (1968) to measure recovery of applied N by corn from various solid N sources and solutions. Crop recovery was low from sources containing NH_2-N, such as urea, formamide, hexamine, and oxamide surface-applied to both acid and alkaline soils. Much higher recovery resulted from mixing these fertilizers with the soils. Greatest recovery (least loss, presumably by NH_3 volatilization) resulted from NH_4NO_3 and $NH_4H_2PO_4$. Similar results were obtained with solid and solution sources (Table III). Volatilization loss as NH_3 increased in the laboratory to a

TABLE III
NH$_3$-N Losses in 41 Days from Fluid Fertilizers Surface-Applied to Hartsells Fine Sandy Loam (pH 6.3)[a]

N source	Total N (%)	NH$_3$-N volatized (mg) Aeration at 100% RH	Aeration at 70% RH
Solutions			
AN	20.5	1	1
75% AN, 25% urea	19.5	2	2
25% AN, 75% urea	20.2	16	28
Urea	20.5	45	55
Suspension			
UAN	36.8	20	22

[a] From Terman *et al.* (1968).

maximum of 55 mg of N over a 41-day period at 24°C with an increase in proportion of urea in solutions, and was less with moist than with drier (70% RH) aeration. Much greater loss was indicated by lower N recovery by corn grown after a 1-week drying period (24°–32°C) prior to planting.

Apparent N recovery by barley (*Hordeum vulgare* L.) in a greenhouse pot experiment from surface-applied N was found by Fenn and Escarzaga (1976) to be well correlated with nonvolatilized N. Losses of N as NH$_3$ from AN and AS decreased greatly with higher soil moisture content. However, N losses are dependent on sufficient H$_2$O to dissolve the applied N compounds. That is, little or no loss occurs from dry soil, as was reported earlier for a sandy Florida soil (Volk, 1966).

D. FIELD RESULTS

Crowther (1941) reported higher cotton yields with CN than with AS on alkaline soils of the Sudan Gezira. Yields increased with rate of CN but not of AS, which gave better results with soil incorporation. Van Schreven (1950) observed that crop yields on Dutch soils containing 6–16% CaCO$_3$ were better with nitrate than with ammonium sources of N applied as topdressings. Loss of N as NH$_3$ was reduced if applications were made during or just before rainy weather.

Several investigators have reported severe losses of N from surface-applied urea. Burton and DeVane (1952) compared N sources topdressed for bermudagrass over a 3-year period in Georgia. Average relative yields were 100, 102, 86, and 77 for NaNO$_3$, AN, cyanamid, and urea. Further analysis (not in the published paper) of the results from this experiment indicated that two or four split applications of urea during the growing season resulted in less recovery of N than was obtained from a single total application made in March. Split applications of AN, however, were more effective than the single application. This indicates that losses of N from urea applied during higher temperatures of the growing period were greater than was the case with application during lower March temperatures. However, Mays and Terman (1969) did not find this result at Muscle Shoals, Alabama.

Volk (1959, 1961) measured losses of ammonia directly by absorption by pads of glass wool saturated with dilute H_2SO_4. Comparable losses from 112 kg of N per hectare applied to four grass sods in Florida were 29% from crystalline urea, 11.5% from urea–AN solution, and 0.3% from pelleted AN (1959). Losses from application of 112 kg of solid urea N per hectare to bahiagrass sod (*Paspalum notatum* Flugge) averaged 29% on unlimed (pH 5.8) Leon fine sand and 39% from sod on this soil that was limed 4 months previously (1961). Corresponding losses were 0.4% and 19.7% from AS, and 0.3% and 3.4% from AN. Losses from urea applied to bare sandy soil averaged 25%; the loss from AN was negligible.

Burton and Jackson (1962) reported the following relative N recoveries of topdressed N on bermudagrass in Georgia over a 5-year period: AN—100; AS—98; AN solution—95; urea–AN solution—86; and urea—74. Half of the N was applied in March, and half after the second cutting. Applications of urea and AN (224 kg of N per hectare) 15 cm below the surface of Coastal bermudagrass sod growing on Tifton loamy sand resulted in equal yields (Jackson and Burton, 1962). Surface applications of urea resulted in 86% as much forage yield on unburned stubble and 97% as much on burned stubble as were found after similar applications of AN.

Meyer *et al.* (1961) found lower yields of irrigated corn in Nebraska with 100 and 180 kg of N applied as urea than when it was applied as AN. Yields decreased as follows: applied when corn was 30–60 cm tall, in the spring before planting, in the fall. Lowest yields were obtained with urea surface-applied in the fall. Losses of N, especially of that applied in the fall, probably include leaching as well as volatilization losses. Urea was slightly less effective in 1956–1957 tests than AN–lime when topdressed on grassland in England (Templeman, 1961). No difference in effectiveness was found for wheat (*Triticum aestivum* L.) in 1956 or for wheat, grass, and other crops in earlier field tests. Laughlin (1963) found that AN, AS, and CN were equally effective as N fertilizers for bromegrass (*Bromus inermis* Leyss) in Alaska, but urea was usually less effective. This was especially true with fall applications and at high N levels.

Devine and Holmes (1963) compared AN, AS, and urea in ten experiments on grassland in England and Scotland. Nitrogen from each source was applied at rates of 34 and 67 kg/ha in the spring and after each cutting except the last.

Similar forage yields resulted with AN and AS at all cuttings in six experiments on soils with pH 5.5–7.0. Yields were lower with AS than with AN at one or more cuttings on four calcareous soils, pH 7.4–8.2. Urea gave lower yields than AN at one or more cuttings in eight of the ten experiments, and was less effective relative to AN as the season advanced.

Topdressed AN and AS at rates of 39–67 kg of N per hectare were compared in 36 experiments by Devine and Holmes (1964) on winter wheat and at rates of 34–112 kg in 89 experiments on grassland. Forage yields were lower with AS than with AN on highly calcareous soils. On other soils yields with the two sources were similar, on both sandy and heavy soils and in various regions of Great Britain. Gasser (1964a) pointed out that NH_3 was lost most rapidly from AS immediately after it was applied to calcareous soil. With urea there was a brief delay before NH_3 volatilized most rapidly. Thus, if rain occurs soon after application of the N, unhydrolyzed urea may be washed into the soil and loss of NH_3 decreased. This is unlikely with AS, which reacts rapidly with calcium compounds in the soil.

Johansson and Johnson (1964) summarized nearly four hundred N experiments on grassland in Sweden. As a topdressing, urea produced slightly lower yields than did AN–lime. Urea applied as a preseeding treatment for barley and spring wheat was also less effective but more economical than AN. Urea, AN, AS, and CN (45 and 90 kg of N) were compared by Devine and Holmes (1965) as early winter, late winter, and spring topdressings for grassland in fourteen experiments in various areas of England. Early winter applications gave lower yields than did late winter or spring applications, and the inferiority increased with winter rainfall. Lower yields were obtained with CN than with the other sources applied in winter, but yields with urea averaged lowest with spring application.

Denmead et al. (1976) found rather high NH_3 losses from grazed pasture measured by a chemical–micrometeorological technique, but minor losses from ungrazed pasture. The taller plant cover apparently absorbed the NH_3.

Hargrove et al. (1977) estimated losses of N as NH_3 from 3 to 10% for AN, from 36 to 45% for granular AS, and from 25 to 55% for AS solution field-applied at rates of 140–280 kg of N per hectare to Houston black clay in Texas. The directly measured losses were lower during lower temperatures in spring than in summer. Consistent diurnal fluctuations in NH_3 losses occurred at both times.

It has been determined from personal contacts that a large volume of unpublished data has been obtained on field comparisons of surface-applied urea and other N fertilizers. There is much variability in the results from year to year, among sources, application times, and crops. Results obtained with no-till farming are often different from those obtained with conventional tillage. Such variability makes it difficult to summarize the results and to derive firm conclusions. Some results show equal effectiveness of surface-applied urea and other N sources. It is obvious that much research still needs to be done to forecast a given NH_3 loss resulting from urea fertilizer application under field conditions.

IV. NH₃ Losses from Urea

Urea requires contact with urease for hydrolysis to $(NH_4)_2CO_3$. Consequently, surface-applied urea may be potentially less subject to initial loss of NH_3 than are NH_4-N sources applied to alkaline soils. If little soil urease is present, rain or irrigation water may move urea deeper into the soil before hydrolysis occurs. Gasser (1964c) reported a typical example of such effects (Fig. 7). Losses of NH_3 occurred immediately from AS applied to a chalky loam (pH 8.0), but losses from urea were delayed a few days until hydrolysis took place. Mixing urea with this soil was not effective in reducing NH_3 losses over surface placement. Temperature is also important for urea hydrolysis, as indicated by hydrolysis being only 50% complete at 10°C but 85% complete at 20°C and 30°C in Hermitage silt loam after 2 weeks (Fisher and Parks, 1958). Broadbent et al. (1958) found complete hydrolysis of urea in California soils at 7°C in 7 days; at 24°C no urea remained after 2 or 3 days.

Chin and Kroontje (1963) observed that chemical hydrolysis of urea is very slow in comparison with biochemical (urease) hydrolysis. Conrad (1942) and others had recognized the role of urease much earlier.

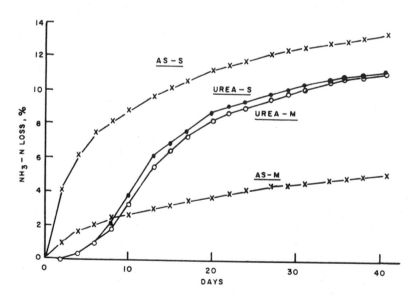

FIG. 7.   Cumulative percentage losses of NH_3 from 112 kg of N/per hectare as urea and AS surface-applied (S) or mixed (M) with chalky loam. Adapted from Gasser (1964c).

Sufficient water must also be present for the urea to dissolve so that hydrolysis can occur. Volk (1966) reported data indicating slow dissolution of urea surface applied to Florida sandy soils. Over 80% of prilled urea field-applied to the surface of air-dry Leon fine sand did not hydrolyze in 14 days, and 42–72% persisted on initially moist sand, even though dews formed nightly. On continually moist sand, urea hydrolyzed completely in 7 days and high volatilization loss resulted from the dissolved urea. These differences in dissolution rates explain some erratic responses to surface-applied urea.

As previously indicated, loss of N as NH_3 from surface-applied urea occurs from both acid and alkaline soils. Losses of dissolved urea depend not only on the usual factors of drying conditions, soil water content, soil CEC, and temperature, but also on urease activity.

A. UREASE ACTIVITY IN SOILS AND EFFECTS OF INHIBITORS

Urease activity in 100 Australian surface soils was found by McGarity and Myers (1967) to be highly correlated with organic carbon content and poorly correlated with soil pH. Overrein and Moe (1967) found that urea hydrolysis at 28°C was directly proportional to rate of urea applied to Chalmers silt loam and Plainfield sand (Indiana). Ammonia volatilization increased exponentially with application rate. Aeration rate had no significant effect on urea hydrolysis, but NH_3 volatilized increased with an increase in gas exchange rate over the soil.

Tabatabai and Bremner (1972) developed a simple soil urease assay technique involving determination of NH_4-N released when soil is incubated with tris(hydroxymethyl)aminomethane (THAM) buffer, urea solution, and toluene at 37°C for 2 hours.

Dalal (1975) found that urease activity was highly correlated with organic carbon or CEC in 15 Trinidad soils. Bremner and Zantua (1975) detected urease, phosphatase, and sulfatase activities in soils at −10°C and −20°C. This was attributed to interaction in unfrozen water at soil particle surfaces.

Addition of urea to Iowa soils did not induce urease activity (Zantua and Bremner, 1976, 1977). However, addition of glucose and other organic materials that stimulated biological activity also increased urease activity. Persistence of urease activity varied among soils, but activity in all soils eventually returned to the level of an unamended soil. Apparently, soil constituents protect urease against microbial degradation and other processes leading to enzyme inactivation, so that each soil has a stable level of urease activity, depending on the soil constituents.

Zantua et al. (1977) found high correlations of urease activity with organic carbon, clay content, and CEC in 21 Iowa soils. Organic matter accounted for most of the variation in urease activity. Tabatabai (1977) found that urease activity decreased with soil depth, in proportion to a decrease in organic carbon content. Thus, a given soil apparently has a stable urease level and temporary levels, depending primarily on soil organic matter content.

Numerous compounds have been found effective in inactivating urease in solution, but relatively few are effective in soil. Waid and Pugh (1967) showed that acetohydroxamic acid inhibited hydrolysis of urea in an acid loamy sand and delayed loss of NH_3. Pugh and Waid (1969a,b) found that acetohydroxamic acid was more effective in retarding NH_3 loss than was benzo-, capro-, or salicylohydroxamic acid.

Bremner and Douglas (1971b) evaluated more than 100 compounds as urease inhibitors in soils. Dihydric phenols and quinones were the most effective organic compounds, and silver and mercury salts the most effective inorganics. In later studies (1973) they found that of eight urease inhibitors 2,5-dimethyl-p-benzoquinone was most effective in reducing urea hydrolysis and NH_3 loss from soil. This compound reduced NH_3 loss from urea applied to a sandy soil from 63 to 0.3%.

Methyl-, chloro-, bromo-, and fluoro-substituted p-benzoquinones markedly inhibited urease activity in soils, whereas phenyl-, t-butyl-, and hydroxy-substituted p-benzoquinones had little, if any, effect (Bundy and Bremner, 1973). Various nitrification inhibitors retarded nitrification of NH_4-N but had little effect on urease activity (Bundy and Bremner, 1974).

Bremner and Bundy (1976) found that potassium azide (KN_3) retarded urea hydrolysis in soils, but did not retard nitrification of NH_4-N. Tabatabai (1977) reported that numerous metals inhibited urease activity in eight Iowa soils in the order $Ag \geqslant Hg > Cu > Cd > Zn > Sn > Mn$. Additional trace elements also inhibited urease.

Inhibition of soil urease by several heterocyclic mercaptans was investigated by Gould et al. (1978). Several quinones, catechol, and 1,3,4-thiodiazole-2,5-dithiol were most effective.

B. ADDITIVES TO UREA IN RELATION TO NH_3 LOSSES

Chemical amendments are applied not only to inhibit urease activity, but also to have other direct effects on potential NH_3 losses from urea. These include (a) reduction in pH, (b) coating of urea granule surfaces, and (c) reduction in NH_3 loss through precipitation of Ca and Mg carbonates.

Urea forms addition compounds, or adducts with numerous acids and salts. For example, urea and $Ca(NO_3)_2$ form an adduct, as follows:

$$4CO(NH_2)_2 + Ca(NO_3)_2 \rightarrow 4CO(NH_2)_2 \cdot Ca(NO_3)_2$$

This product, known as Calurea (34% N), was manufactured in Germany and exported in the 1930s. It was a usable product, but often was in poor physical condition. Other adducts having industrial and other uses (Terman and Fleming, 1968) include urea nitrate [$CO(NH_2)_2 \cdot HNO_3$] and urea phosphate [UP, $CO(NH_2)_2 \cdot H_3PO_4$]. Both form strongly acid solutions.

Gasser and Penny (1967) found that urea nitrate applied to barley and grassland produced lower yields than did AN, likely as a result of toxicity of the resulting extremely acid solution. Urea phospate and UP plus urea were superior to AN as sources of N; however, P was not equalized in the experiments, and urea alone was not included.

Bremner and Douglas (1971a) found that H_3PO_4 formed from decomposition of UP in soils retarded enzymatic hydrolysis of the urea by urease. This resulted in reduction of NH_3 loss from urea. Loss of N as NH_3 from urea applied to Iowa soils was 5–61%, but only 0.1–1% from UP. Paulson and Kurtz (1969) found that urea N in a pelleted urea-aliphatic acid complex plus p-benzoquinone inhibitor released NH_4-N slightly more slowly than did urea alone.

Watkins et al. (1972) measured lower NH_3 losses from mixtures of urea and NH_4Cl than from crystalline or pelleted urea alone (50% less, with 25% of the N as NH_4Cl). Terman and Hunt (1964) were apparently the first to report higher crop recoveries from surface-applied SCU than from uncoated urea. Recoveries in pot experiments with corn were also higher (less NH_3 loss) from granules containing urea + phosphate (UAP) than from urea alone. The lower NH_3 losses with SCU probably occur as a result of slower dissolution and hydrolysis of urea in SCU and from some oxidation of S, which lowers the pH at the granule sites. In the case of UAP, the presence of MAP or NH_4-polyphosphates tends to reduce the fertilizer solution pH and thus restricts NH_3 loss. The presence of DAP in the granules would not be effective in reducing the solution pH, since DAP solution is alkaline (pH 8.0). However, Phipps (1964) found that NH_3 losses from fertilizers applied to Nebraska soils were in the order urea > UAP > DAP, and they decreased with a decrease in soil pH from 8.2 to 5.5. Losses from AN and APP were negligible.

Matocha (1976) reported that <1% of AN, AS, and SCU (20% and 30% initial dissolution in water) surface-applied to unlimed Darco fine sand (pH 6.0) was lost in 14 days as NH_3, whereas 18.5% of uncoated urea N was lost. Topdressing lime with N increased N loss more initially from AS than from urea, but cumulative losses in 14 days were 52%, 23%, 9%, and 2% from urea, AS, SCU-30, and SCU-20, respectively.

Urea–AN (UAN) solutions and suspensions are used widely to supply N for a wide range of crops. The question frequently arises as to whether AN with urea reduces NH_3 losses from topdressed UAN and whether such solutions are more effective than solid urea, especially under no-till.

Kresge and Satchell (1960) showed that higher AN/urea ratios of fertilizer solutions (158 ppm of total N) surface-applied to two Pennsylvania soils (pH 6.3) decreased NH_3 volatilization. However, it is not clear (Table IV) whether the reduced NH_3 losses were due to less urea N applied or to the effects of increasing

G. L. TERMAN

TABLE IV
Volatilization of NH$_3$ in 11 Days, as Affected by AN/Urea Ratio of Solutions Surface-Applied to Two Soils[a]

| AN/urea ratio | N applied as AN (ppm) | N applied as urea (ppm) | NH$_3$-N volatilized | | | |
| | | | Morrison sandy loam | | Hagerstown silt loam | |
			% of urea N	% of total N	% of urea N	% of total N
0/100	0	158	77	77	79	79
36/65	45	113	42	30	32	23
45/55	60	98	37	23	28	17
56/44	77	81	36	18	25	13
70/30	101	57	34	12	23	8

[a] Calculated from Kresge and Satchell (1960).

the proportions of AN, or to both. It was shown in other studies that NH$_3$ losses from AN or AS alone were negligible.

Results obtained in Mississippi (Table V) also were not definitive in regard to possible effects of AN on NH$_3$ losses from urea. Negligible NH$_3$ losses occurred from AN surface-applied to an acid (pH 5.9) sandy loam. Losses were low (5–13%) from urea or AN + urea. Percentage losses of NH$_3$ decreased with an increase in the amount of AN if expressed on the basis of total N applied, but not if expressed on the basis of urea N applied. Losses of NH$_3$-N were 6–9% from AN surface-applied to an alkaline (pH 7.8) sandy loam, and much higher from urea or urea + AN. Percentage losses again decreased with increasing amounts of AN if expressed on the basis of total N applied, but increased greatly with amounts of AN if expressed on the basis of urea N applied. Thus, susceptibility of AN to NH$_3$ losses on the two soils and method of expressing losses greatly affected interpretation of the results. This experiment was also not adequate to determine effects of AN on NH$_3$ losses from urea. Adequate answers to comparative effectiveness of urea + AN solutions versus urea alone await more definitive results.

Terman *et al.* (1968) found the usual increases in NH$_3$-N losses with an increase in urea content of fluid fertilizers. Negligible losses occurred from AN and from solutions of 75% AN, 25% urea. Higher losses occurred with drier than with moist aeration.

V. NH$_3$ Losses from Anhydrous NH$_3$ and NH$_4$OH

Intensive studies on equipment for applying anhydrous NH$_3$ and on crop response were carried out in Mississippi in 1945–1947 (Andrews *et al.*, 1948).

Use has grown rapidly since then. Jackson and Chang (1947) found little loss of NH_3 from anhydrous NH_3 injected 2.5–5.0 cm into soil varying widely in physical properties.

Laboratory studies of aqua ammonia surface-applied to Hawaiian soils by Humbert and Ayres (1957) showed losses of up to 15% from surface application to acid soils and 50% or more from alkaline soils, even with injection 10 cm into the soils. With injection into irrigation water, about 20% volatilization loss was found from water flowing about 60 m in irrigation furrows.

McDowell and Smith (1958) found negligible loss of NH_3 from anhydrous NH_3 injected only 7 cm deep in a calcareous clay soil. Losses were much higher from injections 15 cm deep in an acid sandy soil.

Robertson and Hansen (1959) reported that NH_3 losses from low-pressure N solutions dribbled on the surface of Kalamazoo sandy loam in Michigan increased with rate of application, were greatest immediately after application, and decreased with time. No appreciable loss occurred with injection to a depth of 5 cm. Baker et al. (1959) also reported negligible losses from anhydrous NH_3 with injection 10 cm or deeper at practical rates of application. In contrast, Henderson et al. (1955) reported NH_3 losses as high as 60% when anhydrous NH_3 was applied through jet sprinklers at NH_3 concentrations greater than 100 ppm.

Chao and Kroontje (1964) observed a linear relationship between rate of application of NH_4OH to soils and rate of NH_3 loss during drying. Rates of NH_3

TABLE V

Volatilization of NH_3 in 18 Days from AN and Urea Surface-Applied to Acid and Alkaline Soils[a]

N as AN applied with urea (mg)	Urea applied and NH_3-N loss			Urea applied and NH_3-N loss			
	100 mg	200 mg	300 mg	None	100 mg	200 mg	300 mg
	% of urea N			% of total N			
Sandy loam, pH 5.9							
0	5	8	12	—	5	8	12
50	5	8	13	—	4	6	11
100	6	9	13	0.3	3	6	10
200	6	9	12	0.6	2	4	7
300	7	9	11	0.5	2	4	6
Sandy loam, pH 7.8							
0	18	18	16	—	18	18	16
50	28	20	16	9	16	16	14
100	33	21	19	7	16	14	14
200	42	24	22	8	14	12	13
300	47	31	25	6	12	12	13

[a] Calculated from unpublished Mississippi Agricultural Experiment Station data reported to the TVA in 1966.

and H_2O losses followed different functions. Losses of NH_3 from finer textured soils were proportional to their original pH values.

DuPlessis and Kroontje (1966) showed that CO_2 increased and stabilized retention of NH_3 applied as NH_4OH to base-saturated bentonite. It was postulated that Ca and Mg exchanged by NH_4-N were precipitated as carbonates of low solubility, thus increasing NH_4-N retention by the bentonite.

Miyamoto *et al.* (1975) observed that volatilization loss of NH_3 from irrigation water was directly proportional to concentrations of NH_4OH. Applied H_2SO_4 reduced NH_3 losses largely by lowering water pH. Simultaneous injection of NH_4OH and H_2SO_4 into the water at equivalent rates reduced losses by 50%. Earlier, van Dijk had reported (1943, cited by Vlek and Stumpe, 1978) NH_3 losses from fertilized irrigation water of about 50 kg of N per hectare in a 4-day period.

Bock (1978) measured NH_3 volatilization losses from the nonevaporated portion of sprinkler-applied irrigation water. Losses from water initially containing 200 ppm of NH_4-N added as NH_3 (pH 9.6–9.8) were about 35%. Reducing the initial pH with H_2SO_4 to 8.0 reduced the losses to <5%. Losses were diminished still more by acid additions in amounts equivalent to NH_3 added to the water.

VI. NH_3 Losses from Organic Amendments and Crop Residues

Usually the chief concern with volatilization losses of NH_3 from organic material is the loss of fertilizer value, particularly of its N content, although air and water pollution effects may also be important. Heck (1931) found initial rates of NH_3 loss from stable manure of 0.5- to 2.0-day half-lives. A second stage of loss proceeded at a slower rate.

Losses of NH_3 are perhaps most noticeable from piles of stable manure or other wastes, which tend to heat and from which NH_3 can be detected by smell. This was recognized by Midgley and Weiser (1937), who recommended application of superphosphate to the manure to reduce NH_3 losses. They found that freezing also resulted in NH_3 losses of 50% in 2 days. Salter and Schollenberger (1938) also reported NH_3 losses up to 50% from stable manure following spreading.

Hutchinson and Viets (1969) measured NH_3 absorption rates by acid traps near feedlots as much as twenty times those at points remote from feedlots in northeastern Colorado. This atmospheric NH_3 volatilized from feedlots contributes to N enrichment of nearby surface waters and soils. Elliott *et al.* (1971) similarly measured 148 kg of atmospheric NH_3 per hectare per year (including organic N) near a feedlot and 16 kg/ha/year 0.8 km away. Feedlot disturbances increased NH_3 losses. Luebs *et al.* (1973) also measured high atmospheric NH_3 concentrations near dairies in California.

Stewart (1970) simulated feedlot wetting and drying conditions in a laboratory study and demonstrated about 80% loss of NH_3 from urine in 3 days. Mosier *et al.* (1973) and others have identified aliphatic amines and other compounds in addition to NH_3 volatilized from feedlots. Pronounced diurnal fluctuations in NH_3 losses from dairies were found by Luebs *et al.* (1974) and from urine by McGarity and Rajaratnam (1973).

Hooker *et al.* (1973) did not measure appreciable losses of NH_3 from Duroc silt loam (pH 7.3) during 160 days under conditions simulating breaking the soil from native sod.

Lauer *et al.* (1976) measured NH_3 losses from surface-applied dairy manure at Ithaca, New York, by determining NH_4-N contents of manure and of the soil beneath with time. Losses of 61–99% of NH_4-N applied in the manure in summer were measured, with mean half-lives of 1.9–3.4 days for low and high rates of application. In winter, NH_3 losses were precluded by subfreezing temperatures, snow cover, or rapid thaws. Beauchamp *et al.* (1978) measured NH_3 losses from newly applied anaerobically digested sewage sludge to a field in Ontario by an aerodynamic method for 5- and 7-day periods in May and October. Fluxes of NH_3 followed a diurnal pattern, with maxima occurring about midday, and decreased exponentially with time. Losses were estimated at 56–60% of the NH_4-N applied in the sludge. Unpublished 1977 results of Beauchamp *et al.* showed similar NH_3 losses from liquid cattle manure.

In cropping experiments in greenhouse pots, Terman *et al.* (1968) found much lower crop recovery of applied N (presumably higher loss as NH_3) from several NH_2-N compounds surface-applied to Hartsells fine sandy loam than was obtained from mixed application. The compounds included urea, formamide, hexamine, and oxamide. Engelstad *et al.* (1964) and Allen *et al.* (1971) also concluded that NH_3 was lost from surface-applied oxamide. Since NH_2-N compounds are present in plants, it follows that serious volatilization losses of NH_3 might occur from drying and decomposing crop residues.

Mayland (1968) found in drying alfalfa (*Medicago sativa* L.) hay that carbon and dry matter were lost more readily than N. However, in research conducted at TVA (Terman, unpublished, 1968), N uptake by corn was lower from alfalfa surface-applied to Hartsells fine sandy loam than it was from the same amount of residue mixed with this soil. Evidently some N volatilization occurred. In other TVA research, losses of dry matter and N from clippings of green alfalfa, low and high N Coastal bermudagrass, and red clover (*Trifolium pratense* L.) placed in the field, on greenhouse benches, and in an NH_3 recovery train were measured from July 6 to 15. Losses of N from samples in the field were relatively greater than losses of dry matter in the first 2 days (Fig. 8), and then both declined more slowly during a 7-day sampling period. In the greenhouse, the samples dried rapidly, and there was no appreciable weight loss in dry matter from the continuously dry samples. Losses of dry matter and N began when samples were wetted.

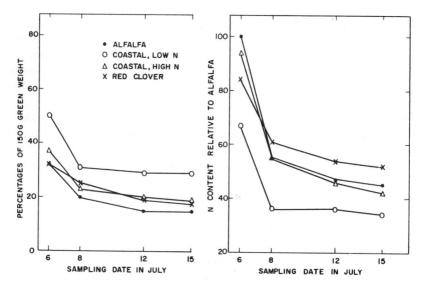

FIG. 8.　Losses in weight and N content of alfalfa residue during drying for 9 days (TVA data).

Losses of N as NH_3 were not measurable during rapid drying of fresh samples, but were measurable when water was added to the samples and decomposition began. Percentages of N in field-placed samples decreased at first, but then increased as decomposition narrowed the C/N ratio. The same was true for wetted samples in the greenhouse.

Several investigators have postulated N volatilization losses from maturing plants, but apparently no definitive results have been reported that show volatilization losses during decline in N concentrations in addition to losses in plant parts.

VII. NH₃ Losses from Flooded Soils

Low recovery of applied N by flooded rice is usually assumed to result largely from nitrification–denitrification reactions and the consequent loss of N as N_2O and N_2 to the atmosphere. Loss may also occur by leaching from coarse-textured soils. However, research results have demonstrated severe losses of topdressed N as NH_3 from flooded soils under some conditions. The extent of loss is a function of soil and fertilizer NH_4-N concentration, water pH, wind movement, and temperature.

Willis and Sturgis (1944) found large volatilization losses of NH_3 at 38°C from Crowley clay loam (pH 4.6) fertilized with 200 ppm of N as AS in laboratory studies. Losses were higher (Fig. 9) from flooded than from nonflooded soil and were much higher from limed soil. Since more N was lost than was applied, it is obvious that NH_3 was being volatilized from organic matter, as well as from AS. MacRae and Ancajas (1970) in laboratory studies found NH_3 losses of 3–19% from urea surface-applied to moist soil followed by flooding. Losses from incorporated urea and AS were only 3–5%. The pH of the four soils studied remained < 7.5.

Ventura and Yoshida (1977) found that most of the NH_3 losses from flooded soil occurred during the first 9 days after N application. Losses increased greatly with an increase in soil pH and were greater from urea than from AS, from which losses below pH 7.5 were very small. Mixing the N with puddled soil reduced losses. Losses were greater from flooded soil; drying reduced the duration and amount of loss. Wetselaar et al. (1977) reported NH_3 volatilization losses from AS surface-applied for flooded rice in Thailand (soil pH 7.0–7.5) up to 12%. Loss was less from AS incorporated into the soil.

Vlek and Stumpe (1978) found in laboratory studies with open containers that the NH_3 loss capacity of flooded soil was equivalent to its alkalinity [NH_3(aq), HCO_3^-]. In nonalkaline solutions, loss of NH_3 from AS was low, but loss from $(NH_4)_2CO_3$ was almost quantitative. The rate of NH_3 loss was directly related to the concentration of aqueous NH_3 and thus to the concentration of NH_4-N and the

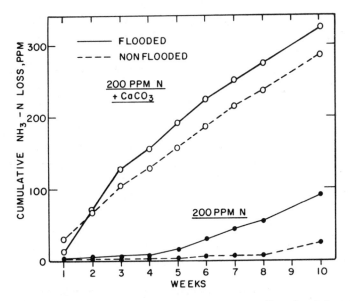

FIG. 9.   Cumulative losses of NH_3-N from Crowley clay loam over a 10-week period, as affected by flooding and 5000 ppm of applied $CaCO_3$. From Willis and Sturgis (1944).

pH. Volatilization of NH_3 increased with water turbulence and exponentially with higher temperatures.

Loss of NH_3 from AS surface-applied to flooded rice at the International Rice Research Institute was found (IRRI, 1976; Sahrawat, 1978) to be related to growth of algae in the floodwater. As a result of depletion of CO_2 in the water by algal growth, the pH rose as high as 9.0 by mid-afternoon, and as high as 9.7 with AS + DAP applied to the floodwater. As much as 50% of the applied N was lost in 5 days at these high water pH levels. Both N and P contribute to algal blooms, accompanied by higher water pH. Deep placement of NH_4-N greatly reduces NH_3 losses from flooded soils. In other studies, NH_3-N losses were least from acid soil (pH 5.7), intermediate from a soil at pH 6.6, and highest from a sodic soil (pH 8.6). Results of NH_3-N loss determined by an aerodynamic method agreed with losses from flat dishes and open cans placed in the paddy field.

Mikkelsen et al. (1978) and Mikkelsen and DeDatta (1978) concluded that pH of floodwater, and thus NH_3-N losses, are greatly influenced by the total respiration activity and gross photosynthesis of the organisms present. Losses from fertile Maahas clay were as high as 20%, but those from acid Luisiana clay, where floodwater did not support algae growth, was <1% of the applied N.

Volatilization losses of NH_3 at IRRI from open dishes filled with rice paddy floodwater fertilized with 60 kg of N per hectare as AS were found by Bouldin and Alimagno (1976, quoted by Vlek and Stumpe, 1978) to range from 10 to 24 kg/ha over the first 2–3 days. Depletion of CO_3 by photosynthetic activity of algae appeared to be a major factor in raising the pH of floodwater to 8.0 and above, which increases NH_3 losses greatly. D.R. Bouldin (private communication) believes that real losses of NH_3 are not measured in any studies involving any sort of small container. Losses are greatly affected by wind movement and can be estimated better from larger areas through micrometeorological techniques.

The results of Bouldin et al. (1974) are of interest in relation to NH_3 losses from flooded soils. After application of AN to six New York ponds, they found that loss of NH_4-N from solution ranged from 2 to 38% per day. Biological immobilization of N was low; reductions in NO_3-N ranged from 7 to 15% per day. They concluded that NH_3-N loss to the atmosphere was a major cause of reductions in NH_4-N concentrations in solution.

VIII. NH₃ Losses in Forest Fertilization

Studies concerning NH_3 loss from forest soils have been fewer than those from agricultural soils, and reported losses have usually been rather low. The same general factors affect NH_3 loss in both situations. For example, Watkins *et al.* (1972) found that NH_3 loss from urea applied to mineral soils and forest floors increased with increasing air movement, with higher temperatures, and with higher pH. Losses with an airflow of 3 liters/min at 19°C ranged from 6 to 30% from mineral soils and from 27 to 46% from forest floors.

Overrein (1968) reported lower NH_3 losses (maximum of 3.5%) from an application of 500 kg of N per hectare with 5 mm of water to Scotch pine (*Pinus* sp.) humus. Losses (1969) were similarly low from urea applications of up to 1000 kg of N per hectare. Urea hydrolysis was largely complete in 3 days.

Losses of NH_3-N reported by Watson *et al.* (1962) from urea applied to bare clay soil in Malaya was negligible from dry soil, 10% from soil at 20% moisture, and 24% at 4% moisture. Losses from urea applied to rubber tree (*Hevea* sp.) leaf litter was 10–18%. Losses occurred largely during the first 4 days.

Volk (1970) found that NH_3 losses from prilled urea surface-applied (100 kg of N per hectare) to undisturbed organic residue over moist soil under slash pine averaged 4% in 7 days, as compared with 2% from an area control-burned 5 weeks previously. Loss from an area from which loose debris had been removed was 5%.

Nommik (1973a) and Johnson (1977) found in Sweden that NH_3 losses from large-pellet urea were markedly less than were losses from small-pellet urea during the first 2 weeks of exposure. Differences were less after 4 weeks. The large pellets apparently dissolved more slowly. Results in a 5-year experiment on pines showed best growth response to AN (150 kg of N per hectare), better response to large than to small urea pellets, and no effect of adding H_3PO_4 to the urea.

Raison and McGarity (1978) determined the effects of plant ash on volatilization of NH_3 from applied N fertilizers. In moist ash alone, urea and KNO_3 were unaffected, but 63–70% of the N was lost from AS over a 4-day period from autoclaved and nonautoclaved ashes. Losses of NH_3 were greater from urea or AS applied to soil treated with a surface layer of ash than they were after application to soil (pH 6.1) alone. The ash increased surface soil pH, saturated exchange sites with basic cations, and apparently increased soil urease activity. These results are of interest in relation to NH_3 losses from urine excretions by wildlife, and in relation to timing of N fertilizer applications.

Numerous field comparisons of urea and AN for forest fertilization have been made, especially in Europe, as summarized by Bengtson (1976). Results of comparisons of urea and AN have been highly variable. Various theories have been proposed to explain the results, but none is universally applicable. Poorer results with urea are usually explained by more volatilization loss of NH_3 from urea surface-applied on acid forest soils or forest floors. A second possible explanation is the more rapid immobilization of urea by humus. Poorer results with AN have usually been explained by the more rapid leaching of NO_3-N from AN.

Urea was formerly favored over AN in forest fertilization because of its higher analysis and lower pollution potential, during both manufacture and use. The shift away from urea to AN in Sweden starting in 1968 (Möller, 1974) has prompted renewed research to resolve older, anomalous results.

IX. NH_3 Sorption by Soils and Vegetation

It is abundantly clear that NH_3 is volatilized to the atmosphere from soils having insufficient sorption capacity to hold NH_4-N released from fertilizers. However, soils also sorb NH_3 from the atmosphere. This sorption capacity is, of course, greater for acid, fine-textured soils, which also have higher base exchange capacities and greater capacities for retaining NH_4-N from applied fertilizers.

As cited by Hanawalt (1969), Liebig stated as early as 1847 that soil colloids absorb NH_3 from the atmosphere. The first estimate of quantity absorbed, 46 kg of NH_3-N per hectare per year, was published by Bretschneider in 1872. Schloesing in 1875 estimated an absorption of 40 kg of NH_3-N per hectare per year. Rossi in 1945 reported that thin layers of soil absorbed 20–25 kg of NH_3-N per hectare per year.

Malo and Purvis (1964) observed NH_3 levels in New Jersey of three times the normal atmospheric levels and measured sorption rates of 22–78 kg of NH_3-N per hectare per year by soil exposed under field conditions. Hanawalt (1969) measured NH_3-N sorption of 55–74 kg/ha/year by six New Jersey soils during 4-day exposures in an enclosed chamber. Sorption of NH_3 increased greatly with higher NH_3 concentrations in the air, higher temperatures, and greater velocities of airflow. Coffee and Bartholomew (1964b) also found that adsorption by soil increased with NH_3 concentations in the air, with lower soil pH, and with higher clay content. NH_3 did not penetrate more than 10–20 mm into the soil.

The above results are in apparent conflict with increasing NH_3 volatilization losses with higher temperatures and air velocities, as measured by other workers. However, these discrepancies are apparently related to the degree of saturation of the soil sorption capacity by NH_3.

Ingram (1950) reported sorption rates of several organic materials for NH_3-N. Coffee and Bartholomew (1964a) found that retention of atmospheric NH_3 by both moist and dry plant residue samples increased with NH_3 concentration. Physical retention by moist samples was higher.

No measurements have apparently been reported, but living vegetation and bodies of water constitute tremendous sinks for atmospheric NH_3. The various results on NH_3-N sorption by soils and vegetation indicate that this source of N for crop growth is quite important under some conditions, such as near urban areas (Malo and Purvis, 1964), near feedlots (Hutchinson and Viets, 1969), and near power plants and other industrial plants. Contributions of atmospheric NH_3 should be considered in N balance studies. However, Hoeft *et al.* (1972) and Rodgers (1978) found dry deposition of NH_3-N to contribute less than 5 kg of N per hectare per year. Rodgers found that dry deposition of NH_3 was inversely related to rainfall.

X. Conclusions

A large amount of research has been carried out to study NH_3-N losses from soils, organic residues and amendments, and water. Much of the research has been done under laboratory conditions, for which numerous factors affecting losses can be more closely controlled and separated. The results show that the most severe loss problems are concerned with surface-applied urea on both acid and alkaline soils and with surface-applied NH_4-N sources on neutral to alkaline soils. More loss tends to occur with NH_4-N sources (such as AS or DAP), which tend to react with $CaCO_3$ to form reaction products of low solubility, than from sources (such as AN) that form soluble salts. Losses of NH_3-N are essentially eliminated in acid soils by covering urea 5 cm or more deep, as are losses from urea placed deeply in flooded rice soils. Losses are reduced by covering in alkaline soils but may not be reduced to zero.

Research results almost invariably show that NH_3-N losses from fertilizers surface-applied to moist soils increase with intensity of drying conditions (higher temperatures, lower humidity) and with decrease in soil sorption capacity for NH_4-N (coarser texture, lower CEC, higher pH, lower water content). Results from laboratory, greenhouse pot, and field plot experiments agree in general concerning the various factors contributing to NH_3-N losses.

Results from field experiments tend to be highly variable among seasons, crops, soils, years, and various weather conditions. The major need in further research on NH_3-N losses seems to be for field experiments in which more loss-contributing factors are quantified. Apparently only then can we hope to predict accurately a given loss result from a surface-applied N fertilizer or organic amendment.

Part of the variability in results from laboratory, greenhouse pot, and field studies is also due to variation in application patterns for solid granules and liquid droplets. This should be made uniform for comparable results.

G. L. TERMAN

REFERENCES

Allen, S.E., Terman, G.L., and Hunt, C.M. 1971. *J. Agric. Sci.* **77**, 397–404.
Allison, F.E. 1955. *Adv. Agron.* **7**, 213–250.
Allison, F.E. 1964. *In* "Soil and Fertilizer Nitrogen Research." TVA, Muscle Shoals, Alabama, pp. 1–17.
Andrews, W.B., Edwards, F.E., and Hammons, J.G. 1948. *Miss. Agric. Exp. Sta. Bull.* 451.
Avnimelech, Y., and Laher, M. 1977. *Soil Sci. Soc. Am. J.* **41**, 1080–1084.
Baker, J.H., Peech, M., and Musgrave, R.B. 1959. *Agron. J.* **51**, 361–362.
Beauchamp, E.G., Kidd, G.E., and Thurtell, G. 1978. *J. Environ. Qual.* **7**, 141–146.
Bengtson, G.W. 1976. Presented to working group on forest fertilization. IUFRO World Congr. *16th, Oslo*.
Bock, B. 1978. Ph.D. Thesis, Univ. of Nebraska, Lincoln.
Bouldin, D.R., Johnson, R.L., Burda, C., and Kao, C.W. 1974. *J. Environ. Qual.* **3**, 107–114.
Bremner, J.W., and Bundy, L.G. 1976. *Soil Biol. Biochem.* **8**, 131–133.
Bremner, J.W., and Douglas, L.A. 1971a. *Soil Sci. Soc. Am. Proc.* **35**, 575–578.
Bremner, J.W., and Douglas, L.A. 1971b. *Soil Biol. Biochem.* **3**, 297–307.
Bremner, J.M., and Douglas, L.A. 1973. *Soil Sci Soc. Am. Proc.* **37**, 225–226.
Bremner, J.W., and Zantua, M.I. 1975. *Soil Biol. Biochem.* **7**, 383–387.
Broadbent, F.E., Hill, G.N., and Tyler, K.B. 1958. *Soil Sci. Soc. Am. Proc.* **22**, 303–307.
Bundy, L.G., and Bremner, J.M. 1973. *Soil Biol. Biochem.* **5**, 847–853.
Bundy, L.G., and Bremner, J.M. 1974. *Soil Biol. Biochem.* **6**, 369–376.
Burton, G.W., and DeVane, E.H. 1952. *Agron. J.* **44**, 128–132.
Burton, G.W., and Jackson, J.E. 1962. *Agron. J.* **54**, 40–43.
Carter, J.N., and Allison, F.E. 1961. *Soil Sci. Soc. Am. Proc.* **25**, 484–486.
Chao, T.T., and Kroontje, W. 1964. *Soil Sci. Soc. Am. Proc.* **28**, 393–395.
Chin, W., and Kroontje, W. 1963. *Soil Sci. Soc. Am. Proc.* **27**, 316–318.
Coffee, R.C., and Bartholomew, W.V. 1964a. *Soil Sci. Soc. Am. Proc.* **28**, 482–485.
Coffee, R.C., and Bartholomew, W.V. 1964b. *Soil Sci. Soc. Am. Proc.* **28**, 485–490.
Conrad, J.P. 1942. *Soil Sci.* **54**, 367–380.
Cooke, G.W. 1964. *Fert. Soc. (London) Proc.* **80**.
Court, M.N., Stephen, R.C., and Waid, J.S. 1962. *Nature (London)* **194**, 1263–1265.
Crowther, F. 1941. *Emp. J. Exp. Agric.* **9**, 125–136.
Dalal, R.C. 1975. *Soil Biol. Biochem.* **7**, 5–8.
Denmead, O.T., Simpson, J.R., and Freney, J.R. 1974. *Science* **185**, 609–610.
Denmead, O.T., Simpson, J.R., and Freney, J.R. 1976. *Soil Biol. Biochem.* **8**, 161–164.
Denmead, O.T., Simpson, J.R., and Freney, J.R. 1977. *Soil Sci. Soc. Am. J.* **41**, 1001–1004.
Devine, J.R., and Holmes, M.R.J. 1963. *J. Agric. Sci.* **60**, 297–304.
Devine, J.R., and Holmes, M.R.J. 1964. *J. Agric. Sci.* **62**, 377–379.
Devine, J.R., and Holjes, M.R.J. 1965. *J. Agric. Sci.* **64**, 101–107.
Douglas, L.A., and Bremner, J.M. 1971. *Soil Biol. Biochem.* **3**, 309–315.
DuPlessis, M.C.F., and Kroontje, W. 1964. *Soil Sci. Soc. Am. Proc.* **28**, 751–754.
DuPlessis, M.C.F., and Kroontje, W. 1966. *Soil Sci. Soc. Am. Proc.* **30**, 693–696.
Elliott, L.F., Schuman, G.E., and Viets, F.G., Jr. 1971. *Soil Sci. Soc. Am. Proc.* **35**, 752–755.
Engelstad, O.P., Hunt, C.M., and Terman, G.L. 1964. *Agron. J.* **56**, 579–582.
Ernst, J.W., and Massey, H.F. 1960. *Soil Sci. Soc. Am. Proc.* **24**, 87–90.
Feagley, S.E., and Hossner, L.R. 1978. *Soil Sci. Soc. Am. J.* **42**, 364–467.
Fenn, L.B. 1975. *Soil Sci. Soc. Am. Proc.* **38**, 366–369.
Fenn, L.B., and Escarzaga, R. 1976. *Soil Sci. Soc. Am. J.* **40**, 537–541.
Fenn, L.B., and Escarzaga, R. 1977. *Soil Sci. Soc. Am. J.* **41**, 358–363.

Fenn, L.B., and Kissel, D.E. 1973. *Soil Sci. Soc. Am. Proc.* **37** 855-859.

Fenn, L.B., and Kissel, D.E. 1974. *Soil Sci. Soc. Am. Proc.* **38**, 606-610.

Fenn, L.B., and Kissel, D.E. 1975. *Soil Sci. Soc. Am. Proc.* **39**, 631-633.

Fenn, L.B., and Kissel, D.E. 1976. *Soil Sci. Soc. Am. Proc.* **40**, 394-398.

Fenn, L.B., Taylor, R.M., and Matocha, J.E. 1979. *Soil Sci. Soc. Am. J.* **43**, (in press).

Fisher, W.B., Jr., and Parks, W.L. 1958. *Soil Sci. Soc. Am. Proc.* **22**, 247-248.

Forster, I., and Lippold, H. 1975. *Arch. Acker Pflanzenbau Bodenkd.* **19**, 619-630.

Gandhi, A.P., and Paliwal, K.V. 1976. *Plant Soil* **45**, 247-255.

Gasser, J.K.R. 1964a. *Soils Fert.* **27**, 175-180.

Gasser, J.K.R. 1964b. *World Crops* March, 25-32.

Gasser, J.K.R. 1964c. *J. Soil Sci.* **15**, 258-272.

Gasser, J.K.R., and Penny, A. 1967. *J. Agric. Sci.* **69**, 139-148.

Gould, W.D., Cook, F.D., and Bulat, J.A. 1978. *Soil Sci. Soc. Am. J.* **42**, 66-72.

Hanawalt, R.B. 1969. *Soil Sci. Soc. Am. Proc.* **33**, 231-234.

Harding, R.B., Embleton, T.W., Jones, W.W., and Ryan, T.M. 1963. *Agron. J.* **55**, 515-518.

Hargrove, W.L., Kissel, D.E., and Fenn, L.B. 1977. *Agron. J.* **69**, 473-476.

Hauck, R.D., and Stephenson, H.F. 1965. *J. Agric. Food Chem.* **13**, 486-492.

Heck, A.F. 1931. *Soil Sci.* **31**, 467-481.

Henderson, D.W., Bianchi, W.C., and Doneen, L.D. 1955. *Agric. Eng.* **36** 398-399.

Hoeft, R.G., Keeney, D.R., and Walsh, L.M. 1972. *J. Environ. Qual.* **1**, 203-208.

Hooker, M.L., Peterson, G.A., and Sander, D.H. 1973. *Soil Sci. Soc. Am. Proc.* **37**, 247-249.

Humbert, R.P., and Ayres, A.S. 1957. *Soil Sci. Soc. Am. Proc.* **21**, 312-316.

Hutchinson, G.L., and Viets, F.G., Jr. 1969. *Science* **166**, 514-515.

Ingram, G.J. 1950. *Soil Sci.* **70**, 205-212.

International Rice Research Institute 1977. *Ann. Rep. 1976* 236-237, Los Baños, Philippines.

Jackson, J.E., and Burton, G.W. 1962. *Agron. J.* **54**, 47-49.

Jackson, M.L., and Chang, S.C. 1947. *J. Am. Soc. Agron.* **39**, 623-633.

Jewitt, T.N. 1942. *Soil Sci.* **54**, 401-409.

Johansson, O., and Jonson, L. 1964. *Vaxt. Nar. Mytt.* **20**(3), 23-25.

Johnsson, S. 1977. *Inst. For. Improv. Fert. Inf.* **I.**

Khan, D.H., and Haque, M.Z. 1965. *J. Sci. Food Agric.* **16**, 725-729.

Kissel, D.E., Brewer, H.L., and Arkin, G.F. 1977. *Soil Sci. Soc. Am. Proc.* **41**, 1133-1138.

Kresge, C.B., and Satchell, D.P. 1960. *Agron. J.* **52**, 104-107.

Larsen, S., and Gunary, D. 1962. *J. Sci. Food Agric.* **13**, 566-572.

Lauer, D.A., Bouldin, D.R., and Klausner, S.D. 1976. *J. Environ. Qual.* **5**, 134-141.

Laughlin, W.M. 1963. *Agron. J.* **55**, 60-62.

Lehr, J.J., and van Wesemael, J.C. 1961. *Landbowk. Tydschr.* **73**, 1156-1168.

Low, A.J., and Piper, F.J. 1961. *J. Agric. Sci.* **57**, 249-255.

Luebs, R.E., Davis, K.R., and Laag, A.E. 1973. *J. Environ. Qual.* **2**, 137-141.

Luebs, R.E., Davis, K.R., and Laag, A.E. 1974. *J. Environ. Qual.* **3**, 265-269.

MacRae, I.C., and Ancajas, R. 1970. *Plant Soil* **33**, 97-103.

McDowell, L., and Smith, G.E. 1958. *Soil Sci. Soc. Am. Proc.* **22**, 38-42.

McGarity, J.W., and Myers, M.G. 1967. *Plant Soil* **27**, 217-238.

McGarity, J.W., and Rajaratnam, J.A. 1973. *Soil Biol. Biochem.* **5**, 121-131.

Malo, B.A., and Purvis, E.R. 1964. *Soil Sci.* **97** 242-247.

Martin, J.P., and Chapman, H.D. 1951. *Soil Sci.* **71**, 25-34.

Matocha, J.E. 1976. *Soil Sci. Soc. Am. J.* **40**, 597-601.

Mayland, H.F. 1968. *Agron. J.* **60**, 658-659.

Mays, D.A., and Terman, G.L. 1969. *Sulphur Inst. J.* **5** (Autumn), 7-10.

Meyer, R.D., Olson, R.A., and Rhoades, H.F. 1961. *Agron. J.* **53**, 241-244.

Midgley, A.R., and Weiser, V.L. 1937. *Vermont Agric. Exp. Sta. Bull.* 419.

Mikkelsen, D.S., and DeDatta, S.K. 1968. Paper presented at Symposium of Nitrogen and Rice, IRRI, Los Baños, Philippines, September.

Mikkelsen, D.S., DeDatta S.K., and Obcemea, W.N. 1978. *Soil Sci. Soc. Am. J.* **42**, 725-730.

Mitsui, S., Ozaki, K., and Moriyama, M. 1954. *J. Soil Sci. (Tokyo)* **25**, 17-19. (*Chem. Abstr.* **48**, 11702.)

Miyamoto, S., Ryan, J., and Stroehlin, J.L. 1975. *Soil Sci. Soc. Am. Proc.* **39**, 544-548.

Möller, G. 1974. *Phosphore Agr.* No. 68, 33-48.

Mosier, A.R., Andre, C.E., and Viets, F.G., Jr. 1973. *Environ. Sci. Technol.* **7**, 642-644.

Nommik, H. 1973a. *Plant Soil* **38**, 589-603.

Nommik, H. 1973b. *Plant Soil* **39**, 309-318.

Okuda, A., Takahashi, E., and Yoshida, M. 1960. *Nippon Dojo Hiryogaku Zasshi* **31**, 273-278.

Overrein, L.N. 1968. *Soil Sci.* **106**, 280-290.

Overrein, L.N. 1969. *Soil Sci.* **107**, 149-159.

Overrein, L.N., and Moe, P.G. 1967. *Soil Sci. Soc. Am. Proc.* **31**, 57-61.

Paulson, K.N., and Kurtz, L.T. 1969. *Soil Sci. Soc. Am. Proc.* **33**, 973.

Phipps, R.L. 1964. M.S. Thesis, Univ. of Nebraska, Lincoln.

Pugh, K.B., and Waid, J.S. 1969a. *Soil Biol. Biochem.* **1**, 195-206.

Pugh, K.B., and Waid, J.S. 1969b. *Soil Biol. Biochem.* **1**, 207-217.

Raison, R.J., and McGarity, J.W. 1978. *Soil Sci. Soc. Am. J.* **42**, 140-143.

Rashid, G.H. 1977. *Plant Soil* **48**, 549-556.

Robertson, L.S., and Hansen, C.M. 1959. *Mich. Agric. Exp. Sta. Q. Bull.* **42**(1), 47-51.

Rodgers, G.A. 1978. *J. Agric. Sci.* **90**, 537-542.

Sahrawat, K.L. 1978. *Int. Rice Res. Newslett.* **3**(2), 16.

Salter, R.M., and Schollenberger, C.J. 1938. *In* "Soils and Men," pp. 445-461. Yearbook of Agric., U.S. Govt. Printing Office, Washington, D.C.

Stangel, P.J. 1977. Presented at Int. Rice Res. Conf., IRRI, Los Baños, Philippines. April.

Steenbjerg, F. 1944. *Tids. Planteavl.* **48**, 516-543. (*Chem. Abstr.* **41**, 4878-9).

Stewart, B.A. 1970. *Environ. Sci. Technol.* **4**, 579-582.

Tabatabai, M.A. 1977. *Soil Biol. Biochem.* **9**, 9-13.

Tabatabai, M.A., and Bremner, J.M. 1972. *Soil Biol. Biochem.* **4**, 479-487.

Templeman, W.G. 1961. *J. Agric. Sci.* **57**, 237-240.

Terman, G.L. 1965. *Agrichem. West* **8**(12), 8-9 and 13-14.

Terman, G.L., and Fleming, J.D. 1968. *In* "New Fertilizers Materials," pp. 320-326. Noyes Data Corp. Park Ridge, N.J.

Terman, G.L., and Hunt, C.M. 1964. *Soil Sci. Soc. Am. Proc.* **28**, 667-672.

Terman, G.L., Parr, J.F., and Allen, S.E. 1968. *J. Agric. Food Chem.* **16**, 685-690.

Trickey, N.G., and Smith, G.E. 1955. *Soil Sci. Soc. Am. Proc.* **19**, 222-224.

Van Schreven, D.A. 1950. *Trans. Int. Congr. Soil Sci., 4th, Amsterdam* **1**, 259-261.

Van Schreven, D.A. 1956. *Trans. Int. Congr. Soil Sci. 6th* **D**, 65-73.

Ventura, W.B., and Yoshida, T. 1977. *Plant Soil* **46**, 521-523.

Vlek, P.L.G., and Stumpe, J.M. 1978. *Soil Sci. Soc. Am. J.* **42**, 416-421.

Volk, G.M. 1959. *Agron. J.* **51**, 746-749.

Volk, G.M. 1961. *J. Agric. Food Chem.* **9**, 280-283.

Volk, G.M. 1966. *Agron. J.* **58**, 249-252.

Volk, G.M. 1970. *Soil Sci. Soc. Am. Proc.* **34**, 513-516.

Wagner, G.H., and Smith, G.E. 1958. *Soil Sci.* **85**, 125-129.

Wahab, A., Randhawa, M.S., and Alam, S.Q. 1957. *Soil Sci.* **84**, 249-255.

Waid, J.S., and Pugh, K.B. 1967. *Chem. Ind. (London) No. 2*, 71-73.

Watkins, S.H., Strand, R.F., DeBell, D.S., and Esch, J., Jr. 1972. *Soil Sci. Soc. Am. Proc.* **36,** 354–357.
Watson, G.A., Tsoy, C.T., and Weng, W.P. 1962. *J. Rubber Res. Inst. Malaya* **17,** 77–90.
Wetselaar, R., Shaw, T., Firth, P., Oupathum, J., and Thitipoca, H. 1977. *Proc. Int. Seminar SEFMIA. Soc. Sci. Soil Manure, Tokyo* pp. 282–288.
Willis, W.H., and Sturgis, M.B. 1944. *Soil Sci. Soc. Am. Proc.* **9,** 106–113.
Zantua, M.I., and Bremner, J.M. 1976. *Soil Biol. Biochem.* **8,** 369–374.
Zantua, M.I., and Bremner, J.M. 1977. *Soil Biol. Biochem.* **9,** 135–140.
Zantua, M.I., Dumenil, L.C., and Bremner, J.M. 1977. *Soil Sci. Soc. Am. J.* **41,** 350–352.

ADVANCES IN AGRONOMY, VOL. 31

DIFFUSION OF IONS AND UNCHARGED SOLUTES IN SOILS AND SOIL CLAYS

P.H. Nye

Soil Science Laboratory, Department of Agricultural and Forest Sciences, University of Oxford, Oxford, England

I. The Diffusion Process and Its Range

The rates of many important soil processes, such as release of ions by weathering of minerals, the exchange of cations on clay surfaces, the delivery of nutrient ions to plant roots, and the spread of pesticides applied to the soil, depend wholly or in part on the rates at which these solutes diffuse.

There has been no previous comprehensive review of diffusion of solutes in soils and clays; this one concentrates on the process rather than its many applications. Diffusion processes in the inorganic nutrition of soil-grown plants have been reviewed by Nye and Tinker (1977), Barley (1970), and Olsen and Kemper (1968); in the movement of organic materials by Hamaker (1972), of pesticides by Letey and Farmer (1974), and of herbicides by Hartley (1976); in the move-

225

ment of nitrogen by Gardner (1965); and in the movement of gases in soil respiration by Currie (1970).

Diffusion results from the random thermal motion of ions or molecules. If we imagine a column of unit cross section enclosing a solution whose concentration is greater at section A than it is at section B, a distance x away from it, then, on balance, more molecules of solute will tend to move by random motion from A to B than from B to A. The net amount crossing unit section in unit time, the flux, is given by a relation known as Fick's first law:

$$F = -D(dC/dx) \tag{1}$$

where F is the flux, and dC/dx is the concentration gradient across a particular section. The minus sign enters because net movement is from high to low concentration. This equation *defines D*, the diffusion coefficient, which is thus a proportionality coefficient between two terms, F and dC/dx, that can be measured experimentally. So defined, it may be applied to ions as well as to molecules. Although for molecules in simple systems like dilute solutions D may be nearly constant over a range of concentrations, for ions in complex systems like soils and clays D will usually depend on the concentration of the ion, and on that of other ions present as well.

Fick's first law may be derived from thermodynamic principles in ideal systems, but in such a complex medium as soil Eq. (1) may be regarded as giving an operational definition of the diffusion coefficient.

The self-diffusion coefficient of an ion—its diffusion coefficient when it is interchanging with its own isotope—is related to its mobility by the Nernst–Einstein equation (Jost, 1960), $D = ukT$, where the absolute mobility, u, is the velocity attained under unit force, k is the Boltzman constant, and T is the thermodynamic temperature. Because of this relation the diffusion coefficients of ions in homoionic clays may be deduced from their conductivity, which is related to the absolute mobility by the equation $u = N\lambda_i / \mid z \mid F^2$ (Robinson and Stokes, 1959), where λ_i is the equivalent ionic conductivity, F is the Faraday, N is Avogadro's number, and z is the ionic charge. Values of D in systems of interest range from about 10^{-8} m²/sec for H in water to 10^{-27} m²/sec for K in an illitic clay. The physical significance of these values becomes more real if we imagine one of the ions to be marked. Then, on the average, as a result of its thermal motion it would have moved in a linear system, after time t, a distance $(2Dt)^{1/2}$ from its starting point (Jost, 1960, p.25). Thus, the H would move 1 mm in 50 sec, and the K would move only 1 nm in 5×10^8 sec (16 years).

Some examples of the orders of magnitude of diffusion coefficients in soil may be given. The release of ions by gradual weathering of minerals over thousands of years is marked by apparent diffusion coefficients of less than 10^{-24} m²/sec. The release of K from hydrous mica, fast enough to meet the needs of a crop, involves diffusion coefficients of the order 10^{-19} m²/sec. The exchange of cations between the interlayer positions in clays and the external solution, nearly complete in minutes, is governed by diffusion coefficients of the order of 10^{-13} m²/sec. In moist soil, ions such as Cl and NO_3, which are usually only in the soil solution where they are mobile, have diffusion coefficients of about 10^{-10} m²/ sec. On the other hand, cations, and anions such as H_2PO_4, which are adsorbed by the soil solid and so spend much of their time in a state less mobile than in solution, have correspondingly lower diffusion coefficients, 10^{-11}–10^{-12} m²/sec. In dry soil, near the wilting point, their mobility may be less by a factor of 10 to 100. In free solution all the simple cations and anions of interest, as well as simple molecules of molecular weight less than 200, have diffusion coefficients in the range 0.5–2.0 × 10^{-9} m²/sec at 25°C, except for the hydrogen ion, 9.3 × 10^{-9} m²/sec, and the hydroxyl ion, 5.3 × 10^{-9} m²/sec, which are carried along faster by a chain mechanism involving rotation of water molecules (Glasstone *et al.*, 1941, p.560).

Finally, molecules of volatile solutes are approximately 10,000 times as mobile in the vapor phase as they are in the aqueous phase. A full account of diffusion of gases in porous media is given by Barrer (1951), and of the kinetic theory of gas diffusion by Chapman and Cowling (1951).

II. The Mechanism of Ion Movement

We have seen that the mobility of an ion or molecule varies widely with its situation. In soil an ion may be in solution, adsorbed on a charged surface, or within a crystal lattice. Of the adsorbed ions, some, remote from the charged surface, are normally hydrated; others, closer to it, are surrounded by water molecules whose arrangement is modified by the charged surface; the remainder, closer still, are dehydrated and are held by ionic bonds of varying strength in the surface of the lattice. Thus, the explanation of the mobility of adsorbed ions is complicated. Therefore, before discussing experimental data it is worth considering some of the salient features of their movement in the solid state (Jost, 1960) and in free solution (Robinson and Stokes, 1959).

In a crystal lattice an ion cannot diffuse simply by exchanging places with an adjacent ion of like charge. The activation energy required is too high. Movement is possible because lattices have defects. Thus, in Fig. 1 the cation at a could move into the vacancy b, but could not exchange directly with the cation at c. It is found experimentally that the variation of the diffusion coefficient with temperature can usually be expressed by the relation $D = \text{const.} \times e^{-E/kT}$. The cation at a occupies a position of minimum potential energy and has to pass an energy barrier, U, to reach position b. The proportion of ions with the necessary energy may be expected to vary as $e^{-U/kT}$. The rate of diffusion will also vary with the concentration of vacancies in the lattice, which will also be proportional to a term $e^{-B/kT}$, where B is the energy of disorder required to create a vacancy. Hence the experimental activation energy, E, may not be entirely due to the energy barrier between adjacent sites. For example, in a NaCl crystal Mott and Littleton (1938) calculate that the total activation energy for diffusion of a positive ion vacancy is 1.44 eV/ion (138 kJ/mole), of which 0.51 eV/ion is contributed by the energy barrier and 0.93 eV/ion by the energy of disorder. For diffusion in ionic crystals, values of E are often in the range 40–200 kJ/mole. For example, the self-diffusion of Pb in PbS can be represented as (Anderson and Richards, 1946)

$$D = 1.3 \times 10^{-4} \exp -176,000/RT \text{ m}^2/\text{sec}$$

(See also Section III,B,3.)

In addition to lattice vacancies, crystals have fractures. Diffusion along such surfaces is more rapid than it is through the body of the crystal; this is called grain boundary diffusion.

In aqueous solution ions diffuse through a medium that has a loose structure formed by H bonds. The ionic charge distorts this structure, and the water dipoles tend to be oriented around the ion. In a simple model, the ion, together with a sheath of strongly attached water molecules, is treated as a spherical particle moving through an infinite homogeneous medium. By Stokes' law its diffusion coefficient should then be given by (Robinson and Stokes, 1959, p. 44)

$$D = kT/6\pi\eta r \tag{2}$$

where η is the dynamic viscosity of the medium, and r is the effective radius. The diffusion coefficient is found to decrease as the unhydrated crystallographic radius decreases (see Table II), so that the small cations must have a larger effective radius because they are more strongly hydrated. Calculation of the extent of hydration is difficult, since Stokes' law is not strictly applicable to

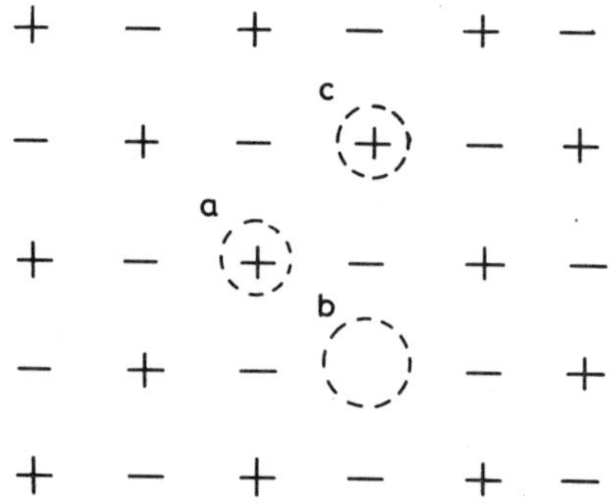

FIG. 1.   Movement of ions in solids. The cation at a can move into vacancy b, but cannot readily exchange with c.

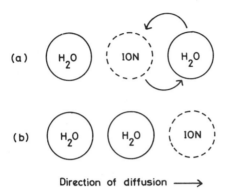

Direction of diffusion ⟶

FIG. 2.   Movement of ions in solution. State a leads to state b.

particles whose size is comparable to the molecules of the medium (Robinson and Stokes, 1959, Chapter 6).

The other point of interest in Eq. (2) is that the diffusion coefficient should be inversely proportional to the viscosity—that is, directly proportional to the fluidity of the medium—a fact that has been verified experimentally (Waldens Rule). Glasstone *et al.* (1941) point out that the activation energies for diffusion of most ions in a given solvent are nearly the same, and the values are almost identical with the activation energy of the viscous flow of solvent. In water these activation energies are 16–18 kJ/mole. These observations strongly suggest that the rate of migration of ions in aqueous solution is determined by the movement of the water molecules from one site to an adjoining one, possibly by a mechanism such as that shown in Fig. 2. It is clear that in order to move through solution an ion does not depend upon the formation of a molecular vacancy in the water structure, since this would involve an activation energy of 42 kJ/mole.

Following this brief survey of diffusion in liquids and solids, we may now consider what has been learned about the mobility of adsorbed ions in well-defined minerals important in soils, before considering the complexity of soils themselves.

III. Diffusion of Adsorbed Ions in Soil Clays and Clay-Type Minerals

A. SELF-DIFFUSION COEFFICIENTS AND ACTIVATION ENERGIES

Our knowledge is derived mainly from studies of the self-diffusion of ions in well-crystallized specimens of micas and vermiculites, and on the diffusion and conductivity of ions in homoionic clay pastes.

1. Micas and Vermiculites

The self-diffusion coefficients of ions in mica and vermiculite minerals are shown in Table I. In these experiments the ion exchanges with its own isotope, and the interlayer spacing does not change during the exchange. Many experiments have also been made on the release of ions from micas and vermiculites. Although these reactions are undoubtedly controlled in part by the rate of diffusion of the interlayer ions in exchange for different ions in the surrounding solution, the c-axis spacing changes simultaneously, and the interpretation of the measurements is uncertain.

Clearly, the main factor influencing the diffusion coefficient is the interlayer spacing. For the unexpanded potassium illite studied by de Haan et al. (1965), the value of $D \sim 10^{-27}$ m²/sec corresponds to a penetration of only about 0.3 nm, or the diameter of one ion in a year. Of the different cation forms of a dioctahedral vermiculite studied by Graf et al. (1968), the Rb and Cs forms are unexpanded, and their self-diffusion coefficients were too low to measure. The Ba and Sr forms, with a c-axis spacing of 1.24–1.23 nm, corresponding to one molecular layer of water between the sheets, yielded values of $D \sim 10^{-17}$–10^{-16} m²/sec. The Ca and Mg forms with a c-axis spacing of 1.5–1.4 nm, corresponding to two molecular layers of water, gave values of $D \sim 10^{-15}$–10^{-16} m²/sec. For other vermiculites containing two water layers, greater mobilities have been reported. Keay and Wild (1961) found $D = (1.3–4.5) \times 10^{-11}$ m²/sec for the self-diffusion of Ba in six different minerals, and Lai and Mortland (1968) found $D = 6.1 \times 10^{-13}$ m²/sec for Na vermiculite. The difference in these results may be caused by the structure of the octahedral layer. In the trioctahedral forms the O-H bond near the center of the hexagonal network of tetrahedrally coordinated silicon atoms is directed along the c-axis. In the dioctahedral forms the direction of the O-H group is altered; consequently, a cation in the interlayer will be less shielded by the proton from the negative O atom, and will occupy a more stable position. Rausell-Colom et al. (1965) have offered this explanation for the relative stability of dioctahedral micas to weathering.

The interlayer water in expanded (1.4–1.5 nm) vermiculite has an ordered structure (Walker, 1956; Barshad, 1949; Bradley and Serratosa, 1960) in which it surrounds the interlayer ions between the aluminosilicate layers. It is therefore to be expected that the activation energy for diffusion will be greater than that for ions in solution, which is about 17 kJ/mole. Keay and Wild (1961) determined activation energies of 46 kJ/mole for Na self-diffusion in Na vermiculite. When Na replaced divalent (Mg, Ca, Sr, Ba) cations, the activation energy was 37 kJ/mole, but for the reverse reaction the activation energy was 56 kJ/mole. In both forms the spacing corresponded with two molecular layers of water, and the difference is the increase in enthalpy in the change from the Na to the divalent form. Walker (1959) determined an activation energy of 50 kJ/mole for diffusion

TABLE I

Self-Diffusion Coefficients of Interlayer Ions in Mica and Vermiculite

System	Ion	Temperature	c-axis spacing (nm)	D (m²/sec)	Activation energy (kJ/mole)	References
Illitic subsoil (< 2-μm fraction)	K	Laboratory		10^{-27}		de Haan et al. (1965)
Dioctahedral vermiculite 5–20 μm CEC 2.29 mq/g	Rb	25°C	1.01	Very low		Graf et al. (1968)
	Cs		1.06	Very low		
	Ba		1.24	8.9×10^{-18}		
	Sr		1.23	1.8×10^{-16}		
	Ca		1.48	$<4 \times 10^{-15}$		
	Mg		1.43	$(7.2 \pm 2.9) \times 10^{-17}$		
Vermiculites 3.6-mm diameter	Ba	25°C	1.53	$(1.3–4.5) \times 10^{-11}$		Keay and Wild (1961)
CEC 1.1–1.6 mq/g	Na		1.51	Not determined	46	
Montana vermiculite (< 2 μm)	Na	20°C	1.5	6.1×10^{-13}		Lai and Mortland (1968)

of Sr into Mg vermiculite, during which there was a slight expansion of the lattice from 1.45 to 1.50 nm.

2. Montmorillonites and Kaolins

In the studies with micas and vermiculites the movement of the ion in the plane of the aluminosilicate layers into a surrounding solution has been measured. In the work on clay pastes, which will now be described, the particles are, as a rule, oriented randomly, and movement through the whole mass is studied. The pathway for the movement of the ions is thus tortuous, and the observed mobility includes a "geometry" effect. Most of the work has been done on montmorillonite, in which the number of water molecules between the aluminosilicate layers is greater than in vermiculite, and the surface charge density is less.

A great variety of experimental conditions have been used. In some examples the mobility of the ions has been determined from the electrical conductivity of the homoionic pastes, prepared in a variety of ways; in others, from the self-diffusion of the ions, determined by a variety of methods. The proportion of clay to solution has ranged from a dilute suspension to a stiff paste, and solutions of varying electrolyte concentration have been used. The resulting measurements have been interpreted in a number of ways. Table II attempts to order them by comparing the apparent mobilities of the ions in "salt-free" gels with their mobilities in solution at infinite dilution. Although many gels that are claimed to be salt-free are in fact hydrolyzed, in all instances the mobility cited can be attributed to the overwhelming majority of cations that are satisfying negative charges on clay lattices rather than on free anions in solution.

In spite of the inclusion of geometry effects, all the ions have greater mobility in the montmonillonites and kaolins than in the vermiculites. The values are also consistent in showing that the ion with the largest unhydrated radius suffers the greatest reduction in mobility. This is particularly well shown by Cremers (1968) for the alkali cations on both montmorillonite and kaolinite, and by Gast (1962) for the alkali and alkaline earth cations on montmorillonite. Large cations are more polarized in an electric field than small ones, and they interact more strongly with the negatively charged O atoms that form the clay surfaces (Bolt $et\ al.$, 1967). (See Section III,B,2.) There is some indication that the divalent ions have lower mobility than monovalent ions of comparable unhydrated size (cf. Na^+, $r = 0.095$ nm, and Ca^{2+}, $r = 0.099$ nm), although Hoekstra (1965) found that Na and Ca had similar mobilities in frozen Na and Ca bentonite pastes, where in each case the unfrozen water amounted to 56%.

The mobilities of the monovalent ions on kaolinite are similar to their mobilites on montmorillonite, in spite of the fact that the ions on kaolinite are adsorbed on external surfaces, whereas on montmorillonite they will be mainly in interlayer positions.

TABLE II

Mobility of Ions in Salt-Free Clay Gels Relative to Solution at 25°C

Method	Clay concentration	Ion								References
		Li	Na	K	Rb	Cs	Ca	Sr	Ba	
Montmorillonites										
Conduction	0.1–10 g clay/100 ml soln		0.37				0.13			van Olphen (1957)
Diffusion	15–28%		0.27			0.19	0.22			Lai and Mortland (1962)
Diffusion	4–6 g clay/100 g gel		0.37							Bloksma (1957)
Diffusion	4.9–10.6 g clay/100 g gel						0.08			Fletcher and Slabough (1960)
Diffusion	3 g clay/100 ml soln		0.25	0.23		0.06	0.08	0.09	0.04	Gast (1962)
Conduction	60 g clay/100 g gel		0.13	0.06	0.02	0.01	0.08			Cremers (1968)
Kaolinites										
Diffusion	56 g clay/100 g gel		0.28	0.08						Bloksma (1957)
Conduction	70 g clay/100 g gel	0.19	0.18			0.03				Cremers (1968)
Conduction	31–35 g clay/100 ml gel		0.14					0.05		Gast (1966)
Self-diffusion coefficients in water at infinite dilution (m²/sec × 10⁹)		1.04	1.35	1.98	2.07	2.11	0.78	0.78	0.84	

TABLE III
Diffusion Coefficients of Ions Diffusing into Water-Saturated Clay Films in Ca Form[a]

Clay	g H_2O/g oven-dry clay	CEC (me/100 g)	$D \times 10^9 (m^2/sec)$				
			Cu	Mn	Zn	Fe^{2+}	Fe^{3+}
Kaolinite	49	6.5	0.22	0.25	0.32		
Montmorillonite	44	70	0.02	0.05	0.06	0.055	0.01
Illite	39	20	0.09	0.12	0.12		
Vermiculite	30	80	0.01	0.02	0.04		
Diffusion coefficient in free solution $\times 10^9$ m²/sec			1.09	1.04	0.73		

[a]From Ellis et al. (1970a,b).

Ellis et al. (1970a,b) measured the diffusion of heavy metal ions into water-saturated films of Ca clay deposited on a slide. The clay particles were highly oriented so that hindrance caused by tortuosity should be slight. Since they measured the concentration profile of the penetrating ions by X-ray flourescence, they were able to determine the diffusion coefficient at varying proportions of the metal ion in the exchange complex. At low proportions nearly all the ion is in the exchangeable form and little is in the solution, so that the corresponding diffusion coefficient is a measure of the mobility of the ion in the adsorbed state. Table III shows the values obtained.

The order of mobility was kaolinite > illite > montmorillonite > vermiculite. In kaolinite all ions will be on external surfaces. In illite and montmorillonite most exchangeable ions are in interlayer positions. However, when the heavy metal ion is diffusing into the clay, it may diffuse preferentially on external surfaces, and this may explain its greater mobility in illite than in montmorillonite, which has a wider interlayer spacing. In vermiculite the c-axis spacing is 1.5 nm, which accounts for the low mobility. The mobility of the heavy metal ions is in the order Zn > Mn > Cu, which does not accord with either their mobility in solution or their unhydrated radii.

The mobility of Fe^{3+} is about a fifth of that of Fe^{2+}, although it will be noted that Fe^{3+} has appreciable mobility. The mobility of the divalent heavy metal ions is of the same order of magnitude as for the Ca, Sr, Ba series on kaolinite and montmorillonite in Table II.

The important feature of the activation energies for diffusion in clay gels shown in Table IV is that they nearly all lie in the range 17–25 kJ/mole. In the lower range this is close to the activation energy for diffusion in solution, and it is considerably lower than the value of about 42 kJ/mole found for diffusion in the structures having two water layers.

Davey and Low (1968) have shown that the activation energies reported for Na bentonite may be up to 4.2 kJ/mole too high because of the formation of hydrous aluminum oxide on the surface during preparation of the Na-saturated clay. Street *et al.* (1968) and Miller and Brown (1969) also hold aluminum oxide responsible for the variability in activation energies reported for Li, Na, and K bentonites.

These general findings have been illuminated by a number of detailed studies. It would be expected that the mobility of ions in clays would be reduced by geometry effects, by electrostatic attraction between the ions and the clay lattice, and by changes in the structure of the water near the clay surfaces. These effects will now be discussed.

B. FACTORS AFFECTING THE DIFFUSION COEFFICIENTS

1. Geometry Effects

Cremers and Thomas (1966), Cremers and Laudelout (1965), and Thomas and Cremers (1970) have measured the conductance of the sodium forms of montmorillonite, illite, and kaolinite clays suspended in sodium chloride solutions. Both the proportions of clay and the concentration of electrolyte were varied over a wide range. Figure 3 illustrates their results. They relate the

TABLE IV
Activation Energies for Self-Diffusion or Conduction in Salt-Free Clay Gels[a]

Method	Li	Na	K	Rb	Cs	Ca	Sr	References
				Ion				
			Montmorillonite					
Diffusion		16				36		Lai and Mortland (1962)
		25				45		
Conduction	18	18	17					Low (1958)
Conduction		19	20	22	23	18		Cremers (1968)
Conduction	24		18					Street *et al.* (1968)
Conduction		14–16					16	Gast (1966)
			Kaolinite					
Conduction	18	17	21	25				Cremers (1968)
Conduction		16					16	Gast (1966)

[a] Values given in kilojoules per mole.

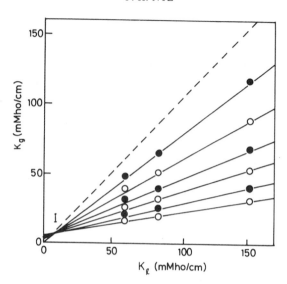

FIG. 3.   Electrical conductivity (mMho/cm) of Na montmorillonite gels, K_g, versus conductivity of equilibrium NaCl solutions, K_l, for various porosities. I is the isoconductivity. From top to bottom: $\theta = 1$ (broken line); 0.96, 0.91, 0.86, 0.80, 0.73, 0.64. After Cremers (1968).

specific conductance of the gel, K_g, to the specific conductance of the solution in the pores, K_l, by the equation

$$K_g = K_l/\mathfrak{F} + K_\sigma \qquad (3)$$

K_σ is the excess specific conductance created by the mobility of the exchangeable ions. At high electrolyte concentration K_σ will be small in comparison with K_l. The "formation factor," $\mathfrak{F}$, is then a measure of the extent to which the solid particles reduce the specific conductance of the gel in relation to the solution by reducing the cross section for the passage of the ions, and by increasing the tortuosity and viscosity of their pathway.

They found that the experimental values obtained for $\mathfrak{F}$ in dilute gels (porosity of 1.0–0.65 for montmorillonites and 1.0–0.5 for kaolinites) are well described by a theoretical equation due to Fricke (1924):

$$\mathfrak{F} = 1 + (1 + 0.21n)(1 - \theta)/\theta \qquad (n > 10) \qquad (4)$$

Here θ is the volume fraction of the liquid, and n is the ratio of the diameter of the particles to their thickness. A plot of $\mathfrak{F} - 1$ against $(1-\theta)/\theta$ is a straight line passing through the origin, and the slope gives the value of n (see Fig. 4). These values of n, in the range 10–60, for six different clays, agreed very well with independent determinations by electron microscopy or viscosity. In concentrated gels Cremers (1968) found that $\mathfrak{F}$ is better described by a theoretical equation due to Bruggeman and developed by Meredith and Tobias (1962):

$$\mathfrak{F} = (1/\theta)^{(1 + 0.21n)} \qquad (n > 10) \qquad (5)$$

The high values of $\mathfrak{F}$ measured for clays with high axial ratios are quite inconsistent with a cubic model of a concentrated clay gel, which has sometimes been used to deduce geometry effects.

Turning now to the excess conductivity created by the exchangeable ions, we see in Fig. 3 that there is an "isoconductivity" value (I) at which the specific conductivity of the gel is the same as that of the solution—for all concentrations of clay. Dakshinamurti (1960, 1965) had previously noticed this property in a number of clay systems. Cremers (1968) shows that, because there is such an isoconductivity point, the exchangeable ions may be assigned a constant "surface conductance," and that this is governed by the same formation factor as the

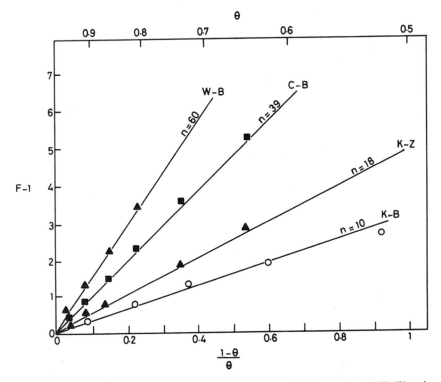

FIG. 4. Formation factors versus porosity according to the Fricke equation. W-B (Wyoming bentonite), C-B (Camp Berteau montmorillonite), K-Z (Zettlitz kaolinite), K-B (Boulvit kaolinite). n is the axial ratio. After Cremers (1968).

ions in solution. Thus, as far as the geometry effect is concerned, the exchangeable ions behave in Eq. (3) as though they were distributed uniformly over the adjacent pore solution—by no means an obvious result. With this knowledge of the geometry effect for the exchangeable ions, Cremers is able to conclude that the conductivity of the gels is "consistent with" 0.5–0.6 of the exchangeable Na in each clay having the same mobility as in solution and the remainder being immobile.

For other cations the experimental data are less complete, but Cremers concludes tentatively that on montmorillonite the following fractions are freely mobile: K 0.3, Cs 0.15, Ca 0.15; and on kaolinite: K 0.15, Cs 0.05.

2. Electrostatic and Viscosity Effects

It is difficult to distinguish experimentally between a proportion of exchangeable ions having the same mobility as in free solution with the rest having none, and a more continuous distribution of mobilities over all the exchangeable ions. That the activation energy for diffusion of Na in clays approximates that in water is some indication that the observed mobility derives from freely mobile ions. The somewhat higher activation energies for K, Rb, and Cs ions (see Table IV) suggest that some ions with modified hydration are contributing to the observed mobility. This problem has been analyzed further by Shainberg and Kemper (1966a) and by van Schaik et al. (1966), who have developed the ideas of Low (1962, 1968). He calculated the variation in electrostatic potential of an ion as it moved parallel to the surface of a clay lattice from one position of stability to the next. He estimated that there would be a negligible potential barrier to surmount if the ion were more than 1.0 nm above the plane of negative charge on the lattice. Shainberg and Kemper (1966a) calculated that if the negative charge is in the octahedral layer (as in montmorillonite) an unhydrated ion adjacent to the surface will require 4.8 kJ/mole more activation energy than an ion separated from the surface by one molecular thickness of water (about 0.25 nm). This effect alone would lead to a sixfold difference in the mobility of the ion. If, in addition, the first molecular layer of water is more viscous than subsequent ones—and there is much evidence for this (Low, 1961; Grim, 1968)—it is reasonable to assume that the mobility of the unhydrated ions can be neglected.

The next step is to consider how much the mobility of the hydrated ions is restricted by reduction in the fluidity of water near the surfaces. By measuring the diffusion of DOH in oriented flakes of bentonite containing varying amounts of water, Kemper et al. (1964) found that in Na bentonite the relative mobility of water in layers one, two, and three molecular thicknesses from each bentonite surface was 0.3, 0.6, and 0.65; in Ca bentonite the corresponding values were 0.05, 0.5, and 0.7.

Against this background, van Schaik *et al.* (1966) have measured the self-diffusion of Na and Ca in oriented, expanded flakes of Wyoming bentonite. Because the ions were diffusing in the plane of the flakes, the tortuosity factor—the factor by which the mobility is multiplied to allow for an increased path length—was high, 0.55, as estimated from diffusion of DOH in similar flakes. They were able to calculate a weighted average fluidity of the water surrounding the hydrated ions, the weighting factors being provided by the relative concentrations of ions in each molecular layer calculated from the theory of the diffuse double layer. The reasonable assumption was made that the relative diffusion coefficient of DOH is a measure of the relative fluidity of water. From the measured diffusion coefficients they concluded that 0.60–0.87 of the Na ions and 0.15–0.47 of the Ca ions are, on a time average, hydrated.

By measurement of conductivity of Li, Na, and K bentonite pastes, Shainberg and Kemper (1966a) have similarly estimated that the hydrated fractions are Li 0.64, Na 0.57, and K 0.39. This work was, however, done with centrifuged pastes, pushed from a glass tube into the conductance cell; and it has been assumed that the tortuosity factor can be taken as 0.67. If the flakes were not well oriented, this value seems likely to be too high. A lower value would have the effect of increasing the fractions hydrated.

Although these estimates of the fraction of hydrated ions are subject to numerous uncertainties—for example, the average fluidity depends on the accuracy of the calculation of the distribution of ions in the diffuse double layer; and the assumption that unhydrated ions are virtually immobile is very sensitive to the value chosen for the dielectric constant near the surface of the lattice—they illustrate well the difficulties that arise in arriving at an exact model of ion movement through clays. They also agree reasonably well with the independent calculations of Shainberg and Kemper (1966b), based on considerations of electrostatic energy, that the fractions of ions hydrated in homoionic bentonites are Li 0.82, Na 0.64, K 0.51. These calculations also are admittedly uncertain, since they neglect terms involving the energy of water-to-water links.

Further insight into the fluidity of interlayer water is provided by neutron-scattering spectroscopy, which has been applied to clays by White, Hunter, and co-workers (Olejnick and White, 1972). Figure 5 shows a plot of $\log_{10} D_{H_2O}$ against the reciprocal of the thickness of the interlayer water in Li and Na montmorillonite and Li and Na vermiculite. The experimental values agree well with the theoretical expression represented by the straight line in Fig. 5:

$$D = D_{bulk} \exp(-2\delta V/dRT)$$

where δ = surface energy of included water plus ions;

V = molar volume of water;

d = interlayer thickness of water.

Here $2\delta V/d$ is the reduction in free energy of water confined between two

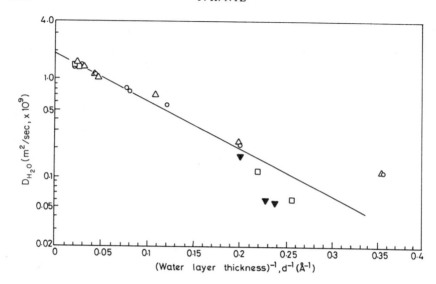

FIG. 5.   Variation of the diffusion coefficient of water in montmorillonite and vermiculite with reciprocal of the interlayer spacing. After Olejnick and White (1972).

parallel plates. The reduction is equal to an increase in the activation energy of diffusion because the water molecules jump from a lower energy level.

These measurements of water mobility are lower than those obtained from diffusion of DOH by a factor of about 4. If correct, they indicate that the fraction of ions hydrated is underestimated by van Schaik *et al.* (1966).

3. Effect of Dehydrating Clays

The work on montmorillonite described in the previous section has been done with clay suspensions or pastes with expanded lattices. Work has also been done on drier montmorillonites in which only one or two molecular layers of water lie between the sheets; in this respect it links with the studies of diffusion in vermiculites.

Mott (1967) has measured the self-diffusion coefficients of Na and Sr in homoionic bentonite at varying degrees of hydration. Figure 6 shows his results. He used oriented flakes in which diffusion in the plane of the flakes was very little reduced by geometry; in fact it proceeded 300 times as fast as it did across the plane of the flakes in specimens with two water layers between the sheets (279 mg of H_2O per gram of dry clay). The mobility of Na over the range 350–200 mg of H_2O per gram of clay (three to two water layers) was also measured by electrical conductivity, with satisfactory agreement between the two methods. In drier clay the conductivity method proved unreliable because of difficulty in

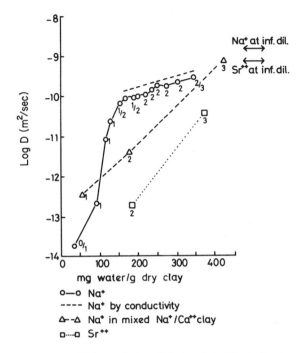

FIG. 6. Self-diffusion coefficients of Na and Sr in oriented bentonite at varying hydration. Integers on curves are the number of molecular thicknesses of water determined by c-axis spacing. After Mott (1967).

obtaining good electrical contact. The sodium is some 1000 times as mobile as the Sr in the clay with two water layers, and is also more mobile than it was in Lai and Mortland's (1968) vermiculite (Table I). On the other hand, the Sr was less mobile than the Ba in Keay and Wild's (1961) vermiculite (Table I).

As might be expected, the activation energy for movement of ions in partially dehydrated bentonite is considerably greater than it is in the fully hydrated form. Oster and Low (1963) determined the following activation energies (kJ/mole) for the conductance of dry homoionic Li, Na, and K bentonites:

Relative humidity (%)	Li	Na	K
48	29 (1–2)	24 (1–2)	31 (1)
73	40 (2)	33 (2)	44 (1)
100	17	18	17

The number of water layers between the sheets, determined by x-ray diffraction, is given in parentheses. The higher activation energy at 73% than at 48% relative humidity is unexpected.

Calvet (1973) has made a detailed study of the diffusion of the exchangeable cations Na, Rb, Cs, and Ca in slightly hydrated montmorillonite (0–250 mg of H_2O per gram of clay) (Table V). He has interpreted the results in terms of geometrical models of the interlayer positions of the cations and water, using the transition-state theory of rate reactions (Glasstone *et al.* 1941). The diffusion coefficient is expressed in terms of the hopping frequency of an ion ω, its hopping distance from one stable position to another δ, and a geometry factor g: $D = g\delta^2\omega$. The hopping frequency is $v \exp(\Delta S/R) \exp(-\Delta E/RT)$, where ΔS is the entropy of activation, and ΔE is the activation energy. Here $\exp(\Delta S/R) \exp(-\Delta E/RT)$ is the probability of the ions being in the activated state, and v is the vibration frequency in that state. Thus,

$$D = g\delta^2 v \exp(\Delta S/R) \exp(-\Delta E/RT)$$

In the moist montmorillonite studied, δ is 0.5 nm, the distance from one hexagonal hole to another. The electrostatic interactions between the cation, the lattice, and the water molecules determine the frequency of vibration and the energy of activation. The frequency is of the order of the Debye frequency for solids, 10^{13}–10^{12} sec^{-1}. The activation energy—the more influential term—consists of two terms, one corresponding to a vertical jump from the plane of the lattice, the other a movement parallel to its surface.

The activation energies calculated from the model system are in reasonable agreement with those measured (Table VI). At a given water content the smallest ion, Na, has the lowest activation energy and the highest diffusion coefficient. For a given ion, the greater the hydration of the clay, the lower will be the energy of the activated state and hence the lower the activation energy.

These activation energies are lower than those reported for vermiculites at comparable degrees of hydration. In the vermiculites the negative charge is

TABLE V

Self-Diffusion Coefficients in Dry Montmorillonite[a]

H_2O (wt %)	D_{self} (m^2/sec), 20°C			
	Na	Rb	Cs	Ca
5	9.2×10^{-13}	2.3×10^{-13}		
10	5.2×10^{-12}	3.3×10^{-12}	6×10^{-16}	
15	2.6×10^{-11}	1.4×10^{-11}	7.5×10^{-15}	1.9×10^{-13}
20				4.9×10^{-12}

[a] After Calvet (1973).

TABLE VI
Activation Energy for Diffusion in Dry
Montmorillonite[a]

% H_2O (wt.)	Activation energy (kJ mole^{-1})			
	Na	Rb	Cs	Ca
5	36	58		
10	25	33		
15	18	21	126	73
20				48

[a] After Calvet (1973).

largely in the tetrahedral layer adjacent to the migrating cation; in bentonite it is largely in the more remote octahedral layer, so that the cation should require less energy of activation.

In the work on clays so far described, no distinction has been drawn between ions adsorbed on internal or external surfaces. In the kaolinite minerals and collapsed 2–1-type layer minerals, the mobility measured over a short time, as in conventional cation exchange, is due to ions on external surfaces. For example, on an illitic clay de Haan *et al.* (1965) found that a fraction of the potassium exchanged completely with ^{40}K within an hour, but they followed the very slow exchange of a further fraction over a period of 16 months. On the other hand, in fully expanded 2–1 lattices the interlayer ions may be nearly as mobile as the ions on external surfaces, and all would be counted as exchangeable ions. When clays are dried, the distinction between external and internal surfaces is not clear-cut, since some of the channels between individual crystallites may be very fine. Lai and Mortland (1968) followed the diffusion of sodium in a sodium vermiculite clay which they oriented and compressed. They found that the mobility of the sodium was five times as great in the fine, intercrystalline channels as it was in the interlayer positions. The rate of exchange was fairly rapid at first, being controlled by ions readily accessible to the channels, but after 24 hours the rate of exchange was reduced and was controlled by the ions that had to diffuse through the interlayers.

4. Effect of Companion Ions in Mixed Ion Systems

a. Self-Diffusion Studies. The effect of one kind of cation on the mobility of another in clays is of special importance because, in practical applications in soils, we are usually interested in the diffusion of a cation in a system dominated by other cations—often Ca. Measurements of the self-diffusion of a particular ion in a mixed ion system are the easiest to interpret. At low moisture levels

Calvet (1973) has studied the self-diffusion of Na and of Ca in Na–Ca montmorillonite systems. Figures 7 and 8 show the contrasting behavior of the two ions. Calvet explains in terms of transition-state theory that "a cation will diffuse less readily to the extent that its affinity for water is lower than that of the other cations in the medium." Hence, the mobility of Na is much reduced in the presence of Ca, which has a much higher affinity for water. The effect is enhanced at low water content (cf. 10% H_2O and 22.5% H_2O in Fig. 7). When Ca satisfies more than about a third of the exchange capacity, the interlayer spacing increases from 1.24 to 1.55 nm, with a corresponding increase in the activation energy for the Na. This accounts for the change in slope of the curves in Fig. 7. Figure 8 shows that the mobility of Ca is raised by the presence of Na. At higher moisture levels Mott (1967) has measured the self-diffusion of Na in oriented Ca

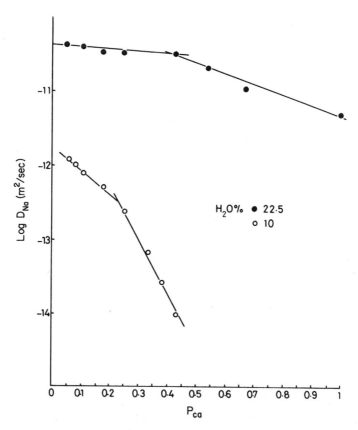

FIG. 7.   Variation of the self-diffusion coefficient of Na in Na–Ca montmorillonite with the degree of Ca saturation. After Calvet (1973).

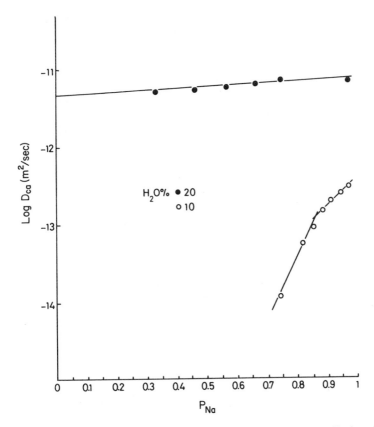

FIG. 8. Variation of the self-diffusion coefficient of Ca in Na–Ca montmorillonite with the degree of Na saturation. After Calvet (1973).

bentonite containing 8.5% Na (by equivalents), with results as shown in Fig. 6. In clay with approximately three water layers, the mobility of the Na was similar in the Na and Ca clays. In the clay with two water layers, the mobility of the Na is much lower in the Ca system than in the Na system, in agreement with Calvet's finding.

The proportion of Na in Ca montmorillonite suspensions in water can be raised to about 15% without changing the interlayer spacing (Shainberg and Otoh, 1968). At higher proportions of Na the tactoids tend to disperse, and the geometry for diffusion changes. The effect of this is not marked: Verdonck (1969) measured Na surface diffusion coefficients of 0.15–0.18×10^{-9} m^2/sec over the range 20–50% Na, followed by a steady increase to 0.3×10^{-9} m^2/sec over the range 50–100% Na. Cremers (1968), however, reports a relatively small, steady decrease in the surface diffusion coefficient of Na with increasing proportion of Na in Na–Cs montmorillonite—from 0.45×10^{-9} m²/sec at 35% Na to 0.25×10^{-9} m²/sec at 100% Na. Clearly, precise interpretation of these small effects is not possible at present.

The fact that the Na ion retains its mobility at low equivalent fractions indicates that the "hopping" distance from one position of stability to the next does not depend on the mean distance between one sodium ion and its neighbor, but rather on the existence of nearby vacant sites. There is likewise no reason to suppose that divalent ions will diffuse more slowly than monovalent ions because they are farther apart in the interlayers.

b. Counter-Diffusion Studies. In these studies, ions of type A counter-diffuse against ions of type B in the same clay system. Such experiments are difficult to interpret for the following reasons. (*a*) As discussed fully and clearly by Helfferich (1962, Chapter 6), the measured counter-diffusion coefficient, D_{AB}, depends on the mobility of ions A and B, and also on their relative concentration, so that the mobility of each ion may be difficult to calculate. (*b*) The interlayer spacing may vary with the ionic proportions. (*c*) Counter-diffusion on external surfaces may differ from that on internal surfaces. (*d*) Water may be redistributed in clay prepared with ion A when ion B diffuses into it. Khafagi *et al.* (1978) note all these effects in a summary of their study of the counter-diffusion of Na and Ca in bentonite.

IV. Diffusion of Ions and Molecules in Soil

A. THE COMPONENTS OF THE DIFFUSION COEFFICIENT

So far we have examined experiments on the mobility of adsorbed or lattice ions in soil minerals. In soil itself the greater proportion of the cations and often the anions—for example, phosphate—and uncharged molecules are adsorbed on minerals and organic matter; but a small, although very important, fraction are in the solution in the pores. In the overall movement of solutes through the soil, both solid and liquid play a part. In so heterogeneous a medium—irregular in the shapes and composition of its solids, and in the sizes of the channels and pores between them—it is impossible to develop a theoretical equation, such as the Fricke or Bruggeman–Meredith equations already cited, to express accurately the overall flux in terms of the mobilities and concentration gradients of the solute in the constituent parts. However, as long as we are concerned with diffusion in volumes large enough to average microscale variations, we may treat the soil as a quasi-homogeneous body, to which Fick's first law, $F = -D(dC/dx)$, may be applied. Here C is the concentration of diffusible solute in the system—that is, all those ions or molecules that are in or pass through a mobile phase during a period of time that is short in comparison with the time of the diffusion process.

It will be shown later that in practical applications only differences between values of C occur, and the absolute value is not required.

This operational definition of the diffusion coefficient in soil has the advantage that it is determined directly by the transient-state methods to be described later, and it makes no assumptions about the mechanism of diffusion or the mobilities of the ions and molecules in their various states in the soil. In addition, the distance over which a solute will diffuse in a given time may at once be appreciated, and the mathematical solutions for diffusion equations—as published, for example, by Crank (1975)—are available without transformation. However, if the boundary conditions of a diffusion system are expressed in terms of the solute concentration in solution, as in many root uptake problems, there are advantages in adopting this frame of reference (see Olsen and Kemper, 1968).

To understand the values of D obtained by experiment, and also to predict its value, we may consider the total rate of transfer of solute through a unit cross section of soil as being due to a flux through the pore solution alone—as though the molecules or exchangeable ions on the solid have no surface mobility—together with an excess flux created by any mobility of the solute on the solid. Hence,

$$F = - D_l \theta f_l (dC_l/dx) + F_E \qquad (6)$$

where D_l is the diffusion coefficient of the solute in free solution;

θ is the fraction of the soil volume occupied by solution, and gives the cross section for diffusion;

f_l is an impedance factor;

C_l is the concentration of solute in the soil solution;

F_E is the excess flux created by the exchangeable ions or adsorbed molecules.

Since by definition $F = -D(dC/dx)$, we obtain by substitution in Eq. (6)

$$D = D_l \theta f_l (dC_l/dC) + D_E \qquad (7)$$

where D_E is an excess term, which is zero when the ions or molecules on the solid have no surface mobility, but represents their extra contribution to the diffusion coefficient if they are mobile.

The concentration of the soil solution is that of an equilibrium dialyzate; θ refers to the water associated with the ions or molecules in the soil solution, and does not include that associated with the exchangeable ions. The product θC_l is thus the amount of solute in solution per unit volume of soil.

The impedance factor, f_l, can readily be determined by experiment. This is particularly easy for nonadsorbed anions, such as Cl and NO_3 in most soils. For these, $\theta dC_l/dC = 1$; hence $D = f_l D_l$. The impedance factor, f_l, takes account primarily of tortuosity. It also allows for the effect of exclusion of anions by negative adsorption from very narrow pores, which may thus cut off larger pores

(Porter *et al.*, 1960). It may include the effect on ionic mobility of the increased viscosity of water near charged clay surfaces, although their influence on the water of the soil solution should be small. All these effects cannot readily be separated experimentally (see Section V,B,1).

Much has been learned about diffusion in soils from self-diffusion experiments, in which one isotope is allowed to counter-diffuse against another of the same species. For instance, a block of soil labeled with a radioactive isotope may be placed in contact with an unlabeled, but otherwise identical, block, and the movement of the labeled isotope into the unlabeled block followed (Schofield and Graham-Brice, 1960). In such an experiment we have $C_l'/C' = C_l''/C'' = C_l/C$, where the primed terms refer to the diffusing isotopes and the unprimed terms refer to the total. Since the distribution of the total number of ions between solution and solid, C_l/C, remains constant, C_l'/C' is also constant, and we may write

$$dC_l'/dC' = C_l'/C' \tag{8}$$

Hence, from Eq. (7), we obtain

$$D_{\text{self}} = D_{l_{\text{self}}} f_l \theta C_l/C + D_{E_{\text{self}}} \tag{9}$$

which is a constant, independent of C_l' and C_l'', for the experiment.

The way in which the term $\theta C_l/C$ affects the self-diffusion coefficient may be more clearly appreciated if it is realized that this is the fraction of their time that the diffusible ions spend in the solution phase.

B. MOBILITY OF ADSORBED SOLUTES IN SOIL

1. Exchangeable Ions

a. Self-Diffusion Experiments. A number of experiments have been made to assess the importance of the excess diffusivity, $D_{E_{\text{self}}}$. The ratio C_l/C is varied, usually by changing the concentration of ion in the soil solution, and D_{self} is measured at each ratio. When D_{self} is plotted against C_l/C, the slope at high values of C_l/C is $D_l\theta f_l$, and the intercept is $D_{E_{\text{self}}}$. An example for the self-diffusion of Sr in a Sr-saturated and moisture-saturated soil is shown in Fig. 9 (Mott and Nye, 1968). The product $D \times C$ is linearly related to C_l, and the value of $D_{E_{\text{self}}} \times C$ is small and can evidently be neglected except at low concentrations in solution. The self-diffusion coefficients of Na in a moist Na-saturated illitic sandy loam, and of H_2PO_4 in a natural sandy loam, were found by Rowell *et al.* (1967) to be proportional to C_l/C, and the values of $D_{E_{\text{self}}}$ were negligible. Similar conclusions for H_2PO_4 may be drawn from the self-diffusion measurements of Lewis and Quirk (1962).

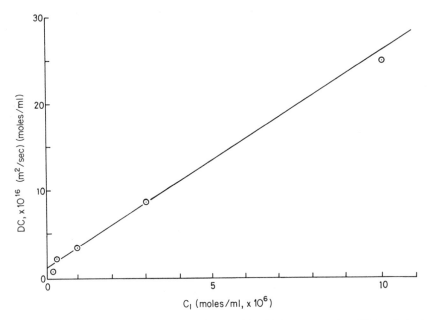

FIG. 9.   Variation of the self-diffusion coefficient of Sr with the concentration of Sr in solution in a Sr-saturated soil. After Mott and Nye (1968).

When the fraction of the total number of ions that is in solution is very small, $D_{E_{self}}$ may be significant. For example, in a soil near the wilting point, Rowell *et al.* (1967) found a value of $D_{E_{self}} = 7.9 \times 10^{-12}$ m²/sec for Na, whereas the value predicted if diffusion occurred only through solution was 1.9×10^{-12} m²/sec. On the other hand, the D_{self} value for Sr was very low at the wilting point, about 2×10^{-13} m²/sec, and there was no evidence for any excess flux.

In moist soils Clarke and Graham (1968) measured self-diffusion coefficients for Zn three to five times as great as expected if only ions in solution diffuse. Their estimate of the proportion of Zn in the soil solution was, however, based on an extrapolation from values at wide solution–soil ratios to a narrow ratio, a range over which the proportion was changing rapidly and nonlinearly; hence, it does not seem that reliable conclusions can be drawn from the data presented.

b. Counter-Diffusion Experiments.   Although self-diffusion experiments are useful in interpretation of diffusion processes, in practical situations an ion is either diffusing against an ion of like charge (counter-diffusion) or with an ion of opposite charge (salt diffusion), and D is given by Eq. (7). Here, for adsorbed ions, dC_l/dC varies as a rule with C, as is shown for example in Fig. 10, for adsorption of H by a Ca–H soil (Farr *et al.*, 1970). Consequently, D also should depend on concentration. This is shown in Fig. 11; and the relation between D and dC_l/dC is shown in Fig. 12. It will be noticed that the intercept D_E is

P. H. NYE

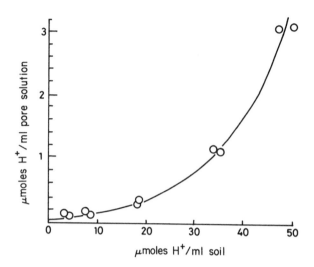

FIG. 10.   Concentration of H⁺ in pore solution at constant chloride concentration (0.02 *M*) versus concentration of H⁺ in the soil. After Farr *et al*. (1970).

negligible. Similar results have been obtained for counter-diffusion of K against Ca in a moist, largely Ca-saturated montmorillonite clay soil (Vaidyanathan *et al*., 1968); for counter-diffusion of K against H in sand, a sandy loam, and a clay loam at a range of moisture levels (Nielsen, 1972); and for counter-diffusion of

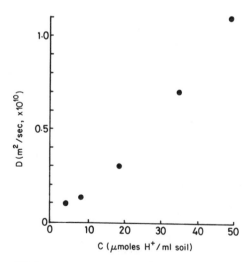

FIG. 11.   Counter-diffusion coefficient H–Ca versus concentration of H⁺ in the soil. After Farr *et al*. (1970).

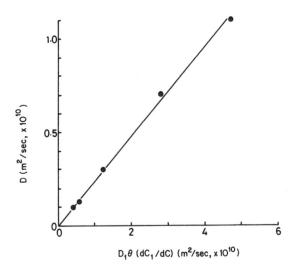

FIG. 12. Counter-diffusion coefficient H–Ca versus concentration derivative dC_l/dC ($\times \; \theta \, D_l$). The slope is f_l. After Farr *et al.* (1970).

H_2PO_4 against Cl in a moist illitic sandy loam (Vaidyanathan and Nye, 1970). In this work with H, K, and H_2PO_4, the ion of interest was in small proportion in the soil solution. Tinker (1969b) has measured the counter-diffusion of Ca against Mg over the whole range of ionic proportions from Ca-saturated to Mg-saturated soil, and found that the variation of D could be accounted for by the term dC_l/dC, and that the exchangeable ions did not contribute appreciably to the diffusive flux.

Warncke and Barber (1973) reported the diffusion coefficient of zinc in four soils to be *less*, by a factor of 5 to 15, than the diffusion coefficient calculated from Eq. (7) using independently measured adsorption isotherms. They measured diffusion coefficients from the uptake of zinc by the H form of cation exchange resin paper, which was placed on the soil surface to act as a sink. It was assumed that the concentration of exchangeable zinc in the soil at the interface would be reduced to zero. If the soils retained an appreciable concentration of zinc there, the discrepancy would be explained. Vaidyanathan and Nye (1970) found that a similar discrepancy in phosphate diffusion coefficients was caused by the erroneous assumption that the phosphate concentration at the resin paper interface had been reduced to zero.

c. The Low Mobility of Exchangeable Cations in Soil—Discussion. At present it is not certain why the exchangeable cations in soil appear to have such low mobility, whereas in clay suspensions and pastes their mobility is, as we have seen, appreciable. The soils used in the experiments cited have contained predominantly 2–1 expanded lattice clays with high Ca content, so that the interlayer spacing should accommodate at least three molecular layers of water. It seems that in natural soils there may be a break in the continuous pathway of water molecules required to enable an ion to diffuse freely. Such a break could be caused by aluminum–oxygen or iron–oxygen groups attached to the edges of the aluminosilicate lattices or deposited in the interlayer. Fripiat and Dondeyne (1960) have suggested that tetrahedral alumina, perhaps as a gel, forms a coating on clay minerals, and Dion (1944) has pointed out that hydrated iron oxide reduces the exchange capacity of clay minerals. The packing of the clay crystallites within microaggregates may also be important, since an occasional lamella arranged normal to a parallel series of interlayers would also be an effective barrier. Organic matter should offer its own exchange sites as a diffusion pathway, but there may be organic molecules adsorbed in interlayer positions with either few or unfavorably disposed exchange sites. More work on the surface mobility of ions over a wider range of soils is clearly required.

Since many treatments of diffusion in soils and clays merely add together expressions for the flux through the solution and through the solid to give the combined flux, it should be particularly noted that in Eq. (6) the excess flux, F_E, does not represent the flux of ions through the solid alone, since ions do not move through solution and solid independently, and it has to take care of the complicated interactions between them. The exceedingly low diffusion coefficients of cations in dry soils show that there is virtually no continuous pathway through the solid alone. For some porous systems with well-defined geometry, F_E may be expressed as a function of the mobility and concentration gradients of the ions in solid and solution, and the porosity and particle shape, as in the work on clay suspensions already described; but for soils such exact treatment is clearly impossible. The formulation of equations of transport through heterogeneous materials has a long history, which has been reviewed by Goring and Churchill (1961) and Meredith and Tobias (1962), and, for diffusion in particular, by Barrer (1968). The subject has been further discussed in the light of this background by Cremers (1968) and Thomas and Cremers (1970) for colloidal suspensions, and by Nye (1968) for soils.

The mobility of adsorbed ions or molecules may conveniently be expressed as the "virtual mobile fraction," α, that, if spread over the adjacent pore solution, would result in a diffusion coefficient equal to that measured. It may be calculated from the relation

$$D = D_l f_l \frac{d[\theta C_l + \alpha(C - \theta C_l)]}{dC}$$

In the experiment illustrated in Fig. 5, $\alpha \simeq 0.001$.

2. Uncharged Solutes

Scott et al. (1974) have measured the diffusion coefficient of Metribuzin over a range of soil water content. They find D_E significantly positive at $\theta = 0.42$,

0.40, 0.38, and 0.36, but negative at 0.34 and 0.32. They determined f_l by chloride diffusion. The negative values calculated for D_E are associated with Cl diffusivities at $\theta = 0.34$ and 0.32, which seem unexpectedly high. Gerstl, Yaron, and Nye (1979) have found that the diffusion coefficient of parathion over a water content ranging from $\theta = 0.18$ to 0.34 could be accounted for if 10% of the adsorbed parathion were mobile. However, the diffusion coefficient of polyethyleneglycol (MW 4000) over a wide range of soil moisture level is satisfactorily explained with the assumption that the adsorbed molecules are immobile (see Section V,B,2). More critical work is required to establish the surface mobility of adsorbed organic solutes.

V. Prediction of Diffusion Coefficients in Soil

In practice it is very convenient that excess diffusivity of ions appears to be small, for when this can be assumed it allows us to predict the concentration-dependent diffusion coefficient by Eq. (7) from a knowledge of D_l, dC_l/dC, θ, and f_l—quantities that can be fairly readily measured. The use of these terms will be discussed in turn.

A. THE DIFFUSION COEFFICIENT IN SOLUTION D_l

1. Ions

Ions, in contrast to neutral molecules, cannot diffuse independently, because every microvolume must remain electrically neutral. This is achieved in counter-diffusion by ions of like charge moving in opposite directions, and in salt diffusion by cations and anions moving together. Hence, ions with greater mobility tend to be slowed down by those with less. The diffusion coefficient of an ion in solution is thus modified by the other ions present. It can be shown (Nye, 1966) that, if the excess diffusion, D_E, is negligible, the flux of ion A is given by

$$F_A = (-\theta f_l D_{l_{\text{self}}} \, dC_l/dx)_A + z_A C_{l_A} D_{l_A} \frac{\Sigma \, z_i D_{l_i} \theta f_l (dC_{l_i}/dx)}{\Sigma \, z_i^2 C_{l_i} D_{l_i}} \quad (10)$$

where z is the charge on the ion, and the subscript i refers to all ions in the soil. It will be seen that the modifying term is small if (a) the ion of interest is a small proportion of the total solution concentration (C_{l_A} small); this is very often the case for H_2PO_4, K, and trace elements; (b) all ions have similar mobilities—in which case $\Sigma z_i D_{l_i}(dC_{l_i}/dx) \sim 0$. As has been seen, the self-diffusion coefficients in solution of the main cations and anions range from 0.7×10^{-9} m²/sec for Mg to 2.03×10^{-9} m²/sec for Cs, so the modifying effect is usually fairly small in practice. If the relative concentration of the very mobile H is high, the modifying effects may be greater—for instance, the counter-diffusion coefficient between H and Ca ion in solution decreased from 8.2×10^{-9} to 4.4×10^{-9} m²/sec as the pH in $M/100$ CaCl$_2$ changed from 4.5 to 2.5 (Farr et $al.$, 1970).

2. Uncharged molecules

The solution diffusion coefficients of organic molecules of interest have rarely been determined. Their mobility depends upon their molecular weight and shape. Tanford (1961, Chapter 6) gives a clear account of this subject. For close-packed spherical molecules, such as globular proteins, the molecular weight is

$$M = \tfrac{4}{3} \rho \pi r^3$$

where ρ is the density, and r is the molecular radius. Consequently the diffusion coefficient varies inversely as the cube root of the molecular weight. Goring (1968) found that the diffusion coefficient at 25°C of a range of lignosulfonates from spruce could be expressed by $D_l = 4 \times 10^{-9}/M^{1/3}$m²/sec. This formula gives a useful guide to the diffusion coefficients of the many herbicides and pesticides with compact molecules in the range of molecular weight 100–10,000.

Long-chain molecules usually form a flexible random coil of roughly spherical shape, with solvent molecules trapped between the elements of the chain. The polycarboxylic acids of the fulvic acid fraction of humus are an example. The radius of the equivalent sphere is $\tfrac{2}{3} R_G$, where R_G is the radius of gyration of the random coil. Since $R_G \propto M^{1/2}$, for such molecules $D \propto 1/M^{1/2}$.

Rigid long-chain molecules undergo Brownian rotation, and the equivalent sphere has a large radius compared with that of a random coil. For thin, rod-shaped molecules, such as collagen, $D \propto 1/M^{0.81}$.

B. THE IMPEDANCE FACTOR f_l

It was noted in Section IV,A that the impedance factor in the liquid phase included the effect of tortuosity, increased viscosity of water near charged surfaces, and restriction of entry to narrow pores. Most measurements of f_l have been interpreted as tortuosity effects, although some evidence for other effects has been obtained.

1. Simple Ions and Molecules

Clearly, as soil dries, the pathway for diffusion will become more tortuous and f_l will decrease. Figure 13 shows the values that have been found in different experiments relating θ and f_l.

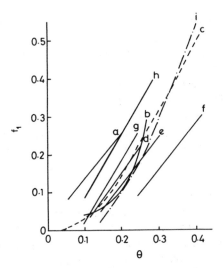

FIG. 13.   Relation between the impedance factor, f_l, and the volume fraction of the soil solution, θ. (a) Wanbi sand (6% clay) (Paul, 1965); (b) Urrbrae loam (19% clay) (Clarke and Barley, 1968); (c) sandy loam (24% clay) (Rowell et al., 1967); (f) Ft. Collins loam (26% clay) (Porter et al., (1960); (e) Apishapa silty clay loam (37% clay) (Porter et al., 1960); (f) Pierre clay (53% clay) (Porter et al., 1960); (g) sand (4% clay) (Nielsen, 1972); (h) sandy loam (15% clay) (Nielsen, 1972); (i) average of six silt loams (Warncke and Barber, 1972).

It will be seen that in very dry soil f_l is very low. Rowell et al. (1967) found f_l = 2×10^{-4} at -100 bar and 10^{-2} at -15 bar water potential. At water potentials between about -1.0 bar and zero, f_l increases approximately linearly with moisture content. Thus, over the field moisture range -0.1 to -10 bar, the product θf_l may change by a factor of as much as 100. Figure 13 also shows that at a given moisture content clay soils have a lower value of f_l than sandy soils. Porter et al. (1960) found that f_l tends to zero at a moisture content somewhat above that corresponding to the formation of a monolayer of water molecules on the surfaces, an observation that can be explained by the high viscosity of the monolayer (Section III,B,2). At a given water potential clay soils usually have a higher value of f_l than sandy soils because they hold more water and offer more continuous pathways.

In saturated soils values of f_l between 0.4 and 0.7 have been obtained (Dakshinamurti, 1959; Mott and Nye, 1968; Farr et al., 1970). Such values accord with the theoretical derivation by Bruggeman of the impedance factor for a mixture of different-sized spherical particles: $f_l = \theta^{0.5}$ (Cremers, 1968).

The high value of f_l in nearly saturated soil shows that the micropores as well as the macropores are readily available for diffusion by small ions. For an ion such as Cl, which is only in the soil solution, relatively slow diffusion out of aggregates into interaggregate pores has not been detected in diffusion experiments (but see Section VIII,C).

The exact distribution of water in the pores depends on whether a given content was attained by wetting or drying. Phillips and Brown (1965) found the self-diffusion coefficients of Rb and Sr in a moist soil to be reduced by about one-half if it was saturated with water and then drained to the original moisture content. Since this procedure may reduce the concentration of the soil solution, which should be finite even though the soils were prepared "salt-free," this, rather than the distribution of the water, could also explain the observed effect.

Although the point has not been critically tested, it seems that the value of f_l is little influenced by the type of ion—at any rate, for simple ions in moist soils. In drier soils a greater proportion of the ions in solution are near charged surfaces where their exact distribution might be important; for example, the diffuse layer thickness for exclusion of a monovalent anion in a solution of 0.003 M $CaCl_2$ is approximately 3.0 nm, which is comparable to the thickness of the water films joining aggregates at -15 bar water potential (Collis-George and Bozeman, 1970; Kemper and Rollins, 1966). Thus, the anion might be unable to enter some pores through which a cation could pass more freely.

It seems that compacting the soil may increase or decrease the value of f_l at a given θ. Graham-Bryce (1965, Fig. 2) found that the self-diffusion coefficient of Rb in a Ca soil was increased from 0.35 to 1.2 $\times$ 10^{-11} m²/sec when the bulk density increased from 1.35 to 1.95 g/ml and θ was 0.25. Since the value of C_l was held constant, and compaction should decrease C_l/C, it follows that compaction must have raised f_l. However, Warncke and Barber (1972) found that f_l values for Cl diffusion in five silt loams decreased two- to threefold with increase of density from 1.3 to 1.6 g/ml. Clearly more critical studies of compaction are needed, in which all the terms in Eq. (9), particularly the pore solution concentration, are measured.

2. Macromolecules

Two effects will reduce the mobility of a molecule in pores of diameter less than ten times the molecular diameter. The cross section of the pore available to the molecule is only π(pore radius)² $-$ π(molecular radius)²; and the viscous drag on a moving particle increases near the wall of the pore by a factor of $(1 - 2.09x + 2.14x^3 - 0.95x^5)$, where x is the ratio of the molecular to the pore radius (Faxen, 1922). The drag factor modifies Stokes' law when the medium is finite [see discussion of Eq. (2)] and is not to be confused with any effect due to an increase in viscosity of one or two molecular layers of water near charged surfaces (see Section III,B,2). Renkin (1954) has confirmed these effects in experiments with cellulose membranes, and Beck and Schultz (1970) in mica made porous by bombarding it with ^{235}U fission products. When x is 1/10, the reduction in diffusion coefficient is nearly 40%.

In soil, Williams et al. (1966, 1967) found that polyvinyl alcohol (PVA) penetrated aggregates with pores of maximum diameter 6 nm more slowly as its molecular weight increased from 25,000 to 100,000. PVA is adsorbed on the soil surfaces, thus restricting the pores further, and there was little penetration of pores less than 3 nm across by PVA of MW 70,000. Saxena *et al.* (1974) report that the mobility of 2,4-dichlorophenoxyacetic acid is reduced by about 50% by glass beads of average pore radius 2 μm in comparison with beads of radius 7 μm. The pore sizes seems large for such an effect. Barraclough and Nye (1979) measured the self-diffusion coefficients of Cl ion, polyethylene glycol 4000, and polyvinyl pyrollidone 40,000 in a sandy loam over a wide range of water content. These solutes have effective radii of 0.18, 1.9, and 18.3 nm, respectively. The impedance factors of PEG 4000 and Cl are similar (Figs. 14 and 15). In moist

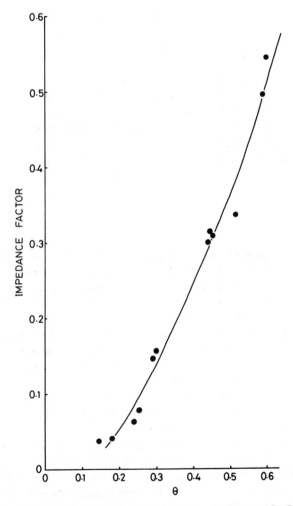

FIG. 14.   Impedance factor for Cl in a sandy loam at a range of moistures. After Barraclough and Nye (1979).

soil it appeared that PEG did not diffuse rapidly into 0.085 of the soil volume, whereas there was no evidence of exclusion of Cl. PVP 40,000 did not diffuse into 0.28 of the soil volume, which corresponded to the intra-aggregate pore space. In dry soil its f_l value (Fig. 16) was correspondingly small, but in moist soil its f_l value exceeded that of Cl, probably because the interaggregate pores to which it was confined offered a more direct pathway. The ratio of the impedance factors of PVP and Cl in dry soil was consistent with Renkin's (1954) explanation of hindered diffusion in narrow pores.

C. THE DERIVATIVE dC_l/dC

1. Uncharged Solutes

At low concentrations the adsorption isotherms of a great range of herbicides and pesticides are approximately linear (Hamaker, 1972), so the diffusion coefficients, by Eq. (7), should be independent of concentration. Scott and Phillips (1972) have found that the variation in the diffusion coefficients of seven non-volatile herbicides can be accounted for by variation in their solid–liquid distribution coefficients. Hamaker (1972) and Letey and Farmer (1974) cite many more examples of the dominant influence of adsorption on the mobility of uncharged solutes in the soil.

2. Ions

Change in the proportion of diffusible ion in the soil solution explains many variations in diffusion coefficients. For example, the increase in solution concentration following addition of chelating ions satisfactorily accounts for the increased self-diffusion coefficient of Zn with EDTA (Elgawhary et al., 1970), and of Fe with EDDHA (O'Connor et al., 1971). Peaslee and Phillips (1970) have found that the effect of salts on the self-diffusion coefficient of phosphate in kaolinite is proportional to their effect on the concentration of phosphate in the equilibrium solution. Prokhorov and Frid (1972) found that increased levels of humus decrease the diffusion coefficient of ^{90}Sr.

The correct determination of the relation between C and C_l is not as easy as it might appear, since it must reproduce exactly the same conditions as occur in the diffusion process. The following points arise:

a. The True Pore Solution Concentration. Methods of determining pore solution concentrations are described by Moss (1969), who finds the alcohol displacement and "null point" quantity–intensity methods satisfactory. The value will be influenced by the moisture content, and by the concentrations of the other ions in solution. Hence, concentrations measured on saturation extracts are

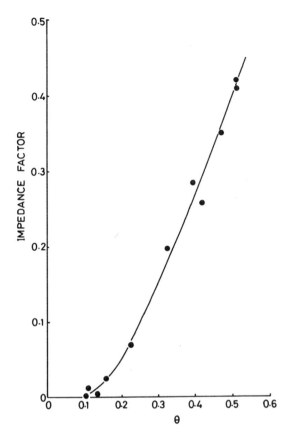

FIG. 15. Impedance factor for PEG 4000 in a sandy loam at a range of moistures. After Barra-clough and Nye (1979).

not usually sufficient. Nor is it possible to prepare unsterilized soils with electrolyte-free pore solutions by washing with distilled water, and even sterilized soils will have HCO_3 in solution.

b. The Choice of the Exchanging Ion. It is particularly important that the exchanging ion should be correctly chosen. For example, in a diffusion process in which phosphate is being desorbed from the soil, the relation between C_l and C is very different in a solution containing an indifferent anion, such as Cl or NO_3, from one containing a specifically adsorbed ion, such as HCO_3 or citrate (Nagarajah *et al.*, 1968; Vaidyanathan and Nye, 1970). If the exchanging ion is an isotope, dC_l/dC will be constant, C_l/C, as we have seen. It is clear from Fig. 10 that dC_l/dC is often very different from C_l/C. Hence the self-diffusion coefficient cannot usually be used as the effective coefficient over a given range

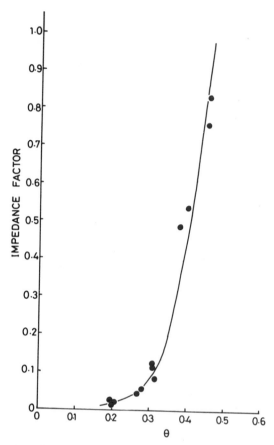

FIG. 16. Impedance factor for PVP 40,000 in a sandy loam at a range of moistures. After Barraclough and Nye (1979).

of concentrations. Further discussion of the effect of the derivative dC_l/dC has been given by Olsen and Kemper (1968), Nye (1968), and Tinker (1970).

In practical applications it has not proved necessary to measure the absolute amounts of diffusible ions. This would be a difficult task if C and C_l are related as shown in Fig. 10, since at very low solution concentrations the amount that will desorb is indefinite, and is often affected by release of ions that are only slowly exchangeable. In practice, one is always concerned with diffusion between certain concentration limits, and hence with a difference, ΔC. If, as is frequently the case, the concentration limits are expressed in terms of the solution concentration, then ΔC becomes the change in the total diffusible ions over the change between the specified limits of solution concentration.

c. *The Rate of Equilibration between the Solution and the Solid.* In Eq. (7) dC_l/dC, and hence D, will be independent of time only if there is virtually instantaneous equilibrium between the ions on exchange sites and the adjoining solution, so that, for a small change in solution concentration, δC_l, there is a definite change, δC, in the total concentration. If, for a given change δC_l, δC changes with time, then D will also be time-dependent, and this complicates its application. Fortunately, cation exchange is usually rapid. Sometimes, however, in addition to a rapid exchange there is a slow exchange with relatively inaccessible sites, so that the final equilibrium is attained slowly—if at all. If this reaction can be approximately described by reversible first-order kinetics, there is a useful, rough method of deciding whether it limits the amount of ion diffusing (Crank, 1975). If the half-time for the slow reaction is the same as the half-time for the diffusion process, the amount diffusing is approximately the same as it would be if the reaction were infinitely rapid. If the reaction is slower, then it will significantly influence the process.

The effect of slow reaction has been noted in measurements of the self-diffusion coefficient of P in soil using ^{32}P (Rowell *et al.*, 1967); and Phillips (1969) has described how the relation between concentration of isotope and distance is affected in such experiments.

If the reaction cannot be regarded as effectively instantaneous, the system is one of diffusion with simultaneous reaction, considered in Section VIII,B.

d. Hysteresis and Relaxation. The relation between C_l and C, and hence the value of dC_l/dC, may differ between an adsorption and a desorption process. Examples have been given for K by Arnold (1970), for SO_4 by Tinker (1970), and for P by Muljadi *et al.* (1966). It may be that insufficient time has been allowed for true equilibration between the solution and the solid, in which case the phenomenon is described as relaxation (Everett and Smith, 1954). Or the difference may be stable—the phenomenon of hysteresis—and be caused, for example, by alteration of the lattice spacing or geometrical rearrangement of the particles as the concentration of an adsorbed ion alters. Whatever the reason for the irreversibility, it is essential that, for a desorption process such as diffusion of ions from a soil to plant roots, or for an adsorption process such as diffusion of an ion from a fertilizer pellet in the soil, the appropriate value of dC_l/dC should be chosen.

VI. Volatile Solutes

The diffusive flux of a volatile solute may be expressed as

$$F = -[D_g v_a f_g (dC_g/dx) + D_l \theta f_l (dC_l/dx)] + F_E \tag{11}$$

The first term accounts for diffusion through the gaseous pathway; it is analogous to the second term, which accounts for the diffusion through the liquid pathway [see Eq. (7)]. These pathways are usually continuous and act in parallel.

The term F_E is the additional flux that arises from cooperation between the gas and liquid phases.

Values of f_g have been reviewed by Currie (1970). They are illustrated (Fig. 17) in the theoretical model developed by Millington and Shearer (1971), which accords with Currie's experimental data on moist soil crumbs. In the aggregated soil, the gradual fall in f_g as the water content increases, and v_g decreases, corresponds to filling of intra-aggregate pores. The value of f_g falls more steeply when interaggregate pores are being filled.

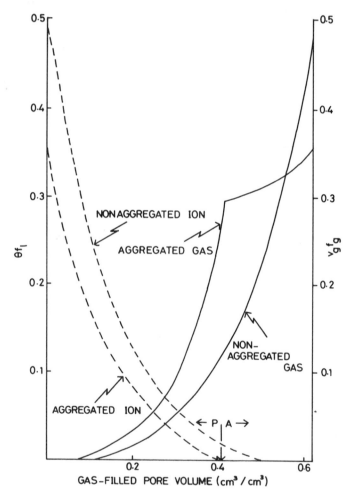

FIG. 17.   Relation between θf_l for solute and $v_g f_g$ for gaseous diffusion with varying proportion of gas and water in soil space. After Millington and Shearer (1971).

According to Henry's law, $C_l/C_g = \beta$ represents the coefficient of solubility. Hence

$$F = -[(D_g/\beta)v_g f_g + D_l \theta f_l](dC_l/dx) + F_E \tag{12}$$

We have seen that $D_g/D_l \sim 10,000$. If $\beta \sim 10,000$, $D_g/\beta \sim D_l$. In this instance the steady-state flux in a nonswelling soil should not be affected if soil water displaces soil air. In agreement with this, Graham-Bryce (1969) has shown that the diffusion coefficient of disulfoton—a volatile insecticide with $\beta = 5500$—changed little when θ ranged from 0.08 to 0.41.

No model calculation of the cooperative term F_E has been made. Its importance is indicated by considering the flux predicted for a compound such as disulfoton ($\beta \sim 10,000$) in a nonaggregated soil, according to the model of Millington and Shearer. The flux predicted (Fig. 17) when the soil is dry ($\theta = 0$) or saturated ($\theta = 0.61$) is about $0.5D_l(dC_l/dx)$. When $\theta = 0.3$, the gas pathway and the liquid pathway alone contribute $0.05D_l(dC_l/dx)$ each, leaving $0.4D_l(dC_l/dx)$ as the predicted contribution of the cooperative pathway.

A small addition of water to a dry soil may greatly increase the vapor pressure of volatile solutes, such as the organochlorine insecticides, adsorbed on it. The water displaces solute molecules from the solid, and when sufficient water to form a monolayer has been added the concentration of the vapor phase increases abruptly (Spencer et al., 1969). There is a correspondingly sharp rise in the diffusion coefficient (Ehlers et al., 1969). Shearer et al. (1973) have noted that the diffusion coefficient of lindane in moist soil is greater than that expected from separate liquid and vapor pathways. They speculate that movement of lindane held at the water–air and the water–solid interfaces accounts for the discrepancy. No supporting evidence is adduced, and they do not allow for the combination pathway in their speculations.

VII. Methods of Measurement of Ion Diffusion Coefficients in Soil

Ideally, a method will reveal how D varies with C and with C_l, and also with time; and it should be possible to use it at any desired moisture level and other imposed condition—for example, salt concentration. A review of methods has been given by Tinker (1970).

A. TOTAL TRANSFER METHODS

In these methods the amount of material crossing a section of soil in a given time is measured, and an effective D between imposed concentration limits is calculated.

1. Steady State

An experiment made by Olsen *et al.* (1965) will illustrate this method. They placed a block of soil, thickness Δx, between two porous plates so that the ends of the block were in equilibrium with solutions of differing concentration, $C_{l_1} > C_{l_2}$, held at the same tension as the soil moisture. When a steady state was reached, the flux, F, across the block was measured. Since $F = D(C_1 - C_2)/\Delta x$, if the value of C at C_{l_1} and $C_{l_2}(C_1, C_2)$ can be determined, $\bar{D}$, the average value of D between C_1 and C_2, can be found.

If it is known that there is no solid excess flux, then by Eq. (6)

$$F = D_l \theta f_l (C_{l_1} - C_{l_2})/\Delta x$$

and f_l can be determined without the need to find C. This seems the main advantage of the method, which is otherwise rather tedious, because of the need to ensure that a steady state has been reached.

2. Transient State

In a typical experiment a block of soil is placed in contact with a sink, and the movement of ion into or out of the block is followed. The flux and the concentrations in the block vary with time—hence, "transient state." The method is particularly convenient for studies of self-diffusion in which the block is labeled, and the sink is provided by an unlabeled block (Schofield and Graham-Bryce, 1960). In this instance

$$M_t = C_i (D_{\text{self}} \, t/\pi)^{1/2}$$

where M_t is the amount of labeled isotope crossing unit area at the junction between the two blocks in time t, and C_i is its initial concentration in the labeled block. The method may be used over a wide range of moisture contents, short of saturation, and is probably the most suitable method for examining the influence of moisture content on f_l. For this purpose a nonadsorbed ion such as Cl is used (Porter *et al.*, 1960), so that $D = D_l f_l$ (p. 247). Mott and Nye (1968) used a stirred unlabeled solution as sink. This has the advantage that the initial concentration of the solution in the soil pores can be known accurately, but it can be used only with saturated oil.

For studies of counter-diffusion Vaidyanathan and Nye (1966) used an ion exchange resin paper sink. This is useful for making quick comparisons over a range of ions and moisture levels. It has the disadvantage that the concentration at the boundary is not known precisely, and cannot readily be varied at will. For precise work it is necessary to use a stirred solution as the sink so that the concentration in the block and the boundary concentration in solution can be accurately controlled. The method can be adapted to moist soils if the sink solution is held under tension. The effective value of D over the required concentration range is obtained directly. If the value of D at each concentration is needed, a series of runs at different initial sink concentrations must be made (Vaidyanathan *et al.*, 1968; Vaidyanathan and Nye, 1972; Nielsen, 1972; Lindstrom, *et al.*, 1968).

B. CONCENTRATION-DISTANCE METHODS

In these methods, the diffusion coefficient is calculated from the variation of concentration with distance within the soil block.

1. Steady State

An experiment made by Tinker (1969) illustrates the principles involved. He placed a Ca-saturated ion exchange resin paper at one end of a block of soil, and a Mg-saturated paper at the other; and established a steady counter-diffusion of Ca and Mg across the block. He then sectioned the block and determined the exchangeable plus solution ions in each section. He thus measured dC/dx directly for different values of C and, knowing the constant flux, could determine the variation of D with C in one experiment. This method has the additional advantage that, if the surface diffusion is zero, the solution concentration in each slice is easily calculated, since in the steady state dC_l/dx is constant.

2. Transient State

As an example, the concentration–distance curve obtained in measurement of the self-diffusion coefficient of Cl is shown in Fig. 18. In this instance the diffusion coefficient is constant, and the experimental points fall on the theoretical curve

$$C/C_0 = \tfrac{1}{2}\,(1 - \mathrm{erf}\ x/2\sqrt{Dt})$$

When D varies with C, the C–x curve can be analyzed by a simple algebraic procedure (Matano's method) (Crank, 1975), and the variation of D with C determined. This method has been developed particularly by Phillips (Brown *et al.* 1964; Phillips and Brown, 1964), who introduced the idea of freezing the soil at the end of a run and cutting it into thin sections with a microtome. Since sections only 10 nm thick can be cut, the method is particularly useful in following the movement of slowly diffusing ions such as H_2PO_4. It has been adapted by Farr *et al.* (1969) for studying the diffusion of ions near plant roots.

If the ion of interest is suitably labeled, its concentration may be determined by autoradiography. The appearance of the plates often gives a useful qualitative indication of the concentration gradient (Place and Barber, 1964); but for quantitative work it is essential that the photographic density be calibrated against

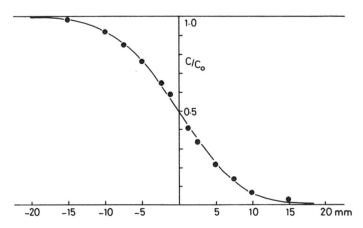

FIG. 18.   Concentrations of ^{36}Cl versus distance. The experimental points lie close to the theoretical curve $C/C_0 = \frac{1}{2} (1 - \mathrm{erf}\ x/2\sqrt{Dt})$, which is drawn. Data of R.J. Dunham.

concentration, and a low-energy isotope used if high resolution is required (Bhat and Nye, 1973). The method can show up irregularities in packing, adsorption, moisture distribution, and contact at boundaries. In the autoradiographic method exposure times of several hours are often required. Wray and Tinker (1969) have described a device for scanning the radioactive surface, which is particularly valuable in following changes in the concentration–distance relation with time.

In all the transient state concentration–distance methods the overall concentration of ion is measured, and the samples are usually too small for direct determination of the solution concentration.

VIII. Diffusion in Practice

In the laboratory it is usually possible to simplify the diffusion system by sieving, packing, and moistening the soil uniformly, and by controlling the other ions present - for example, by ensuring that the soil solution is predominantly $CaCl_2$. With all such variables under control, diffusion coefficients can be measured to an accuracy of around 10%. Under natural conditions, however, diffusion occurs in a coarse, structured soil with irregular moisture content; reactions such as mineralization can increase the concentration of diffusible ions and can alter the partial pressure of CO_2 and hence the ionic strength of the soil solution; many ions move simultaneously, and they move by mass flow of the solution as well as by diffusion.

Many of these sources of variation can be dealt with by intelligent averaging. We may consider some of them in more detail.

A. MULTIPLE ION DIFFUSION

The simultaneous movement of several cations all competing for positions on the exchange complex gives rise to problems requring the simultaneous solution of a number of differential equations. A brief discussion of these questions is given by Helfferich (1962, p. 228) and Olsen and Kemper (1968, p. 124), but their full implications in soil problems have not yet been developed. Fortunately, many diffusion processes in soils take place in a solution of relatively steady salt concentration dominated by Ca and Mg cations and Cl, NO_3, and SO_4 anions, so that other ions of interest can be considered as minor components in Eq. (10). In other processes there may be a gradient in the total salt concentration—for example, because of uptake of anions near a root. This will in itself affect the concentration of cations in the soil solution, and hence dC_i/dC. An example of the calculations involved is given by Olsen and Kemper (1968, p. 126), who describe the diffusion of cations following the application of a salt to the soil. Nye (1972) has noted the many acid–base conjugate pairs that may diffuse between adjacent portions of soil differing in pH. Bar-Yosef et al. (1975) have shown that the concentration profile of zinc diffusing on goethite can be explained if the pH profile induced is also taken into account.

B. SIMULTANEOUS DIFFUSION AND SLOW REACTION

Adsorption was assumed to be effectively instantaneous in Section IV,A. Few instances of diffusion with simultaneous slow reaction have been thoroughly studied in soil, although theory is well developed. Crank (1975, Chapter 14) deals fully with irreversible and reversible first-order reactions. Lindstrom and Boersma (1970) have in addition considered a more complex reversible rate equation, which Fava and Eyring (1956) have introduced to represent the kinetics of adsorbtion of detergents. The main problem with systems involving diffusion with simultaneous reaction in soil lies in characterizing the heterogeneous reaction. Often its rate is controlled by a diffusion process—for example, penetration of a reactant into an aggregate, or into the interlamellar regions of clays. The rate will not be a unique function of the concentration of reactants in solution and solid, but will depend on the history of their diffusion-controlled movement through the solid. Such problems have been investigated by Helfferich (1962, Chapter 6) in ion exchange materials.

Ramzan and Nye (1978) and Nye and Ramzan (1979) have treated the neturalization of a block of acid soil by a source of bicarbonate ion as a process of coupled diffusion and slow reaction. The reaction rate was determined independently, and fitted to an empirical equation.

Reactions that depend on microbiological processes are likewise difficult to characterize in detail. Gerstl, Nye and Yaron (1979) have developed a model that incorporates the growth of microbial activity and successfully predicts the diffusion and simultaneous microbiological degradation of parathion. Many less complete studies are cited by Letey and Farmer (1974).

C. SIMULTANEOUS DIFFUSION AND MASS FLOW

Diffusion of ions often occurs simultaneously with mass flow of the soil solution. This may be caused, for example, by transpiration, evaporation, or drainage. The movement of water through the soil tends to increase the random displacement of a solute that normally occurs by diffusion, because of the irregular pattern of the flow velocity through its pores—a process known as "eddy" or "hydrodynamic" dispersion (Helfferich, 1962, p.486). Consequently, the diffusion coefficient in solution, D_l, has to be replaced by a "longitudinal dispersion coefficient," D_l^*. Nielsen and Biggar (1962) found D_l^*/D_l for Cl in saturated soil to be about 2 when the average pore velocity in the direction of flow, v/θ, was 1.6×10^{-7} m/sec, and it rose to about 60 when the pore velocity increased to 95×10^{-7} m/sec. Frissel and Poelstra (1967) found that the theoretical equation

$$D_l^* = D_l + \lambda dv/f_l\theta \tag{13}$$

expressed the dispersion of Sr as it was percolated through columns of resin–sand and clay–sand mixtures. Here d is the particle diameter, and λ is the packing factor, which is 1 for spheres, but may become as high as 10 when the particle sizes are irregular. In these experiments v ranged from 0.2×10^{-7} to 200×10^{-7} m/sec.

In soil it is not possible to separade d and λ. Frissel et al. (1970) followed the drainage of tritiated water through undisturbed soil columns and found the product, $d\lambda$, to be 0.007, 0.008, and 0.06 m in a sandy, clay, and loess loam soil—over a range of v from 0.06×10^{-7} to 23×10^{-7} m/sec. These data suggest that in undisturbed soil dispersion is significant—even at the low flow rates induced by transpiration, where v at the root surface is of the order of 10^{-8} m/sec (Tinker, 1969a). For example, if in a moist soil $\lambda d = 0.01$ m, $v = 10^{-8}$ m/sec, $\theta f_l = 0.1$, then in Eq. (13) $\lambda dv/\theta f_l = 10^{-9}$ m^2/sec, which is comparable to D_l for most ions. Clearly, for the much greater flow rates that may occur in drainage (1 cm/day $\sim 10^{-7}$ m/sec), eddy dispersion will be far more important than diffusion. In dry soils, with small values of θf_l, the effect of dispersion may be expected to increase, although Eq. (13) has not yet been tested under these conditions.

There is a large literature on the movement of solutes through porous media. Work on the mass flow and dispersion of solutes in soil has been summarized by Nye and Tinker (1977, Chapters 4 and 8), and Rose (1977) has reviewed recent theoretical studies of hydrodynamic dispersion in porous media.

Solutes being displaced from aggregated or unsaturated soils often show unsymmetrical breakthrough curves with marked "tailing." Theories based on a dispersion coefficient account only for symmetrical curves. The unsymmetrical curves can be accounted for if there is slow diffusion of solute between stagnant and moving water in the soil (Gaudet *et al.* 1977; van Genuchten, 1977). Since the slow diffusion is represented by a first-order rate equation, the treatment closely resembles that used for mass flow and diffusion with a slow rate of reaction.

REFERENCES

Anderson, J.S., and Richards, J.R. 1946. *J. Chem. Soc.* 537–541.
Arnold, P.W. 1970. *Proc. Fert. Soc. London* No. 115.
Barley, K.P. 1970. *Adv. Agron.* **22**, 159–201.
Barraclough, D., and Nye, P.H. 1979. *J. Soil Sci.* **30**, 29–42.
Barrer, R.M. 1951. "Diffusion in and through Solids." Cambridge Univ. Press, London and New York.
Barrer, R.M. 1968. *In* "Diffusion in Polymers" (J. Crank and G.S. Park, eds.), Ch. 6. Academic Press, New York.
Barshad, I. 1949. *Am. Mineral.* **34**, 675–484.
Bar-Yosef, B., Posner, A.M., and Quirk, J.P. 1975. *J. Soil Sci.* **26**, 1–21.
Beck, R.E., and Schultz, J.S. 1970. *Science* **170**, 1302–1305.
Bhat, K.K.S., and Nye, P.H. 1973. *Plant Soil* **38**, 161–175.
Bolt, G.H., Shainberg, I., and Kemper, W.D. 1967. *Soil Sci.* **104**, 444–453.
Bloksma, A.H. 1957. *J. Colloid Sci.* **12**, 40–52.
Bradley, W.F., and Serratosa, J.M. 1960. *Proc. Natl. Clay Conf., 7th* Pergamon, New York, pp. 260–270.
Brown, D.A., Fulton, B.E., and Phillips, R.E. 1964. *Soil Sci. Soc. Am. Proc.* **28**, 628–632.
Calvet, R. 1973. *Ann. Agron.* **24**, 77–217.
Chapman, S., and Cowling, T.G. 1951. "Mathematical Theory of Non-Uniform Gases," 2nd ed. Cambridge Univ. Press, London and New York.
Clarke, A.L., and Barley, K.P. 1968. *Aust. J. Soil Res.* **6**, 75–92.
Clarke, A.L., and Graham, E.R. 1968. *Soil Sci.* **105**, 409–418.
Collis-George, N., and Bozeman, J.M. 1970. *Aust. J. Soil Res.* **8**, 239–258.
Crank, J. 1975. "The Mathematics of Diffusion," 2nd ed. Oxford Univ. Press (Clarendon), London and New York.
Cremers, A. 1968. *D.Sc. Thesis,* Univ. Louvain, Belgium.
Cremers, A., and Laudelout, H. 1965. *J. Chim. Phys.* **62**, 1155–1162.
Cremers, A., and Thomas, H.G. 1966. *J. Phys. Chem.* **70**, 3229–3233.
Currie, J.A. 1970. *Soc. Chem. Ind. Monogr.* No. 37, 152–169.
Dakshinamurti, C. 1959. *Soil Sci.* **88**, 209–212.
Dakshinamurti, C. 1960. *Soil Sci.* **90**, 302–305.
Dakshinamurti, C. 1965. *I.A.E.A. Vienna Tech. Rep. Ser.* No. 48, p.57.
Davey, B.G., and Low, P.F. 1968. *Trans. Int. Congr. Soil Sci., 9th, Adelaide* **1**, 607–616.
De Haan, F.A.M., Bolt, G.H., and Pieters, B.G.M. 1965. *Soil Sci. Soc. Am. Proc.* **29**, 528–530.
Dion, H.G. 1944. *Soil Sci.* **58**, 411–424.
Ehlers, W., Farmer, W.J., Spencer, W.F., and Letey, J. 1969. *Soil Sci. Soc. Am. Proc.* **33**, 504–508.
Elgawhary, S.M., Lindsay, W.L., and Kemper, W.D. 1970. *Soil Sci. Soc. Am. Proc.* **34**, 66–70.
Ellis, J.H., Barnhisel, R.I., and Phillips, R.E. 1970a. *Soil Sci. Soc. Am. Proc.* **34**, 866–870.
Ellis, J.H., Phillips, R.E., and Barnhisel, R.I. 1970b. *Soil Sci. Soc. Am. Proc.* **34**, 591–595.

Everett, D.H., and Smith, F.W. 1954. *Trans. Faraday Soc.* **50**, 187–197.

Farr, E., Vaidyanathan, L.V., and Nye, P.H. 1969. *Soil Sci.* **107**, 385–391.

Farr, E., Vaidyanathan, L.V., and Nye, P.H. 1970. *J. Soil Sci.* **21**, 1–14.

Fava, A., and Eyring, H. 1956. *J. Phys. Chem.* **60**, 890–898.

Faxen, H. 1922. *Ann. Phys.* **68**, 89–119.

Fletcher, G.E., and Slabough, W.H. 1960. *J. Colloid Sci.* **15**, 485–488.

Fricke, H. 1924. *Phys. Rev.* **24**, 575–587.

Fripiat, J.J., and Dondeyne, P. 1960. *J. Chim. Phys.* **57**, 543–552.

Frissel, M.J., and Poelstra, P. 1967. *Plant Soil* **26**, 285–302.

Frissel, M.J., Poelstra, P., and Reiniger, P. 1970. *Plant Soil* **33**, 161–176.

Gardner, W.R. 1965. *In* "Soil Nitrogen" (W.V. Bartholomew and F.E. Clark, eds.), pp 550–572. Amer. Soc. Agron. Monogr. 10.

Gast, R.G. 1962. *J. Colloid Sci.* **17**, 492–500.

Gast, R.G. 1966. *Soil Sci. Soc. Am. Proc.* **30**, 48–52.

Gaudet, J.P., Jegat, H., Vachaud, G., and Wierenga, P.J. 1977. *Soil Sci. Soc. Am. J.* **41**, 665–671.

Gerstl, Z., Nye, P.H., and Yaron, B. 1979. *Soil Sci. Soc. Am. J.* (in press).

Gerstl, Z., Yaron, B., and Nye, P.H. 1979. *Soil Sci. Soc. Am. J.* (in press).

Glasstone, S., Laidler, K.J., and Eyring, H. 1941. "The Theory of Rate Processes," McGraw-Hill, New York.

Goring, D.A.I. 1968. *In* "Solution Properties of Natural Polymers." Chem. Soc. London Spec. Pub. No. 27, pp 115–134.

Goring, R.L., and Churchill, S.W. 1961. *Chem. Eng. Prog.* **57**, 53–59.

Graf, H., Reichenbach, V., and Rich, C.I. 1968. *Trans. Int. Cong. Soil Sci. 9th, Adelaide* **1**, 709–719.

Graham-Bryce, I.J. 1965. *I.A.E.A. Vienna Tech. Rep. Ser.* No. 48, pp. 42–56.

Graham-Bryce, I.J. 1969. *J. Sci. Food Agric.* **20**, 489–492.

Grim, R.E. 1968. "Clay Mineralogy," 2nd ed. McGraw Hill, New York.

Hamaker, J.W. 1972. *In* "Organic Chemicals in the Soil Environment" (C.A.I. Goring and J.W. Hamaker, eds.), pp. 341–397. Dekker, New York.

Hartley, C.S. 1976. *In* "Herbicides, Physiology, Biochemistry, Ecology" (L.J. Audus, ed.), pp. 1–28. Academic Press, New York.

Helfferich, F. 1962. "Ion Exchange." McGraw Hill, New York.

Hoekstra, P. 1965. *Soil Sci. Soc. Am. Proc.* **29**, 519–521.

Jost, W. 1960. "Diffusion in Solids, Liquids, Gases." Academic Press, New York.

Keay, J., and Wild, A. 1961. *Soil Sci.* **92**, 54–60.

Kemper, W.D., and Rollins, J.B. 1966. *Soil Sci. Soc. Am. Proc.* **30**, 529–534.

Kemper, W.D., Maasland, D.E.L., and Porter, L.K. 1964. *Soil Sci. Soc. Am. Proc.* **28**, 164–167.

Khafagi, M.S.E., Tinker, P.B., and Townsend, W.N. 1978. *Int. Congr. Soil Sci., 11th,* Edmonton **1**, 193. (Abstr.)

Lai, T.M., and Mortland, M.M. 1962. *Clays Clay Minerals* **9**, 229–247.

Lai, T.M., and Mortland, M.M. 1968. *Soil Sci. Soc. Am. Proc.* **32**, 56–61.

Letey, J., and Farmer, W.T. 1974. *In* "Pesticides in Soil and Water" (W.D. Guenzi, ed.), pp. 67–97. Soil Sci. Soc. Am., Madison, Wisconsin.

Lewis, D.G., and Quirk, J.P. 1962. *Int. Soc. Soil Sci. Trans. Commun. IV, V, Palmerston N.Z.* pp. 132–138.

Lindstrom, F.T., and Boersma, L. 1970. *Soil Sci.* **110**, 1–9.

Lindstrom, F.T., Boersma, L., Gardiner, H. 1968. *Soil Sci.* **106**, 107–113.

Low, P.F. 1958. *Soil Sci. Soc. Am. Proc.* **22**, 395–398.

Low, P.F. 1961. *Adv. Agron.* **13**, 269–327.

Low, P.F. 1962. *Clays Clay Minerals* **9**, 219–228.

Low, P.F. 1968. *Isr. J. Chem.* **6**, 325–336.

Meredith, R.E., and Tobias, C.W. 1962. *Adv. Electrochem. Electrochem. Eng.* **2,** 15–47.
Miller, R.J., and Brown, D.S. 1969. *Soil Sci. Soc. Am. Proc.* **33,** 373–378.
Millington, R.J., and Shearer, R.C. 1971. *Soil Sci.* **111,** 372–378.
Moss, P. 1969. *J. Soil Sci.* **20,** 297–306.
Mott, C.J.B. 1967. *D. Philos. Thesis,* Univ. Oxford.
Mott, C.J.B., and Nye, P.H. 1968. *Soil Sci.* **105,** 18–23.
Mott, N.F., and Littleton, M.J. 1938. *Trans. Faraday Soc.* **34,** 485–499.
Muljadi, D., Posner, A.M., and Quirk, J.P. 1966. *J. Soil Sci.* **17,** 212–229.
Nagarajah, S., Posner, A.M., and Quirk, J.P. 1968. *Soil Sci. Soc. Am. Proc.* **32,** 507–510.
Nielsen, C. 1972. *Yearb. R. Vet. Agric. Univ. Copenhagen* pp. 142–159.
Nielsen, D.R., and Biggar, J.W. 1962. *Soil Sci. Soc. Am. Proc.* **26,** 216–221.
Nye, P.H. 1966. *J. Soil Sci.* **17,** 16–23.
Nye, P.H. 1968. *Trans. Int. Congr. Soil Sci., 9th, Adelaide* **1,** 117–126.
Nye, P.H. 1972. *J. Soil Sci.* **23,** 82–92
Nye, P.H., and Ramzan, M. 1979. *J. Soil Sci.* **30,** 43–52.
Nye, P.H., and Tinker, P.B. 1977. "Solute Movement in the Root-Soil System." Blackwell, Oxford.
O'Connor, G.A., Lindsay, W.L., and Olsen, S.R. 1971. *Soil Sci. Soc. Am. Proc.* **35,** 407–410.
Olejnck, S., and White, J.W. 1972. *Nature (London) Phys. Sci.* **236,** 15–16.
Olphen, H., van 1957. *J. Phys. Chem.* **61,** 1276–1280.
Olsen, S.R., and Kemper, W.D. 1968. *Adv. Agron.* **20,** 91–151.
Olsen, S.R., Kemper, W.D., and van Schaik, J.C. 1965. *Soil Sci. Soc. Am. Proc.* **29,** 154–158.
Oster, J.D., and Low, P.F. 1963. *Soil Sci. Soc. Am. Proc.* **27,** 369–373.
Paul, J.L. 1965. *Agrochimica* **9,** 368–379.
Peaslee, D.E., and Phillips, R.E. 1970. *Soil Sci. Soc. Am. Proc.* **34,** 198–201.
Phillips, R.E. 1969. *Soil Sci. Soc. Am. Proc.* **33,** 322–325.
Phillips, R.E., and Brown, D.A. 1964. *Soil Sci. Soc. Am. Proc.* **28,** 758–763.
Phillips, R.E., and Brown, D.A. 1965. *Soil Sci. Soc. Am. Proc.* **29,** 657–661.
Place, G.A., and Barber, S.A. 1964. *Soil Sci. Soc. Am. Proc.* **28,** 239.
Porter, L.K., Kemper, W.D., Jackson, R.D., and Stewart, S.A. 1960. *Soil Sci. Soc. Am. Proc.* **24,** 460–463.
Prokhorov, V.M., and Frid, A.S. 1972. *Pochvovedenie,* No. 6, 86–94.
Ramzan, M., and Nye, P.H. 1978. *J. Soil Sci.* **29,** 184–194.
Rausell-Colom, Sweatman, T.R., Wells, C.B., and Norrish, K. 1965. *In* "Experimental Pedology" (E.G. Hallsworth and D.V. Crawford, eds.), pp. 40–72. Butterworth, London.
Renkin, E.M. 1954. *J. Gen. Physiol.* **38,** 225–243.
Robinson, R.A., and Stokes, R.H. 1959. "Electrolyte Solutions." Butterworth, London.
Rose, D.A. 1977. *Soil Sci.* **123,** 277–283.
Rowell, D.L., Martin, M.W., and Nye, P.H. 1967. *J. Soil Sci.* **18,** 204–222.
Saxena, S.K., Boersma, L., Lindstrom, F.T., and Young, J.L. 1974. *Soil Sci.* **117,** 80–86.
Schofield, R.K., and Graham-Bryce, I.J. 1960. *Nature (London)* **118,** 1048–1049.
Scott, H.D., and Phillips, R.E. 1972. *Soil Sci. Soc. Am. Proc.* **36,** 714–719.
Scott, H.D., Phillips, R.E., and Paetzold, R.F. 1974. *Soil Sci. Soc. Am. Proc.* **38,** 558–562.
Shainberg, I., and Kemper, W.D. 1966a. *Soil Sci. Soc. Am. Proc.* **30,** 700–706.
Shainberg, I., and Kemper, W.D. 1966b. *Soil Sci. Soc. Am. Proc.* **30,** 707–713.
Shainberg, I., and Otoh, H. 1968. *Isr. J. Chem.* **6,** 251–259.
Shearer, R.C., Letey, J., Farmer, W.J., and Klute, A. 1973. *Soil Sci. Soc. Am. Proc.* **37,** 189–193.
Spencer, W.F., Cliath, M.M., and Farmer, W.J. 1969. *Soil Sci. Soc. Am. Proc.* **33,** 509–511.
Street, N., Miller, R.J., and Kook-Nam Han, 1968. *Trans. Int. Congr. Soil Sci., 9th, Adelaide* **1,** 535–545.
Tanford, C. 1961. "Physical Chemistry of Macromolecules." Wiley, New York.

Thomas, H.C., and Cremers, A. 1970. *J. Phys. Chem.* **74**, 1072-1075.

Tinker, P.B. 1969a. *In* "Ecological Aspects of the Mineral Nutrition of Plants" (I.H. Rorison, ed.), pp. 135-147. Blackwell, Oxford.

Tinker, P.B. 1969b. *J. Soil Sci.* **20**, 336-345.

Tinker, P.B. 1970. *Soc. Chem. Ind. Monogr. (London)* **37**, 120-134.

Vaidyanathan, L.V., and Nye, P.H. 1966. *J. Soil Sci.* **17**, 175-183.

Vaidyanathan, L.V., and Nye, P.H. 1970. *J. Soil Sci.* **21**, 15-27.

Vaidyanathan, L.V., and Nye, P.H. 1972. *J. Soil Sci.* **22**, 94-100.

Vaidyanathan, L.V., Drew, M.C., and Nye, P.H. 1968. *J. Soil Sci.* **19**, 94-107.

Van Genuchten, M. Th., Wierenga, P., and O'Connor, G.A. 1977. *Soil Sci. Soc. Am. Proc.* **41**, 278-285.

Van Schaik, J.C., Kemper, W.D., and Olsen, S.R. 1966. *Soil Sci. Soc. Am. Proc.* **30**, 17-22.

Verdonk, P. 1969. *D.Sc. Thesis,* Univ. Louvain, Belgium.

Walker, G.F. 1956. *Natl. Acad. Sci. U.S.A.* **456**, 101.

Walker, G.F. 1959. *Nature (London)* **184**, 1392-1393.

Warncke, D.D., and Barber, S.A. 1972. *Soil Sci. Soc. Am. Proc.* **36**, 39-46.

Warncke, D.D., and Barber, S.A., 1973. *Soil Sci. Soc. Am. Proc.* **37**, 355-358.

Williams, B.G., Greenland, D.J., and Quirk, J.P. 1966. *Aust. J. Soil Res.* **4**, 131-143.

Williams, B.G., Greenland, D.J., and Quirk, J.P. 1967. *Aust. J. Soil Res.* **5**, 77-83.

Wray, F.J., and Tinker, P.B. 1969. *J. Sci. Instr. (J. Phys. E)* **2**, 343-346.

BORON NUTRITION OF CROPS

Umesh C. Gupta

Research Branch, Research Station, Agriculture Canada, Charlottetown
Prince Edward Island, Canada

I. Introduction

Boron is one of the seven recognized essential micronutrients required for the normal growth of most plants. The essentiality of B, as it affected the growth of maize (*Zea mays* L.) plants, was first mentioned by Maze (1914) in France. However, it was the work of Warington (1923) in England that provided firm knowledge of the B requirement for a variety of crops. Since that time the importance of B as an agricultural chemical has grown very rapidly. In the last 50 years there have been hundreds of reports dealing with the essentiality of B for a large number of agricultural crops in countries from every continent of the world.

Of the known essential micronutrients, B deficiency in plants is most widespread. The deficiency of B has been reported for one or more crops in 43 states of the United States (Sparr, 1970), almost all provinces of Canada, and many other countries of the world. Some of the most severe disorders caused by a deficiency of B include brown-heart of rutabaga or turnips (*Brassica napobrassica*, Mill), cracked stem of celery (*Apium graveolens* L.), heart rot of beets (*Beta vulgaris* L.), brown-heart of cauliflower (*Brassica oleracea* var. *botrytis* L.), and internal brown-spot of sweet potatoes (*Ipomoea batatas* (L.) Lam.).

Boron is unique among the essential mineral nutrients because it is the only element that is normally present in soil solution as a non-ionized molecule over the pH range suitable for plant growth, as Oertli and Grgurevic (1975) have shown. According to their results, boric acid is the form of B that plant roots absorb most efficiently. Alt and Schwarz (1973) hypothesized that B is absorbed as the molecule and that B, at least in high supply, is passively distributed with the transpiration stream. Vlamis and Williams (1970) found that B did not accumulate in barley roots in response to changes in temperature or external B concentration. They suggested that boric acid is present in the medium largely in molecular form (not the ionized form) and does not participate in the metabolic activity associated with the ion uptake and accumulation in roots.

Soils formed from marine sediments are likely to contain more B than those formed from igneous rocks; soils on the average have a higher content of B than rocks (Norrish, 1975). The original source of B in most well-drained soils is tourmaline. Tourmaline (3–4% B) is present in soils formed from acid rocks and metamorphosed sediments. Boron can substitute for tetrahedrally coordinated Si in some minerals. It is likely that much of the B in rocks and soils is dispersed in the silicate minerals in this way and would be available only after long periods of weathering (Norrish, 1975). Most of the B in soil that is available to plants is derived from sediments or plant material (Bowen, 1977).

Because of its non-ionic nature, once B is released from soil minerals it can be leached from the soil fairly rapidly. This explains why soils in high rainfall areas are often deficient in B. Results of Gupta and Cutcliffe (1978) on the podzol soils of eastern Canada have shown that up to 62% of applied B was not recovered from the surface 15 cm of soil 5 months after broadcast application. The amount recovered being referred to is that fraction which was recovered by hot-water extraction of the soil. On the other hand, the availability of B also decreases sharply under drought conditions. This has been attributed partly to the reduced number of microorganisms that can release B from the parent materials (Bowen, 1977). Also, moisture is not available to dissolve B from tourmaline. Lack of soil moisture reduces the mobility of B, thus restricting its uptake by plant roots via mass flow mechanism.

The total B content of most soils varies from 20 to 200 ppm (Berger and Pratt, 1963). Gupta (1968), working on a number of soils from eastern Canada, found that total B ranged from 45 to 124 ppm, whereas hot-water-soluble (hws) B ranged from 0.38 to 4.67 ppm. This indicated that only a small fraction of total B occurred in an available form. Generally, less than 5% of total B is found in an available form, but one report by Kick (1963) showed that hws B averaged 15% of the total B in some Egyptian soils. Very little is known about the mineral forms of B in soils (Lindsay, 1972). The adsorption of B on oxides of Fe and Al is believed to be an important mechanism governing B solubility in soils and clays (Sims and Bingham, 1968).

The purpose of this chapter is to review and report recent information on the B nutrition of agricultural crops in the light of factors such as availability of soil B to plants and factors affecting it, B-containing fertilizers and methods of application, symptoms and levels of B deficiency and toxicity in plants, and the physiological role of B in plants.

II. Boron-Containing Fertilizers

Sodium salts are the most common forms of B fertilizers. Some of the commonly used forms with their chemical formulas and percentages of B as described by Diamond (1972), Chesnin (1972), and Morrill et al. (1977) are given in Table I. Another recently available source of B is ulexite ($NaCaB_5O_9 \cdot 8H_2O$) containing 49.6% B_2O_3. Commercially available ulexite contains 30% B_2O_3 (9.43% B) and is considered to release B slowly as compared with other sources of B. Other kinds of B-containing fertilizers include borated gypsum, calcium carbonate, superphosphate, calcium nitrate, and various mixed fertilizers (Berger and Pratt, 1963). Additional sources include farmyard manure, sewage sludge, compost, and similar materials. The percentage of B in these sources depends on the origin of the materials present in the given source.

The application rates of B vary from 0.3 kg/ha for sensitive crops such as beans (*Phaseolus* spp.), to 3 kg/ha for B-tolerant crops such as rutabaga and

TABLE I
Percentages of Boron and Chemical Formulas of Boron Sources

Boron source	Chemical formula	B (%)
Borax	$Na_2B_4O_7 \cdot 10H_2O$	11
Boric acid	H_3BO_3	17
Boron frits (contained in a moderately soluble glass)	$Na_2B_4 \cdot XH_2O$	10–17
Sodium tetraborate		
Borate-46, Agribor, Tronabor	$Na_2B_4O_7 \cdot 5H_2O$	14
Borate-65	$Na_2B_4O_7$	20
Sodium pentaborate	$Na_2B_{10}O_{16} \cdot 10H_2O$	18
Solubor (partially dehydrated)	$Na_2B_4O_7 \cdot 5H_2O + Na_2B_{10}O_{16} \cdot 10H_2O$	20–21

alfalfa (*Medicago sativa* L.), and they differ according to soil types. Crops grown on peats generally do not show toxicity at the rates that might produce toxicity on mineral soils. For example, no B toxicity was recorded in sweet corn grown on a peat soil even when the hws B level was as high as 10 ppm (Prasad and Byrne, 1975).

The method of application is important in determining the amounts of B to be applied. The rates of applied B for vegetable crops as summarized by Mortvedt (1974) are 0.5–3 kg/ha broadcast, 0.5–1 kg/ha banded, and 0.1–0.5 kg/ha applied as a foliar spray. Vegetables such as rutabaga are very sensitive to a lack of B and require more B than other vegetables to control B deficiency. Gupta and Cutcliffe (1978) stated that this crop may require up to 4 kg B/ha broadcast and 2 kg B/ha applied banded or as foliar spray on some podzol soils of eastern Canada to overcome factors such as high soil pH, improper mixing, and uneven application. In applying B-containing fertilizers, care must be taken to apply them uniformly and to avoid excessive rates, because the range between B deficiency and B toxicity is very narrow. Boron apparently does not react with components of mixed fertilizers, and its availability to plants is related to its distribution and retention in the soil (Mortvedt and Giordano, 1970). The mobility of applied B in soils is probably greater than that of the other micronutrients.

The three most common methods of B application are broadcast, banded, or applied as a foliar spray or dust. In the first two methods of application, an appropriate quantity of the B source is mixed with the bulk N–P–K fertilizer and applied to the soil. Boron has been applied coated on concentrated superphosphate and as a fine granular material mixed with individual plot row fertilizer treatments (Morrill *et al.*, 1977). In many regions, B-containing fertilizers are sold as 0.2, 0.3, or 0.4 B, which means that the bulk NPK fertilizer contains 0.2%, 0.3%, or 0.4% actual B by weight.

For foliar sprays, Solubor and boric acid are most commonly used. The number of sprays used varies from crop to crop and from one region to another. Solubor is also an ideal source of B for addition to liquid fertilizers. It is a specially developed product for speedy and economical correction of B deficiency in fruits, vegetables, and other crops (Turner, 1975). A finely divided powder, it may also be applied directly to plant foliage as a dust.

A detailed discussion on the effects of methods of B application on the plant uptake of B and control of B deficiency will be presented in a subsequent section.

III. Methods of Determining Boron in Plants and Soils

Numerous color reagents are available for determining B in aqueous extracts. Reagents such as chromotropic acid, which form intensely colored complexes with boric acid in aqueous solution, have been developed during recent years.

However, difficulties are experienced in automating this method because both the reagent and the borate–chromotropate complex are sensitive to light (James and King, 1967). Chromogenic reagents, such as quinalizarin and carminic acid, are specific and sensitive for determining B. But their use in automated procedures appears to be limited because they must be used in a concentrated H_2SO_4 medium (Willis, 1970; Lionnel, 1970). The curcumin method has been modified and was found to be rapid and simple (Fiala, 1973). However, reagents such as curcumin suffer from the disadvantage that the values for B determined by such methods vary with the amount and salt content of the material analyzed (Williams and Vlamis, 1970). Other spectrophotometric methods are based on the conversion of sample B into fluoroborate (BF_4^-) in an H_2SO_4-HF system, which is then extracted as a colored complex with Azure-C (or methylene blue) into dichloroethane and determined spectrophotometrically (Weir, 1970).

Procedures other than colorimetric include the determination of B by means of spectrographic and atomic absorption spectrophotometric methods. However, these methods are not as sensitive to B as the color reagents, and therefore their use has been limited.

A new color-developing reagent, azomethine-H, originally used by Russian workers for determining B in organic compounds, was first used by Basson *et al.* (1969) for determining B in plant materials. Since that time the method has achieved prominence for determining B in soil, compost, and manure (Wolf, 1971); it will be dealt with in detail in this section.

A. BORON IN PLANTS

Dry and wet ashing are the mosi common methods for extracting B from plants. The carmine method of Hatcher and Wilcox (1950), also used with some modifications by Gupta (1967a), and the curcumin method of Williams and Vlamis (1970) have been the most common methods for determining B in plant extracts during the past 10 to 20 years. However, with the advent of the azomethine-H method in the late sixties this color reagent has become very popular. Wolf (1971) extended its application in the determination of B from compost and manures using a colorimetric technique. Recently, J. A. Smith and D. A. Tel, University of Guelph, Ontario, Canada, have automated the azomethine-H colorimetric method, using the technique developed by Basson *et al.* (1969). These scientists made some modifications by adding 4 g of NaOH per liter of EDTA reagent to give a pH 4.9 to the waste coming from the AutoAnalyzer system. The only interferences with this method are due to the presence of Al, Cu, and Fe in the plant extracts. Such interference is easily overcome by use of a 0.25 M EDTA (disodium salt) solution (Basson *et al.*, 1969). Sippola and Erviö (1977) reported that the recoveries of added B ranged from 94 to 108% in plant extracts, when determined by the azomethine-H method using spec-

TABLE II
Coefficient of Variability in the Boron Content of Some Plant Leaf Tissues from Various Locations in Prince Edward Island, Canada

Plant species	Location 1		Location 2	
	B (ppm)[a]	Coefficient of variability	B (ppm)	Coefficient of variability
Alfalfa (*Medicago sativa* L.)	37.0	2.7	18.3	1.6
Red clover (*Trifolium pratense* L.)	32.2	1.2	23.7	1.3
Timothy (*Phleum pratense* L.)	14.8	5.4	10.5	3.8
Wheat (boot stage tissue) (*Triticum aestivum* L.)	6.0	3.3	5.8	5.2
Carrots (*Daucus carota* L.)	75.4	2.7	67.3	1.6
Beets (*Beta vulgaris* L.)	102.3	3.0	78.4	1.7
Rutabaga (*Brassica napobrassica*, Mill)	30.9	1.9	94.6	0.8
Brussels sprouts (*Brassica oleracea* var. *gemmifera* Zenker)	50.2	4.9	55.4	6.5
Corn (*Zea mays* L.)	8.7	5.7	9.4	4.3
Tobacco (*Nicotiana tabacum* L.)	29.4	1.0	59.2	1.0

[a] Values (ppm) are averages of four replicate determinations.

trophotometry; these values were similar to those obtained by the carmine method. The azomethine-H method has also been tested by the author; results based on quadruplicate analyses of a number of plant materials indicated good reproducibility based on low coefficient of variability (Table II). This method is very simple and does not require concentrated acids. It is convenient for use on an AutoAnalyzer and is recommended for precise determinations of large numbers of samples on a routine basis.

B. AVAILABLE BORON IN SOIL

Until 10 years ago, the only method for extracting available B in soils involved refluxing the soil with hot water for a period of 5 minutes (Berger and Truog, 1939). However, this method was time-consuming and involved several steps, including the addition of K_2CO_3 to the filtrate, evaporation, and ignition of the residue. The residue, after being dissolved in H_2SO_4, was triturated. Subsequently, Gupta (1967b) developed a rapid and satisfactory method whereby the soils could be extracted by boiling with water directly on a hot plate. The filtrates thus obtained could be analyzed directly, using the then commonly known color-developing reagent, carmine.

Since the development of this simplified method, researchers in various parts of the world (Wolf, 1971; John, 1973) have used direct extraction of soil with hot water. The determination of hws B in soil by using azomethine-H as outlined by Wolf (1971) is probably the most suitable technique. The common interferences due to Cu and Fe in the extract can be eliminated as outlined under the previously described method for plants. Wolf (1974), however, reported that for soils high in Al the use of sufficient tetrasodium salt of EDTA (rather than disodium salt) was necessary for complete elimination of interference due to Al. With this method, precautions should be taken to prevent interference due to organic products in the water and soil extracts by maintaining uniform quantities of charcoal in the soil and in the water. The use of much larger amounts of charcoal than are recommended for soil or water can result in a considerable loss of B from the sample (Wolf, 1974). With the precautions recommended by Wolf, the azomethine-H method should be quite satisfactory in determining B in hot-water extracts of soil.

Azotobacter chroococcum was considered as a possible microbiological indicator for B availability in soils (Gerretsen and de Hoop, 1954). This method is based on the fact that *Azotobacter* requires B, with up to 8 ppm being required in soils for its normal development. However, owing to the complicated behavior of soils, the microbiological determination of B by the *Azotobacter* method has been considered unsatisfactory (Bradford, 1966).

C. TOTAL BORON IN SOIL

There are very few studies on the methods for determining total B in soils. A procedure for extracting total B from a soil fused with Na_2CO_3 was used earlier by Berger and Truog (1939) and subsequently described by Wear (1965). This method is laborious and time-consuming, and it requires an adjustment of the pH of the extract obtained from the fused mass. Large volumes of methyl or ethyl alcohol are required to precipitate sodium sulfate formed in the extract. Furthermore, several evaporations are needed before B can be determined from such extracts. A modified procedure was developed by Gupta (1966) whereby the fused soil mass was extracted with 6 N HCl. The amount of B in the resulting extract could be determined directly by using carmine as the color-developing reagent. No interferences were encountered, the percentage of recovery of B added to the soil was good, and the results were reproducible. For large numbers of determinations the method could be automated by using azomethine-H with some minor modifications.

IV. Role of Boron in Plants

There is perhaps less precise information available on the role of B in plants than for any other essential micronutrient. The functions of B in plants remained almost obscure prior to the mid-1950s. The biochemical role of B is as yet not well understood, and, unlike other generally recognized micronutrients, it has not been shown to be part of an enzyme system (Jackson and Chapman, 1975). The objective of this section is to discuss briefly recent information obtained on the role that B plays in the growth and development of plants.

Some of the chief topics to be discussed here include the effect of B on translocation of sugars, on root extension and meristematic tissues, on the pyrimidine biosynthetic pathway, and on ATPase activity.

Van de Venter and Currier (1977) found that callose accumulates in the tissues of B-deficient bean (*Phaseolus vulgaris* L.) and cotton (*Gossypium hirsutum* L.) plants. Sieve plates in the phloem of B-deficient beans were characterized by heavy plugs of callose, whereas the sieve plates of cotton were essentially unaffected. Since translocation of ^{14}C was drastically reduced in both plants, it was suggested that deposition of callose in B-deficient plants is a secondary effect of cellular damage. Birnbaum et al. (1977) found that cotton ovules callus when B is lacking in the medium.

The research of Sisler et al. (1956) indicated that B enhances uptake and translocation of sugars and is implicated in carbohydrate metabolism. They proposed a micronutrient union with sugars, giving an ionizable sugar–borate complex that moves more readily through cellular membranes than does sugar alone.

Deficient tomato (*Lycopersicon esculentum*, Mill.) plants were found to translocate more sugar when 50 ppm of B was added with sucrose through a cut petiole than when sucrose was applied alone. Subsequent studies by Dugger and Humphreys (1960) implied a direct involvement of B in the enzymatic reactions of sucrose and starch synthesis. It has also been suggested that B deficiency possibly causes reduced synthesis of uridine diphosphate glucose (Birnbaum *et al.*, 1977).

Weiser *et al.* (1964) reported that B does not enhance sugar translocation in plants, but it does enhance the foliar uptake of sucrose applied to the leaves. They concluded that this phenomenon of enhanced foliar uptake of sucrose has given rise in the past to the erroneous conclusion that B enhances sugar translocation.

Zapata (1973) found that sugar cane (*Saccharum officinarum* L.) plants receiving only traces of B suffered growth and quality losses without developing visual B-deficiency symptoms. Lack of B lowered sucrose production in leaves and significantly altered the rates of sugar transport in sugar cane storage tissues (Zapata, 1973). In sugar beets the sucrose content of the storage roots started to decrease at about the same point at which limiting B resulted in a drop in yield (Vlamis and Ulrich, 1971).

The earliest morphological symptoms of B deficiency in mung bean (*Phaseolus aureus* L.) appear to be a slowdown in root extension, followed by a degeneration of meristematic tissue, possibly due to a repressive effect of B deficiency on cell division (Jackson and Chapman, 1975). Results of Robertson and Loughman (1974) indicated that it is unlikely that responses associated with B deficiency are caused by interference with cell division, but they may be related to the role of B in the metabolism, transport, or action of auxin-type hormones in broad beans (*Vicia faba* L.). Whittington (1959) found that B-deficient field bean roots had enlarged apices and fewer cells than the normal B-sufficient roots. Investigations of Kouchi and Kumazawa (1976) on tomato root tips indicated that a lack of B distorted the shape and arrangement of cortical cells and resulted in an abnormal accumulation of a "lipid-like substance." Also, there was an abnormal development of Golgi apparatus, which seemed to be related to the irregular thickening of cell walls.

Cohen and Lepper (1977) established that cessation of root elongation of intact squash (*Cucurbita pepo* L.) plants is an early result of B deficiency. They noted that the ratio of cell length to cell width ranged from a low of 0.8 in B-sufficient root meristems to a high of 3.0 in root meristems grown in a B-deficient nutrient solution for 98 hours. It was concluded that a continuous supply of B is not essential for cell elongation but is required for maintenance of meristematic activity.

Bioassays showed that extracts of substances similar to indoleacetic acid (IAA) taken from B-deficient roots were more inhibitory to the growth of bean-root segments than were those from normal roots (Coke and Whittington, 1968). The IAA treatment and B deficiency together restricted root growth more than did either B deficiency or IAA treatment. Bohnsack and Albert (1977) demonstrated that B deficiency in squash resulted in increased IAA oxidase activity, but root elongation was inhibited. Roots of plants subjected to 12 hours in a B-free medium and then transferred to a medium containing B regained normal elongation rates and oxidase activity within 18 to 20 hours. They further suggested that high levels of IAA under B-deficiency conditions may inhibit cell division and lead to an induction of the IAA oxidase enzyme.

Deficiency of B results in browning of plant tissues, which is thought to be related to the accumulation of polyphenolic compounds; it is also postulated that B is involved in the synthesis of cell-wall components (Slack and Whittington, 1964). Lee and Aronoff (1967) suggested that B combines with 6-phosphogluconic acid to form an enzyme–inhibitor complex, which regulates phenol synthesis, thereby preventing the typical necrosis and ultimate death of B-deficient plants. Birnbaum et al. (1977) found that, when B was lacking in the medium, cotton ovules accumulated brown substances.

Investigations of Jackson and Chapman (1975) with mung beans and broad beans suggest that the earliest known response to the removal of B from a plant culture medium is increased incorporation of precursor into the RNA of the root tip region. These responses to B deficiency are markedly similar to some of the effects of the application to plant tissue of such hormones as auxin, gibberellic acid, and cytokinin.

Any element that is essential for the growth and development of plants must have a direct or indirect influence on N metabolism, including synthesis of proteins. Sherstnev and Kurilenok (1964) found that, in B-deficient sunflower (*Helianthus annuus* L.) plants, the content of many amino acids increased significantly. Such higher content of amino acids was explained by an acceleration in protein decomposition or by a deceleration in protein synthesis. It was suggested that probably both processes occurred, but the deceleration in protein synthesis was probably the dominant process. Studies by Kibalenko et al. (1973) showed that during photosynthesis the rate of $^{14}CO_2$ incorporation into free amino acids was higher in sugar beet (*Beta vulgaris* L.) plants grown on a nutrient medium containing B than in B-deficient plants. However, in leaves of sugar beet and pea (*Pisum sativum* L.) plants, the level of free amino acids was higher and that of protein was lower than in leaves of plants grown in a full nutrient medium. Thus, B deficiency significantly inhibited protein synthesis.

Boron deficiency in tomatoes has been found to be induced by N bases, such as 6-azauracil and 2-thiouracil, whereas uracil antagonized the effect of these bases and prevented the appearance of B-deficiency symptoms (Albert, 1968). Likewise, studies on cotton indicated that B-deficiency-like symptoms were induced by 6-azauracil and 6-azauridine in ovules growing in B-sufficient media (Birnbaum *et al.*, 1977). Other nucleoside–base analogs either reduced or had no effect on overall growth, but did not cause typical B-deficient callus growth of cotton ovules. Orotic acid and uracil countered the effects of 6-azauracil. Similarities between symptoms of B deficiency and 6-azauracil injury, and the ability of uracil to suppress both, suggest that B-deficiency symptoms are related to reduced activity in the pyrimidine biosynthetic pathway and are not related to a reduction in nucleic acid synthesis.

In field experiments on peat and mineral soils, ergot infestation of barley (*Hordeum vulgare* L.) was decreased and grain yields increased by B fertilization (Simojoki, 1969). It was suggested that B deficiency caused structural changes in the plant, resulting in greater susceptibility.

Pollard *et al.* (1977) reported that a deficiency of B in corn and broad beans reduced the capacity for absorption of phosphate. They also found that the B-deficient roots of corn had a reduced ATPase activity, which could be rapidly restored by the addition of H_3BO_3 an hour before extraction of the enzyme. The evidence strongly supports the view that B plays an essential role in the regulation of the functions of higher plant membranes and that the ATPase is a possible component of transport process. The possible mechanisms whereby this control is exercised include direct interaction of B with polyhydroxy components of the membrane and the elevation of endogenous levels of auxins. Gupta and MacLeod (1977) noted that B deficiency produced yellow and violet discoloration in rutabaga leaf tissue. This discoloration may have been due in part to the reduced uptake of P caused by a lack of B, as suggested by Pollard *et al.* (1977) for corn and broad beans.

V. Factors Affecting Boron Requirement and Uptake in Plants

A. SOIL pH, CALCIUM, AND MAGNESIUM

Soil pH is one of the most important factors affecting the availability of B in soil and plants. Generally, B becomes less available to plants with increasing soil pH. Several workers have observed negative correlations between B uptake by plants and soil pH (Bennett and Mathias, 1973; Bartlett and Picarelli, 1973; Gupta, 1972b; Wolf, 1940). However, this relationship is not consistent, and deviations from this effect occur, owing to factors such as crop species (Gupta, 1972a, 1977). Studies by Peterson and Newman (1976) and Gupta and MacLeod (1977) have shown that a negative relationship between soil pH and plant B occurs when soil pH levels are greater than 6.3–6.5. The availability of B to plants decreases sharply at higher pH levels, but the relationship between soil pH and plant B at soil pH values below 6.5 does not show a definite trend. Gupta (1968) did not find any relationship between hws B and pH on 108 soil samples from Eastern Canada ranging in pH from 4.5 to 6.5.

Eck and Campbell (1962) found that liming decreased B uptake when soil B reserves were high. They attributed this effect to a high Ca content. Robertson *et al.* (1975) reported that no close relationship between available soil B and soil pH was found in Michigan soils. However, they reported that soil test levels of B in a calcareous soil decreased rapidly after B application.

Tanaka (1967) reported that B uptake by radish (*Raphanus sativus* L.) was reduced when the Ca content of the medium was increased. Beauchamp and Hussain (1974), in their studies on rutabaga, found that increased Ca concentration in tissue generally increased the incidence of brown-heart. Wolf (1940) found that Mg had a greater effect on B reduction in plants than did Ca, Na, or K, but the differences between Ca and Mg effects were small. However, in the previous work no distinction was made between the effects of soil pH and levels of Ca and/or Mg on B uptake.

Gupta and MacLeod (1977) conducted experiments to distinguish between the effects of soil pH and sources of Ca and Mg on the availability of B to plants. They found that, in the absence of added B, rutabaga roots and tops from Ca and Mg carbonate treatments had more severe brown-heart condition than did roots from the Ca and Mg sulfate treatments. The B concentrations in leaf tissue of rutabaga from treatments with no B were lower at higher soil pH values where Ca and/or Mg were applied as carbonates than they were at lower soil pH where sulfate was used as a source of Ca and/or Mg (Table III). In the presence of added B this trend was not clear, but the levels were well above the deficiency limit. The lower B concentrations in no-B treatments with carbonates than in those with sulfates appear to be related to soil pH differences. It was also noted that the effect of applications of lime on B uptake was not related to the availability of Ca and/or Mg, since equivalent amounts of Ca and/or Mg were applied as sulfates, compared with those added as carbonates; furthermore, Ca and Mg concentrations in the plant tissue were similar (Gupta and MacLeod, 1977). Barber (1971) reported a reduced uptake of B by soybeans [*Glycine max.*(L.) Merr.] as the soil pH increased. However, the author pointed out that the pH effect might be important on some soils and have little effect on others.

Gupta and Cutcliffe (1972) noted an interaction between soil pH and hws B on the severity of brown-heart in rutabaga. The degree of brown-heart was found to be more severe at high soil pH than at low pH. However, at high rates of B, soil pH as high as 6.8 had no effect. Decreased uptake of B with increased soil pH has been reported for alfalfa, soybeans, and barley by Wear and Patterson (1962), Barber (1971), and Gupta (1972b), respectively.

The data by Gupta and MacLeod (1977) showed no differences in B uptake whether the plants were fed with Ca and/or Mg as long as the corresponding

TABLE III

Effects of Ca and Mg Sources and B Levels on Rutabaga (*Brassica napobrassica*, Mill) Plant Tissue B Concentrations, Hot-Water Soluble B, and Soil pH

Treatments		B (μg/g soil)	B (μg/g tissue)	After harvest	
Cation[a]	Anion[a]			Hws B (ppm in soil)	Soil pH
Control		0	33.5d[b]	0.33d[b]	5.6
Ca	CO_3	0	18.4f	0.26de	6.6
Mg	CO_3	0	17.4f	0.27de	6.3
Ca, Mg	CO_3	0	19.9f	0.34d	6.3
Ca	SO_4	0	31.6d	0.24e	4.8
Mg	SO_4	0	26.5e	0.24e	4.9
Ca, Mg	SO_4	0	29.9de	0.26de	4.9
Control		1	112a	0.92a	5.8
Ca	CO_3	1	118a	0.81abc	6.5
Mg	CO_3	1	104ab	0.87ab	6.3
Ca, Mg	CO_3	1	108ab	0.82abc	6.6
Ca	SO_4	1	88c	0.75abc	4.9
Mg	SO_4	1	92bc	0.81abc	5.0
Ca, Mg	SO_4	1	88c	0.76abc	5.0
Means		0	25.3B	0.27B	
Means		1	103A	0.82A	

[a] Treatment consisted of 24 moles/kg of soil either as a Ca or Mg salt or as a mixture in a 1:1 molar ratio of Ca and Mg. Control received 8 millimoles each of $CaCO_3$ and $MgCO_3$ per kilogram of soil.

[b] Values followed by a common letter do not differ significantly at $P = .05$ by Duncan's multiple range test.

anionic components were the same (Table III). Concentrations of Ca and Mg, not shown in the table, were not found to be related to the applications of B. It was further noted that, after the crop was harvested, lower quantities of hws B were found in the soil that received Ca and/or Mg as sulfates than in soil that received Ca and/or Mg as carbonates (Table III).

Until very recently no data had been available on the effects of a wide range of soil pH on the B uptake of crops. Unpublished results of U. C. Gupta and J. A. MacLeod (Research Station, Charlottetown, Prince Edward Island, Canada) on podzol soils with a pH range of 5.4–7.8 showed that liming to pH 7.3–7.5 markedly decreased the B content of pea plant tissue from 117–198 ppm at pH 5.4–5.6 to 36–43 ppm. At values higher than pH 7.3–7.5, even tripling the amount of lime did not affect the B content of plant tissue.

Leaf tissue Ca/B ratios have also been considered as indicators of the B status of crops. The Ca/B ratios of greater than 1370 in barley boot stage tissue (Gupta, 1972b) and of greater than 3300 in rutabaga leaf tissue (Gupta and Cutcliffe,

1972) appeared to be indicators of B deficiency. Drake *et al.* (1941) reported that for tobacco (*Nicotiana tabacum* L.) the boundary between deficiency and optimum Ca/B ratio was quite variable and lay in the range of 1200–1500. The higher Ca/B ratios as indicative of B deficiency are probably related to the higher Ca concentrations in the leaf tissue. Beauchamp and Hussain (1974) noted that the Ca/B ratio decreased significantly as the K concentration of rutabaga roots increased. Likewise, Reeve and Shive (1944) found that the Ca/B ratio in tomato tissue decreased markedly with increasing K concentration in a nutrient solution.

In studies on rutabaga no clear relationship was found between the Ca/B ratio in the leaf blades and the incidence of brown-heart (Beauchamp and Hussain, 1974). They noted that an application of Na increased the Ca concentration in rutabaga tissue, thereby affecting the Ca/B ratio and possibly the incidence of brown-heart. It should be pointed out that the use of the Ca/B ratio in assessing the B status of plants should be viewed in relation to the sufficiency of other nutrients in the growing medium and in the plant. The concentrations of the two elements are also important, as a deficiency or toxicity of one or both of the elements could give a false ratio for determining the nutrient status. Over all, it is the author's opinion that the ratio not be given the same importance as the level of the individual elements.

B. MACRONUTRIENTS AND SULFUR

Among these nutrients, N is of utmost importance in affecting B uptake by plants. Chapman and Vanselow (1955) were among the pioneers in establishing that liberal N applications are sometimes beneficial in controlling excess B in citrus. Jones *et al.* (1963) stated that, under conditions of high B, application of N depresses the level of B in citrus leaves. They further reported that, under conditions of high B, high N depressed the level of B in orange (*Citrus* spp.) leaves from 860 to 696 ppm. Since that time several other investigators have found that large applications of N to the growing medium result in decreased uptake of B by crops.

Although the results of Lancaster *et al.* (1962) on cotton are inconclusive, there was a hint that B deficiency may have been involved in yield reductions with high rates of N. Yamaguchi *et al.* (1958) found that celery plants grown in concentrations of 500 ppm of N were lower in B content than were those grown in 210 ppm of N at 0.1 and 0.25 ppm of applied B. The B concentrations in boot stage tissue of barley and wheat (*Triticum aestivum* L.) increased significantly with increasing rates of compost (Gupta *et al.* (1973). Such increases in B were attributed to a large concentration of B (14 ppm) in the compost. The authors reported that B concentrations decreased with increasing rates of N. Additions of N decreased the severity of B toxicity symptoms, and, at 150 ppm of applied N, the B toxicity was negligible.

In studies by Gupta *et al.* (1976), increasing rates of N applied to initially N-deficient soils significantly decreased the B concentration of boot stage tissue in a greenhouse study, but the field experiments did not show any significant effect of N on concentration of B. The ineffectiveness of N in alleviating B toxicity in cereals under field conditions is due to the fact that N failed to decrease the B concentration in boot stage tissue. Furthermore, the N deficiency was more severe under greenhouse conditions than under field conditions. The decreases in B concentrations were greater with the first level of added N than with the higher rates of added N (Gupta *et al.*, 1976). This may indicate that application of N is helpful in alleviating B toxicity on soils low in available N content. Smithson and Heathcote (1976) found that, where B deficiency occurred in cotton, the application of 250 kg of N per hectare depressed yields. However, with applied B this rate of N produced large increases in yield.

The effects of P, K, and S are less clear than those of N on the availability of B to plants. The first study on this subject, conducted by Reeve and Shive (1943), indicated that the K concentration of the substrate has a definite influence on the accumulation of B in the tissues of tomato and corn plants. They noted that this increased B absorption was especially pronounced at the high B levels. The B-toxicity symptoms on these crops increased in severity with the increase in K concentration in the substrate. However, they noted that, at low levels of B, deficiency of B was progressively itensified with increasing concentrations of K in the growth medium. Nusbaum (1947) reported that, without added B, low rates of K and low rates of P and K together resulted in slight B-deficiency symptoms in sweet potatoes.

The B content of petioles of celery decreased with increasing K in the nutrient solution regardless of the B level in the nutrient solution (Yamaguchi *et al.*, 1958). Bubdine and Guzman (1969) noted that excessive fertilizing with N or K increased symptoms of B deficiency in some celery cultivars, but when N and K were applied together the severity of the symptoms was reduced.

High P increased the severity of B deficiency in tobacco (Stoyanov, 1971). On the other hand, studies of Tanaka (1967) showed that B uptake in radish increased with an increase in P supply. Nusbaum (1947) found that, in the absence of added B, low P fertilizer with optimum rates of N and K resulted in severe B deficiency in sweet potatoes. The results of Reeve and Shive (1944) showed a toxic effect of B only when K in the growing medium was supplied at concentrations in excess of that required for optimal plant growth. Kar and Motiramani (1976), working on various soil types from Madhya Pradesh, India, noted a significant positive relationship between available B and exchangeable K and between available B and the Neubauer value for K. Most recent field studies conducted at the Research Station, Charlottetown, Prince Edward Island, did not reveal a definite effect of K on the B uptake of Brussels sprouts (*Brassica oleracea* var. *gemmifera* Zenker) and cauliflower, although the data did indicate a definite trend toward a slight decrease in the B concentration with applied K.

Tanaka (1967) speculated that there may be a slight effect of sulfate ion on the accumulation of B in plant tissues. The unpublished data of U. C. Gupta from the Research Station, Charlottetown, Prince Edward Island, on a number of crops indicated that the S applications had no effect on the B concentration of peas, cauliflower, timothy (*Phleum pratense* L.), red clover (*Trifolium pratense* L.), and wheat, but they significantly decreased the B content of alfalfa and rutabaga. It is possible that various crops behave differently.

C. METHODS OF APPLICATION

Two principal methods of applying B are by adding it directly to the soil or by foliar spraying. For some elements such as Mo, which plants require in extremely small amounts, seed treatment with a preparation containing Mo is sufficient to overcome a deficiency problem. Because of the comparatively greater requirement for B and because of its toxic effect on the seed or seedlings, seed treatment for B has not received attention.

Soil applications of B made alone or with mixed fertilizers are common, and most data reported on the B uptake have been obtained with B-containing fertilizers added broadcast or in bands. In field studies on rutabaga, Gupta and Cutcliffe (1978) reported that band applications of B resulted in greater B concentrations in leaf tissue than did broadcast applications at five locations. In fact, B applications of 1.12 kg/ha applied in bands resulted in greater B concentrations in leaf tissue than did 2.24 kg/ha applied broadcast. The results of Gupta and Cutcliffe (1978) on rutabaga and of Touchton and Boswell (1975) and Peterson and MacGregor (1966) on corn indicated that band- or foliar-applied B resulted in greater B uptake in plants than did B applied broadcast. Greater uptake when B is applied in bands is likely due to the fact that a large quantity of the available nutrient is concentrated in the immediate root zone. Thus B applied in bands would be concentrated over a small area and would be taken up by the plants very rapidly. Higher quantities of B were required to overcome a B deficiency in rutabaga when B was applied broadcast as compared with B applied in bands or as foliar spray (Gupta and Cutcliffe, 1978).

Foliar sprays are very effective in many areas of California and Arizona where soil applications of micronutrients are ineffective because elements such as Zn, Mn, and Cu are fixed in forms that are not readily available to certain crops (Labanauskas *et al.*, 1969). Foliar applications, besides resulting in higher B uptake, could be used to advantage if a farmer omitted the addition of B in the N–P–K bulk fertilizer or if B deficiency was suspected. Early foliar spray applications result in greater absorption of B than do those applied at later stages of growth (Gupta and Cutcliffe, 1978). Mortvedt (1974) stated that early-morning applications of foliar-applied nutrients may result in increased absorption, as the relative humidity is high, the stomata are open, and photosynthesis is taking place.

D. SOIL TEXTURE

Soil texture is an important factor affecting the availability of B in certain soils (Wear and Patterson, 1962). Gupta (1968), in a study on soils from eastern Canada, found that greater quantities of hws B were found in the fine-textured soils than in the coarse-textured soils. The studies on the recovery of B added to the soil showed that less hws B was recovered in a sandy clay loam than in a sandy loam over a 12-week incubation period. The highest percentage of total B in the hws form occurred in the fine-textured soils. The observed relationship between B and soil texture could be attributed to the fact that some of the B in the soil is adsorbed to clay particles. The lower amounts of B in sandy soils are likely associated with higher leaching of B, which would also explain the lower percentage of total B that occurred in hws form in these soils. For example, Page and Cooper (1955) reported that leaching losses from acid, sandy soils account for as much as 85% of the applied B after addition of 12.5 cm of water. Movement is less rapid in heavy-textured soils because of increased fixation by the clay particles (Reisenauer et al., 1973).

The amount of B adsorbed is significantly influenced by the kind of clay and pH. Hingston (1964) reported that increasing pH resulted in an increase in the monolayer adsorption and a decrease in bonding energy for Kent sand kaolinite and Marchagee montmorillonite and a slight increase in bonding energy for Willalooka illite up to pH 8.5. On a weight basis, illite adsorbed most B over the range of pH values commonly occurring in soils, montmorillonite adsorbed appreciable amounts at higher pH, and kaolinite adsorbed least.

Some workers have shown that fine-textured soils require more B than do the coarse-textured soils to produce similar concentrations of B in plants. Studies of Singh et al. (1976) indicated that B concentrations in solutions of 3.5 ppm in sandy loam and 4.5 ppm in clay loam resulted in similar concentrations of 232, ppm and 221 ppm, respectively, in gram (Cicer arietinum).

E. SOIL ORGANIC MATTER

Organic matter is one of the main sources of B in acid soils, as relatively little B adsorption on the mineral fraction occurs at low pH levels (Okazaki and Chao, 1968). The hws B in soil has been found to be positively related to the organic matter content of the soil (Gupta, 1968). Addition of material such as compost rich in organic matter resulted in large concentrations of B in plant tissues and in phytotoxicity (Purves and MacKenzie, 1973). Berger and Pratt (1963) stated that a large part of the total B in soils is held in the organic matter in tightly bound compounds that have been formed in the growing plants themselves. Boron in organic matter is largely released in available form through the action of microbes (Berger and Pratt, 1963).

Parks and White (1952) suggested that complex formation with dihydroxy compounds in soil organic matter is an important mechanism for B retention. The influence of organic matter on the availability of B in soils is amplified by increases in the pH and the clay content of the soil. The significant interaction between organic matter and pH obtained by Miljkovic et al. (1966) indicates that the increase in hws B associated with an increase in pH is greater in soils with a high rather than a low organic matter content. These findings are contrary to those reported by some European workers, as reviewed by Miljkovic et al. (1966), and are contrary also to observations on the effect of soil pH on hws B as discussed in Section V,A. Little is known of the role of soil organic matter and of the influence of microbial activity on the availability of soil B (Reisenauer et al., 1973).

F. STAGE OF PLANT GROWTH AND PLANT PART SAMPLED

The part of the leaf, its position in the plant, the plant's age, and the plant part are some of the factors that affect the B composition of plants. Studies of Vlamis and Ulrich (1971) showed that young blades of sugar beets contained more B than did the mature and old blades at lower concentrations of B in the Hoagland solution. However, at higher concentrations of B in the solution, no such differences were found. In the case of petioles of sugar beet, no variation was noted in the tissue B concentration at any level of B in the solution. The highest values of B occurred in the older leaves, while the lowest B content occurred in the fibrous and storage roots (Vlamis and Ulrich, 1971). The B concentration of corn leaves increased with age in seedling leaves, but decreased slightly in leaves at higher positions (Clark, 1975a). The uppermost corn leaves had higher concentrations than did leaves at positions below. In the dead bottom corn leaves, B increased to a high of 130 ppm at 74 days before decreasing by over threefold at maturity.

In experiments on corn, leaf B increased with age nearly eightfold and tassel B nearly fivefold, but B in other plant parts remained low and relatively constant (Clark, 1975b). Gupta and Cutcliffe (1973) reported that B levels in leaf tissue of cole crops were generally lower late in the growing season than they were in the early season. Similar results were obtained with rutabaga, where the B content of leaf tissue was greater from early samplings than it was from late samplings (Gupta and Cutcliffe, 1971). Gorsline et al. (1965) noted that B concentration in the whole corn plant decreased during initial growth, remained unchanged during most of the vegetative period, and then decreased after silking. Also, B concentration was higher in the leaves than in the stalks, with the upper leaves higher in B than the lower leaves. Older cucumber (*Cucumis sativus* L.) leaves contained more B than the younger leaves; and within the leaf, B was accumulated in the marginal parts (Alt and Schwarz, 1973). Boron accumulation was greater in the marginal section of corn leaves than in the midrib section (Touchton and Boswell, 1975). Generally, B has a tendency to accumulate in the margin of leaves of plants (Kohl and Oertli, 1961; Jones, 1970). Results of Miller and Smith (1977) on alfalfa showed that the leaves had much higher B content (75–98 ppm) than the tips (47 ppm) or the stems (22–27 ppm).

Supply of B affects the distribution of B in various plant parts. For example, Vlamis and Ulrich (1971) found that in sugar beet plants the blades had a higher B content than the petioles where the B supply was adequate, but this relation was reversed in the B-deficient plants.

G. ENVIRONMENTAL FACTORS

Intensity of light is one of the chief environmental factors affecting the availability of minerals to plants. The faster the plant grows—for example, under high light conditions—the faster it will develop deficiency symptoms in a particular growth period. Observations by Broyer (1971) indicated that deficiencies as well as toxicities are revealed earliest or most intensely in the summer. Experiments conducted with duckweed (*Lemna paucicostata*) showed that reducing light intensity decreased the response to B deficiency and toxicity (Tanaka, 1966). In the absence of B, severe deficiencies were observed in cultures under continuous illumination from a daylight fluorescent lamp at 5500 lux, but not at 1000 lux. Over the range of 0.5–2.5 ppm of B in the culture solution, the uptake of B was reduced with decreasing light intensity. Studies conducted on young tomato plants grown in solution culture showed that B deficiency developed more rapidly at high than at low light intensity (MacInnes and Albert, 1969). Plants supplied with B did not exhibit any B-deficiency symptoms.

Barley leaves grown in Hoagland solution contained more B at 15°C than at 10 or 20°C (Vlamis and Williams, 1970). This effect was consistent on young and mature leaves. However, the B content of roots remained virtually static regardless of temperature.

Moisture appears to affect the availability of B more so than that of some other elements. Studies by Kluge (1971) indicated that B deficiency in plants during drought may be only partially associated with the level of hws B in soil. The reduced soil solution in connection with reduced mass flow, and the reduced diffusion rate as well as limited transpiration flow in the plants during drought periods, may be causative factors of B deficiency in spite of an adequate supply of available B in the soil. Boron deficiencies are generally found in dry soils where summer or winter drought is severe; where there is adequate moisture maintained throughout the summer, deficiency symptoms may not be common (British Columbia Department of Agriculture, 1976). In an experiment on barley, Gupta et al. (1976) found that moisture had a significant effect on plant B uptake when B was applied to the soil. The B concentration of barley, with added B, ranged from 162 to 312 ppm under normal conditions, but only from 87 to 135 ppm when the area near the B fertilizer band was kept dry. Mortvedt and Osborn (1965) likewise reported that movement of B from the fertilizer granules increased with concentration gradient and soil moisture content.

H. PLANT GENOTYPES

Genotypes have affected the uptake of Cu, Fe, and Zn by many plant species. The data on their effect on the B uptake is meager. Susceptibility to B deficiency is controlled by a single recessive gene (Wall and Andrus, 1962), as shown by the tomato varieties T 3238 (B-inefficient) and Rutgers (B-efficient). The data of Wall and Andrus (1962) and Brown et al. (1972) have shown that T 3238 lacks the ability to transport B to the top of the plants and confirms the differential response of T 3238 and Rutgers to a given supply of B. Gorsline et al. (1965) observed that corn hybrids exhibited genetic variability related to B uptake and leaf concentration. One study conducted by E. G. Beauchamp, L. W. Kannenberg, and R. B. Hunter at the University of Guelph, Ontario, indicated that the corn inbred CG 10 was the least efficient, compared with several others, in B uptake as measured by the B content of leaves sampled at the anthesis stage. These researchers also found in a study of eleven hybrids that decreased B uptake was associated with higher stover yield.

VI. Deficient, Sufficient, and Toxic Levels of Boron in Plants

Ulrich and Hills (1967) defined the critical level as that which produces 90% of the maximum yield. The concept is equally valid where crop quality is the main concern rather than yield (Bates, 1971). Rutabaga is an excellent example where a deficiency of B may not affect root yield, but the quality of the roots may be seriously impaired.

The term "critical" level in crops is in my opinion somewhat misleading. Often when one talks about the deficient, sufficient, and toxic levels of nutrients in crops, there is a range in values rather than one definite figure that could be considered as critical. For certain elements the limits between deficiency and sufficiency are narrow, which often results in overlapping of values. A value considered critical by workers in certain areas may not be critical under conditions in other areas. Likewise, the term "optimum" levels of a nutrient, as used in the literature by some researchers to express a relationship to maximum crop yield, is sometimes not clear. Theoretically, such a level for a given nutrient should be sufficient to produce the best possible growth of a crop. Often a single value is published on the "optimum" level when a range of concentrations is equally good. However, in practice there can be no single number or even a very narrow range of numbers to describe this relationship adequately. This indicates that a range of values would be more appropriate to describe the nutrient status of the crop; therefore, for this presentation the term sufficiency will be used, rather than critical or optimum.

The ratio of toxic to adequate levels of B is smaller than that for any other nutrient element (Reisenauer *et al.*, 1973). Thus, both excessive and deficient levels could be encountered in a crop during a single season. This emphasizes the fact that a critical value used to indicate the status of B in crops would be unsuitable. In many cases the values referred to in this section overlap the deficiency and sufficiency ranges.

Boron is among the elements that are not readily translocated from older to younger plant tissue. In a study on cereals, even at very high rates of applied B, very little B was found in the grain of the cereals (Gupta, 1971a).

The sufficiency range varies from one part of the plant to another. Lockman (1972) reported that the sufficiency range for B in sorghum [*Sorghum bicolor* (L.) Moench] was 1–6 ppm at dough stage in the third leaf below the head, 82 to 97 days after planting, whereas it was 1–13 ppm in the whole plant 23 to 39 days after planting.

The deficient, sufficient, and toxic B levels for specific crops as reported by various workers are given in Table IV. The reported deficient and toxic levels of B are associated with plant disorders and/or reductions in the yield of crops.

VII. Deficiency and Toxicity Symptoms of Boron in Plants

As outlined in previous sections of this chapter, B deficiency is more extensive than deficiency of any other micronutrient. This is the principal reason why numerous reports are available on B-deficiency symptoms in plants. Since B is not readily translocated in plants, the deficiency symptoms will generally first appear on the younger leaves at the top of the plants. This is also true of the other micronutrients except Mo, which is readily translocated. In most plants, B deficiency shows up as shortened internodes and arrested top growth. The terminal bud dies and lateral buds produce side shoots; such plants have a bushy or rosette

TABLE IV
Deficient, Sufficient, and Toxic Levels of Boron in Plants

Plant	Part of plant tissue sampled	PPM B in dry matter			References
		Deficient	Sufficient	Toxic	
Rutabaga (*Brassica napobrassica*, Mill)	Leaf tissue at harvest	20–38 <12 severely deficient	38–140	>250	Gupta and Munro (1969)
	Leaf tissue when roots begin to swell	32–40 moderately deficient <12 severely deficient	40	—	Gupta and Cutcliffe (1971, 1972)
	Roots	< 8 severely deficient	13	—	Gupta and Munro (1969)
Sugar beets (*Beta vulgaris* L.)	Blades of recently matured leaves	12–40	35–200	—	Hills and Ulrich (1976)
	Middle fully developed leaf without stem taken at end of June or early July	<20	31–200	>800	Neubert *et al.* (1970)
Cauliflower (*Brassica oleracea* var. *Botrytis* L.)	Whole tops before the appearance of curd	3	12–23	—	Gupta (1971b)
	Leaves	23	36	—	Wallace (1951)
	Leaf tissue when 5% heads formed	4–9	11–97	—	Gupta and Cutcliffe (1973, 1975)
Broccoli (*Brassica oleracea* var. *italica* Plenck.)	Leaves	—	70	—	Wallace (1951)
	Leaf tissue when 5% heads formed	2–9	10–71	—	Gupta and Cutcliffe (1973, 1975)
Brussels sprouts (*Brassica oleracea* var. *gemmifera* Zenker)	Leaf tissue when sprouts begin to form	6–10	13–101	—	Gupta and Cutcliffe (1973, 1975)

Crop	Plant part				Reference
Carrots (*Daucus carota* L.)	Mature leaf lamina	<16	32–103	175–307	Kelly *et al.* (1952)
	Leaves	18	—	—	Smilde and Luit (1970)
Tomatoes (*Lycopersicon esculentum*, Mill)	Plants	14–32	34–96	91–415	Reeve and Shive (1943)
	Mature young leaves from top of the plant	<10	30–75	>200	Neubert *et al.* (1970)
	63-day-old plants	—	—	>125	MacKay *et al.* (1962)
Celery (*Apium graveolens* L.)	Petioles	16	28–75	—	Yamaguchi *et al.* (1958)
Potatoes	Leaflets	20	68–432	720	Eaton (1944)
	32-day-old plants	—	12	>180	MacKay *et al.* (1962)
(*Solanum tuberosum* L.)	Fully developed first leaf at 75 days after planting	<15	21–50	>50**	Neubert *et al.* (1970)
Beans (*Phaseolus* spp.)	43-day-old plants	—	12	>160	MacKay *et al.* (1962)
Dwarf kidney beans (*Phaseolus* spp.)	Plants cut 50 mm above the soil				
	Leaves and stems	—	44	132	Purves and MacKenzie (1973)
	Pods	—	28	43	Purves and MacKenzie (1973)
White pea beans (*Phaseolus* spp.)	Aerial portion of plants 1 month after planting	—	36–94	144	Robertson *et al.* (1975)
Cucumber (*Cucumis sativus* L.)	Mature leaves from center of stem 2 weeks after first picking	<20	40–120	>300	Neubert *et al.* (1970)
Spanish peanuts (*Arachis hypogaea* L.)	Young leaf tissue from 30-day-old plants	—	54–65 / 18–20*	>250	Morrill *et al.* (1977)
Alfalfa (*Medicago sativa* L.)	Whole tops at early bloom	<15	20–40 / 15–20**	200	Meyer and Martin (1976)
	Top one-third of plant shortly before flowering	<20	31–80	>100	Neubert *et al.* (1970)
	Upper stem cuttings in early flower stage	—	30*	—	Melsted *et al.* (1969)

(*continued*)

TABLE IV—(continued)

Plant	Part of plant tissue sampled	PPM B in dry matter			References
		Deficient	Sufficient	Toxic	
Red clover (*Trifolium pratense* L.)	Whole tops	<15	15–20	200	Martin *et al.* (1965)
	Whole tops at 10% bloom	8–12	39–52	> 99	Gupta (1972a)
	Whole tops	<20	—	—	Barber (1957)
	Whole tops at bud stage	12–20	21–45	> 59	Gupta (1971b, 1972a)
Birdsfoot trefoil (*Lotus corniculatus* L.)	Top one-third of plant at bloom	—	20–60	> 60**	Neubert *et al.* (1970)
	Whole tops at bud stage	14	30–45	> 68	Gupta (1972a)
Timothy (*Phleum pratense* L.)	Whole plants at heading stage	—	3–93	>102	Gupta and MacLeod (1973)
Pasture grass (*Graminae*)	Above-ground part at first bloom at first cut	—	10–50	>800	Neubert *et al.* (1970)
Corn (*Zea mays* L.)	Leaf at or opposite and below ear level at tassel stage	—	10*	—	Melsted *et al.* (1969)
	Total above-ground plant material at vegetative stage until ear formation	< 9	15–90	>100	Neubert *et al.* (1970)
Wheat (*Triticum aestivum* L.)	Boot stage tissue	2.1–5.0	8	> 16	Gupta (1971a)
	Straw	4.6–6.0	17	> 34	Gupta (1971a)
Winter wheat (*Triticum aestivum* L.)	Above-ground vegetative plant tissue when plants 40 cm high	< 0.3	2.1–10.1	> 10**	Neubert *et al.* (1970)
Oats (*Avena sativa* L.)	47-day-old plants	—	—	>105	MacKay *et al.* (1962)
	Boot stage tissue	—	15–50	44–400	Jones and Scarseth (1944)
Barley (*Hordeum vulgare* L.)	Boot stage tissue	< 1	8–30	> 30**	Neubert *et al.* (1970)
	Boot stage tissue	1.1–3.5	6–15	> 35	Gupta (1971a)
	Straw	3.5–5.6	14–24	> 50	Gupta (1971a)
	Boot stage tissue	1.9–3.5	10	> 20	Gupta (1971a)
	Straw	7.1–8.6	21	> 46	Gupta (1971a)

*Considered critical.
**Considered high.

appearance. In vegetables such as rutabaga and cauliflower, B deficiency is indicated by dark-brown spots on areas in the storage tissue.

The symptoms of B toxicity are similar on most plants. They consist in a marginal and tip chlorosis, which is quickly followed by a necrosis (Shorrocks, 1974). The pattern of chlorosis and necrosis follows the leaf venation; monocotyledons, for instance, show a tip, and not a marginal necrosis.

As far as B toxicity is concerned, it occurs chiefly under two conditions: owing to its presence in irrigation water or owing to accidental applications of too much B in treating B deficiency. Large additions of materials high in B—for example, compost—also result in B toxicity in crops (Gupta *et al.*, 1973; Purves and MacKenzie, 1973).

Boron toxicity occurs in scattered areas in arid and semiarid regions. It is frequently associated with saline soils, but it most often results from the use of high-B irrigation waters. Some of the main areas of high-B waters in the United States are along the west side of the San Joaquin and Sacramento valleys in California (Branson, 1976). Boron occasionally accumulates in the soil as the result of water's evaporating from the soil surface or from a shallow water table (Wilcox, 1960). The B content of water from a large number of wells is high. This results in a high B content in soil, especially under conditions of poor drainage.

Information on the symptoms of B deficiency and B toxicity in a number of vegetable and field crops is summarized below.

A. BORON-DEFICIENCY SYMPTOMS

1. Vegetable Crops

a. Rutabaga. The B-deficiency disorder in rutabaga is referred to as brown-heart, water core, raan, and blackheart. Upon cutting, the roots show a soft, watery area. Under severe B deficiency the surface of the roots is rough and netted, and often the roots are elongated (Gupta and Cutcliffe, 1971). The roots are tough, fibrous, and bitter and have a corky and somewhat leathery skin (Shorrocks, 1974). Leaves from severely deficient plants show a yellow and violet discoloration (Gupta and MacLeod, 1977).

b. Cauliflower. The chief symptoms are the tardy production of small heads, which display brown "waterlogged" patches, the vertical cracking of the stems, and rotting of the core (Haworth, 1952). When browning is severe, both the outer and the inner portions of the head have a bitter flavor (Purvis and Carolus, 1964). Stems are stiff, with hollow cores, and curd formation is delayed (Mehrotra and Misra, 1974). The roots are rough and dwarfed; lesions appear in the pith, and a loose curd is produced (Bergmann, 1976).

c. Brussels Sprouts (Brassica oleracea L. var. *gemmifera* Zenker). The first signs of B deficiency are swellings on the stem and petioles, which later become suberised. The leaves are curled and rolled, and premature leaf fall of the older leaves may take place (Shorrocks, 1974). The sprouts themselves are very loose instead of being hard and compact, and there is vertical cracking of the stem (Haworth, 1952). Boron deficiency results in purple-colored leaf edges curled inward. There is nonuniform yellowing of most leaves, accompanied by stunted stems (Gupta and Cutcliffe, 1975).

d. Broccoli (Brassica oleracea var. *italica* Plenck). Water-soaked areas occur inside the heads, and callus formation is slower on the cut end of the stems after the heads have been harvested (Inden, 1975). Under severe deficiency the plants show nonuniform yellowing of leaves accompanied by stunted growth; browning of older leaf edges and reduced growth were observed under field conditions (Gupta and Cutcliffe, 1975).

e. Cabbage (Brassica oleracea var. *capitata* L.). Plants develop water-soaked areas in the stems and at the base of the heads, similar to those found in cauliflower (Walker *et al.,* 1941; Purvis and Carolus, 1964). These watersoaked areas dry out and become hollow (Walker *et al.,* 1941).

f. Lettuce (Lactuca sativa L.). Boron deficiency is characterized by spotting and burning of the leaf tips, and death of the growing point of the plant (McHargue and Calfee, 1933). Retardation of growth occurs, and younger leaves are malformed (Purvis and Carolus, 1964).

g. Celery. Boron deficiency results in slight brown checking and cracked stem (Yamaguchi *et al.,* 1958). It also manifests itself by a brownish mottling along the margins of the bud leaves and is accompanied by brittleness of the stem and by brown stripes in the epidermis along the ribs (Purvis and Carolus, 1964). Boron deficiency also results in "brown or hollow heart" of celery (Bergmann, 1976).

h. Radish. Deficiency of B is known as brown-heart, manifested first by dark spots on the roots, usually on the thickest parts (Purvis and Carolus, 1964).

i. Carrots (Daucus carota L.). Boron deficiency results in longitudinal splitting of roots (Smilde and Luit, 1970).

j. Peas. Leaves develop yellow or white veins followed by some changes in interveinal areas; growing points die, and blossoms shed (Piper, 1940). Unpublished data of U. C. Gupta and J. A. MacLeod (Research Station, Charlottetown, Prince Edward Island) indicate that B deficiency in peas results in short internodes and small, shriveled new leaves.

k. Beets. Boron deficiency results in a characteristic corky upper surface of the leaf petiole (Bergmann, 1976).

l. Onions (Allium cepa L.). Deficiency results in the chlorosis of the tip of the leaves, ladder-like rings in the new coming leaves, and stunted growth (Gandhi and Mehta, 1959).

m. Tomatoes. An early symptom of B deficiency is a blackened appearance of the growing point of the stem. The plants look bushy, owing to the growth of new leaves below the growing point. The stems become stunted, and the terminal shoot curls inward, yellows, and dies (Purvis and Carolus, 1964). The growing point is injured; flower injury occurs during the early stages of blossoming, and fruits are imperfectly filled (Inden, 1975). Symptoms of B deficiency in tomato also include the development of secondary roots and axillary buds. Failure to set fruit is common, and the fruit may be ridged, show corky patches, and ripen unevenly. Cork may also be found around the stem end of the fruit (Shorrocks, 1974).

n. Potatoes (Solanum tuberosum L.). Deficiency results in the death of growing points, with short internodes giving the plant a bushy appearance. Leaves thicken, and margins roll upward, a symptom similar to that of potato leaf roll virus (Houghland, 1964).

2. Forage Crops

a. Alfalfa. Symptoms are more severe at the leaf tips, although the lower leaves remain a healthy green color. Flowers fail to form, and buds appear as white or light-brown tissue (Nelson and Barber, 1964). Internodes are short, blossoms drop or do not form, and stems are short (Berger, 1962). Younger leaves turn red or yellow in color (Gupta, 1972a), and "top yellowing" of alfalfa occurs (Bergmann, 1976). Boron-deficiency symptoms in alfalfa must be distinguished from leaf hopper injury, K deficiency, and certain diseases that cause yellowing of both the lower and upper leaves. With B deficiency the yellowing is confined to the upper leaves (Shorrocks, 1974).

b. Clover (Trifolium spp.). Plants are weak, with thick stems that are swollen close to the growing point, and leaf margins often look burnt (Janos, 1963). Symptoms of B deficiency in red and alsike clover may occur as a red coloration on the margins and tips of younger leaves; the coloration gradually spreads over the leaves, and the leaf tips may die (Nelson and Barber, 1964). Boron deficiency in red clover results in poor plants with cupped and shriveled leaves, which are small and yellowish in color (Gupta, 1972a).

c. Birdsfoot trefoil (Lotus corniculatus L.). Except for poor growth, no deficiency symptoms were noted (Gupta, 1972a).

3. Cereal Crops

The typical symptoms of B deficiency in cereals include the development of abnormally thick stems, the death of growing points, and the formation of distorted and imperfect heads (Shorrocks, 1974).

a. Corn. Boron deficiency is seen on the youngest leaves as white, irregularly shaped spots scattered between the veins. With severe deficiency these spots may coalesce, forming white stripes 2.5–5.0 cm long. These stripes appear to be waxy and raised from the leaf surface (Krantz and Melsted, 1964).

b. Wheat. A normal ear forms but fails to flower (Löhnis, 1936). Boron deficiency also results in secondary stalk and root development from the lower nodes (Bergmann, 1976).

c. Barley. No ears are formed (Löhnis, 1936).

d. Oats (Avena sativa L.). Pollen grains are empty (Löhnis, 1936).

4. Other Crops

a. Sugar Beets. Deficiency results in retarded growth, and young leaves curl and turn black (Vlamis and Ulrich, 1971). The old leaves show surface cracking, along with cupping and curling. When the growing point fails completely, it forms a heart rot (Vlamis and Ulrich, 1971). The heart leaves of sugar beet die, the older leaves are wrinkled and wilted, and roots rot at the top or at the side (Janos, 1963).

b. Tobacco. Boron deficiency results in interveinal chlorosis, dark and brittle new, coming leaves, water-soaked areas in leaves, and delayed flowering and formation of seedless pods (Gåndhi and Mehta, 1959). Tissues at the base of the leaf show signs of breakdown, and the stalk toward the top of the plant may show a distorted or twisted type of growth. The death of the terminal bud follows these stages (McMurtrey, 1964).

c. Cotton. The terminal bud often dies, checking linear growth, and short internodes and enlarged nodes give a bushy appearance often referred to as a rosette condition (Donald, 1964). Bolls are deformed and reduced in size. Root growth is severely inhibited, and secondary roots have a stunted appearance (Van de Venter and Currier, 1977). Boron deficiency causes retarded internodal growth (Ohki, 1973).

d. Indian Bean (Dolichos lablab L.). Boron deficiency results in interveinal chlorosis, dark and brittle new, coming leaves, water-soaked areas in leaves, and delaying of flowering and formation of seedless pods (Gandhi and Mehta, 1959).

B. BORON-TOXICITY SYMPTOMS

1. Vegetable Crops

a. Rutabaga. The leaf margins are yellow in color and tend to curl and wrinkle. The symptoms on roots are similar to moderate B-deficiency symptoms—a water-soaked appearance of the tissues in the center of the root

(Gupta and Munro, 1969). Boron toxicity in turnip seedlings results in marginal bleaching of the cotyledons and first leaves (Muller and McSweeney, 1977). However, at harvest no difference is noted among the various B treatments.

b. Peas. Unpublished data of U. C. Gupta and J. A. MacLeod (Research Station, Charlottetown, Prince Edward Island) showed that B toxicity results in burning of the edges of old leaves. The extent of burning in the upper leaves depends on the degree of B toxicity.

c. Onions. Boron toxicity results in burning of the tip of the leaves, gradually increasing up to the base, and no development of bulb (Gandhi and Mehta, 1959).

d. Radishes. Boron excess is characterized by marginal yellowing and burning of lower leaves. The new season's growth of radishes left in the ground over winter turn yellow (Inden, 1975).

e. Tomatoes. Boron toxicity results in dying off of older leaves (Bergmann, 1976).

2. Forage Crops

a. Alfalfa. Boron toxicity is marked by burned edges on the older leaves (Gupta, 1972a).

b. Red Clover. Toxicity symptoms of B include burning and yellowing of older leaf edges (Gupta, 1972a).

c. Birdsfoot trefoil. Growth is poor, and leaves are small and dark with thin stems (Gupta, 1972a).

3. Cereal Crops

a. Wheat. Boron toxicity in wheat appears as light browning of older leaf tips converging into light greenish-blue spots (Gupta, 1971a).

b. Barley. Toxicity of B is characterized by elongated, dark-brown blotches at the tips of the older leaves (Krantz and Melsted, 1964). Severe browning, spotting, and burning of older leaf tips occurs, gradually extending to the middle portion of the leaf (Gupta 1971a, Gupta *et al.*, 1973).

c. Oats. Toxicity of B in oats results in light-yellow bleached leaf tips (Gupta, 1971a).

d. Corn. Leaves show tip burn and marginal burning and yellowing between the veins (Wilcox, 1960; Krantz and Melsted, 1964).

4. Other Crops

a. Tobacco. Boron toxicity results in brown circular spots on the periphery of the leaves and stunted growth (Gandhi and Mehta, 1959).

b. Beans. Excess B causes mottled and necrotic areas on the leaves, especially along the leaf margins (Van de Venter and Currier, 1977).

c. Blackeye Bean (Vigna unguiculata L.). Boron toxicity does not produce any characteristic pattern; yellow or dead areas occur between veins (Wilcox, 1960).

d. Indian Bean. Slight chlorosis is caused, followed by dark-brown spots on the leaves. Marginal scorching is found on the periphery of the leaves, and stunted growth occurs (Gandhi and Mehta, 1959).

VIII. Summary and Future Research Needs

The deficiency of B in crops is more widespread than that for any other micronutrient in many regions of the world. Under deficiency conditions, B applications result in increased crop yields as well as in improved quality of many vegetable crops. Most vegetables and legumes have a higher requirement of B than do other crops. Generally, the rates of applied B to correct B deficiency vary from 0.5 to 3 kg/ha. Lesser amounts are needed when B is applied in bands or as a foliar spray, compared with that applied broadcast. At the same levels of application, banded B and B applied as a foliar spray result in greater absorption of B by most plants.

Although the physiological role of B in plants is less well understood than that for other nutrients, considerable progress has been made in this direction. Some of the observed effects caused by a deficiency of B include slowdown of root extension, inhibition of cell division, abnormal thickening of cell walls, accumulation of callose in the conducting tissues, increased production of indoleacetic acid, and browning of plant tissue as related to the accumulation of polyphenolic compounds. Boron has been demonstrated to unite with various sugars, resulting in an increase in translocation rate of photosynthetic products. Increases in the rate of CO_2 incorporation into the free amino acids have also been shown.

A considerable amount of progress has been made on the methods of determining B in plants and soils. With the advent of the azomethine-H color method, the use of the reagents carmine, quinalizarin, and curcumin requiring concentrated acids has been eliminated. Further, because of the automation made possible by the use of azomethine-H as the color-developing reagent and development of a simplified method of extraction of available B from soils, the determination of B in soil and plants can now be performed satisfactorily with greater rapidity.

With the exception of some environmental factors, specifically under drought conditions where B could be very limiting for plants, soil pH is the most important factor affecting B uptake by plants. Recent studies have shown that increased soil pH plays a significant role in decreasing the uptake of B by plants when the soil pH is greater than 6.3.

In the last 15 years a considerable amount of information has become available on the deficiency and sufficiency levels and deficiency symptoms of B in various crops. However, such data are still lacking for a number of crops, and future studies should be conducted to fill in the gap. The available information on B-toxicity levels and symptoms in plant tissue is scanty, and further investigations need to be done in this regard.

At present, the plant parts analyzed for establishing the optimum and deficiency levels of B in crops vary considerably. An extensive amount of cooperative effort is needed in standardizing plant part and stage of development for a given plant species with respect to determining levels of B in plant tissue. This will facilitate meaningful comparisons of data from different locations and establishment of universally applicable criteria for B deficiency and toxicity in crops.

The information on the levels of hws B (available B) in soil as related to crop production is meager. The sufficiency levels of hws B in soil vary considerably from one crop to another and are often affected by the chemical and physical properties of soil. Interactions of B with various plant nutrients in soils need to be investigated in detail under field conditions.

Future research in the subject areas suggested in this section should result in the correction of problems related to B nutrition of crops.

REFERENCES

Albert, L.S. 1968. *Plant Physiol.* **43**, S 51.
Alt, D., and Schwarz, W. 1973. *Plant Soil* **39**, 277–283.
Barber, S.A. 1957. *Purdue Univ. Agric. Exp. Stn., Lafayette, Indiana, Bull.* 652.
Barber, S.A. 1971. *Proc. Int. Symp. Soil Fertil. Eval. New Delhi* **1**, 249–256.
Bartlett, R.J., and Picarelli, C.J. 1973. *Soil Sci.* **116**, 77–83.
Basson, W.D., Böhmer, R.G., and Stanton, D.A. 1969. *Analyst* **94**, 1135–1141.
Bates, T. 1971. *Soil Sci.* **112**, 116–130.
Beauchamp, E.G., and Hussain, I. 1974. *Can. J. Soil Sci.* **54**, 171–178.
Bennett, O.L., and Mathias, E.L. 1973. *Agron. J.* **65**, 587–591.
Berger, K.C. 1962. *Agric. Food Chem.* **10**, 178–181.
Berger, K.C., and Pratt, P.F. 1963. *In* "Fertilizer Technology and Usage" (M.H. McVickar, G.L. Bridger, and L.B. Nelson, eds.), pp. 287–340. Soil Sci. Soc. Am. Madison, Wisconsin.
Berger, K.C., and Truog, E. 1939. *Ind. Eng. Chem.* **11**, 540–545.
Bergmann, W. 1976. "Ernährungsstörungen bei Kulturpflanzen in Farbbildern." Fischer, Jena pp. 81–113.
Birnbaum, E.H., Dugger, W.M., and Beasley, B.C.A. 1977. *Plant Physiol.* **59**, 1034–1038.
Bohnsack, C.W., and Albert, L.S. 1977. *Plant Physiol.* **59**, 1047–1050.
Bowen, J.E. 1977. *Crops Soils* **29**(9), 12–14.
Bradford, G.R. 1966. *In* "Diagnostic Criteria for Plants and Soils" (H.D. Chapman, ed.), pp. 33–61. Univ. Calif., Riverside.
Branson, R.L. 1976. *In* "Soil and Plant-Tissue Testing in California" (H.M. Reisenauer, ed.), pp. 42–45. Div. of Agr. Sciences, Univ. Calif. Bull. 1879.
British Columbia Department of Agriculture Field Crops Branch 1976. *Boron. Soil Ser.* No. 8, pp. 1–6.

Brown, J.C., Ambler, J.E., Chaney, R.L., and Foy, C.D. 1972. In "Micronutrients in Agriculture" (J.J. Mortvedt, P.M. Giordano, and W.L. Lindsay, eds.), pp. 389–418. Soil Sci. Soc. Am., Madison, Wisconsin.

Broyer, T.C. 1971. Commun. Soil Sci. Plant Anal. 2, 241–248.

Bubdine, H.W., and Guzman, V.L. 1969. Proc. Soil Crop Sci. Soc. Fla. 29, 351–352.

Chapman, H.D., and Vanselow, A.P. 1955. Calif. Citrogr. 40, 455–457.

Chesnin, L. 1972. In "The Fertilizer Handbook" (W.C. White and D.N. Collins, eds.), pp. 65–84. The Fertilizer Institute, Washington, D.C.

Clark, R.B. 1975a. Commun. Soil Sci. Plant Anals. 6, 439–450.

Clark, R.B. 1975b. Commun. Soil Sci. Plant Anal. 6, 451–464.

Cohen, M.S., and Lepper, R., Jr. 1977. Plant Physiol. 59, 884–887.

Coke, L., and Whittington, W.J. 1968. J. Exp. Bot. 19, 295–308.

Diamond, R.B. 1972. Agrichem. Age 15, 12–20.

Donald, L. 1964. In "Hunger Signs in Crops. A symposium" (H.B. Sprague, ed.), 3rd ed., pp. 59–98. McKay, New York.

Drake, M., Sieling, D.H., and Scarseth, G.D. 1941. J. Am. Soc. Agron. 33, 454–462.

Dugger, W.M., Jr., and Humphreys, T.E. 1960. Plant Physiol. 35, 523–530.

Eaton, F.M. 1944. J. Agric. Res. 69, 237–277.

Eck, P., and Campbell, F.J. 1962. Am. Soc. Hortic. Sci. Proc. 81, 510–517.

Fiala, K. 1973. Plant Soil 38, 473–476.

Gandhi, S.C., and Mehta, B.V. 1959. Indian J. Agric. Sci. 29, 63–70.

Gerretsen, F.C., and de Hoop, H. 1954. Plant Soil 5, 349–367.

Gorsline, G.W., Thomas, W.I., and Baker, D.E. 1965. Penn. State Univ. Exp. Stn. Bull. 746.

Gupta, U.C. 1966. Soil Sci. Soc. Am. Proc. 30, 655–656.

Gupta, U.C. 1967a. Plant Soil 26, 202–204.

Gupta, U.C. 1967b. Soil Sci. 103, 424–428.

Gupta, U.C. 1968. Soil Sci. Soc. Am. Proc. 32, 45–48.

Gupta, U.C. 1971a. Can. J. Soil Sci. 51, 415–422.

Gupta, U.C. 1971b. Soil Sci. 112, 280–281.

Gupta, U.C. 1972a. Commun. Soil Sci. Plant Anal. 3, 355–365.

Gupta, U.C. 1972b. Soil Sci. Soc. Am. Proc. 36, 332–334.

Gupta, U.C. 1977. Plant Soil 47, 283–287.

Gupta, U.C., and Cutcliffe, J.A. 1971. Soil Sci. 111, 382–385.

Gupta, U.C., and Cutcliffe, J.A. 1972. Soil Sci. Soc. Am. Proc. 36, 936–939.

Gupta, U.C., and Cutcliffe, J.A. 1973. Can. J. Soil Sci. 53, 275–279.

Gupta, U.C., and Cutcliffe, J.A. 1975. Commun. Soil Sci. Plant Anal. 6, 181–188.

Gupta, U.C., and Cutcliffe, J.A. 1978. Can. J. Plant Sci. 58, 63–68.

Gupta, U.C., and MacLeod, J.A. 1973. Commun. Soil Sci. Plant Anal. 4, 389–395.

Gupta, U.C., and MacLeod, J.A. 1977. Soil Sci. 124, 279–284.

Gupta, U.C., and Munro, D.C. 1969. Soil Sci. Soc. Amer. Proc. 33, 424–426.

Gupta, U.C., Sterling, J.D.E., and Nass, H.G. 1973. Can. J. Soil Sci. 53, 451–456.

Gupta, U.C., MacLeod, J.A., and Sterling, J.D.E. 1976. Soil Sci. Soc. Am. J. 40, 723–726.

Hatcher, J.T., and Wilcox, L.V. 1950. Anal. Chem. 22, 567–569.

Haworth, F. 1952. Q. J. Tea Res. Inst. Ceylon 23, 86.

Hills, F.J., and Ulrich, A. 1976. In "Soil and Plant-Tissue Testing in California (H.M. Reisenauer, ed.) pp. 18–21. Div. Agric. Sci., Univ. of California Bull. 1879.

Hingston, F.J. 1964. Aust. J. Soil Res. 2, 83–95.

Houghland, G.V.C. 1964. In "Hunger Signs in Crops. A symposium" (H.B. Sprague ed.), 3rd ed., pp. 219–244. McKay, New York.

Inden, T. 1975. ASPAC Food Fert. Technol. Cent. (Asian Pac. Council) Ext. Bull. 55, pp. 1–20.

Jackson, J.F., and Chapman, K.S.R. 1975. *In* "Trace Elements in Soil-Plant-Animal systems" (D.J.D. Nicholas and A.R. Egan, eds.), pp. 213–225. Academic Press, New York.

James, H., and King, G.H. 1967. *In* "Automation in Analytical Chemistry," Technicon Symposia, 1966. Medial, New York, p. 123. (as cited by Basson *et al.*, 1969)

Janos, Di G. 1963. *Magy. Tud. Akad. Biol. Orv. Tud. Oszt. Kozl.* **22**, 349–361.

John, M.K. 1973. *Soil Sci. Soc. Am. Proc.* **37**, 332–334.

Jones, H.E., and Scarseth, G.D. 1944. *Soil Sci.* **57**, 15–24.

Jones, J.B., Jr., 1970. *Commun. Soil Sci. Plant Anal.* **1**, 27–34.

Jones, W.W., Embleton, T.W., Boswell, S.B., Steinacker, M.L., Lee, B.W., and Barnhart, E.L. 1963. *Calif. Citrogr.* **48**(107), 128–129.

Kar, S., and Motiramani, D.P. 1976. *Indian Soc. Soil Sci. Bull.* **10**, 99–102.

Kelly, W.C., Somers, G.F., and Ellis, G.H. 1952. *Proc. Am. Soc. Hortic. Sci.* **59**, 352–360.

Kibalenko, A.P., Khomlyak, M.N., and Velikaya, S.L. 1973. *Dopov. Akad. Nauk Ukr. RSR Ser. B* **35**, 457–463.

Kick, H. 1963. *Z. Pflanzenernaehr. Düeng. Bodenkd.* **100**, 102–114.

Kluge, R. 1971. *Arch. Acker-Pflanzenbau Bodenkd.* **15**, 749–754.

Kohl, H.C., Jr., and Oertli, J.J. 1961. *Plant Physiol.* **36**, 420–424.

Kouchi, H., and Kumazawa, K. 1976. *Soil Sci. Plant Nutr.* **22**, 53–71.

Krantz, B.A., and Melsted, S.W. 1964. *In* "Hunger Signs in Crops. A symposium" (H.B. Sprague, ed.), 3rd. ed., pp. 25–37. McKay, New York.

Labanauskas, C.K., Jones, W.W., and Embleton, T.W. 1969. *Proc. Int. Citrus Symp., 1st* **3**, 1535–1542.

Lancaster, J.D., Murphy, B.C., Hurt, B.C., Jr., Arnold, B.L., Coates, R.E., Albritton, R.C., and Walton, L. 1962. *Miss. Agric. Exp. Stn. Bull.* 635, p. 12.

Lee, S., and Aronoff, S. 1967. *Science* **158**, 798–799.

Lindsay, W.L. 1972. *In* "Micronutrients in Agriculture" (J.J. Mortvedt, P.M. Giordano, and W.L. Lindsay, eds.), pp. 41–57. Soil Science Society of America, Madison, Wisconsin.

Lionnel, L.J. 1970. *Analyst* **95**, 194–199.

Lockman, R.B. 1972. *Commun. Soil Sci. Plant Anal.* **3**, 271–282.

Löhnis, M.P. 1936. *Chem. Weekblad* **33**, 59–61. (as cited by Bradford, 1966)

MacInnes, C.B., and Albert, L.S. 1969. *Plant Physiol.* **44**, 965–967.

MacKay, D.C., Langille, W.M., and Chipman, E.W. 1962. *Can. J. Soil Sci.* **42**, 302–310.

Martin, W.E., Rendig, V.V., Haig, A.D., and Berry, L.J. 1965. *Calif. Agric. Exp. Stn. Bull.* 815, pp. 1–35.

Maze, P. 1914. *Ann. Inst. Pasteur* **28**, 21–68.

McHargue, J.S., and Calfee, R.K. 1933. *Plant Physiol.* **8**, 305–313.

McMurtrey, J.E., Jr., 1964. *In* "Hunger Signs in Crops. A symposium" (H.B. Sprague, ed.), 3rd ed., pp. 99–141. McKay, New York.

Mehrotra, O.N., and Misra, P.H. 1974. *Prog. Hortic.* **5**, 33–39.

Melsted, S.W., Motto, H.L., and Peck, T.R. 1969. *Agron. J.* **61**, 17–20.

Meyer, R.D., and Martin, W.E. 1976. *In* "Soil and Plant-Tissue Testing in California" (H.M. Reisenauer, ed.), pp. 26–29. Div. Agric. Sci., Univ. of Calif. Bull. No. 1879.

Miljkovic, N.S., Matthews, B.C., and Miller, M.H. 1966. *Can. J. Soil Sci.* **46**, 133–138.

Miller, D.A., and Smith, R.K. 1977. *Commun. Soil Sci. Plant Anal.* **8**, 465–478.

Morrill, L.G., Hill, W.E., Chrudimsky, W.W., Ashlock, L.O., Tripp, L.D., Tucker, B.B., and Weatherly, L. 1977. *Agric. Exp. Stn.* MP-99, p. 20, Oklahoma State Univ., Stillwater.

Mortvedt, J.J. 1974. *Int. Hortic. Congr., 19th, Warsaw* pp. 497–505.

Mortvedt, J.J., and Giordano, P.M. 1970. *Commun. Soil Sci. Plant Anal.* **1**, 273–286.

Mortvedt, J.J., and Osborn, G. 1965. *Soil Sci. Soc. Am. Proc.* **29**, 187–191.

Muller, F.B., and McSweeney, G. 1977. *N.Z. J. Exp. Agric.* **4**, 451–455.

306 UMESH C. GUPTA

Nelson, W.L., and Barber, S.A. 1964. *In* "Hunger Signs in Crops. A symposium" (H.B. Sprague, ed.), 3rd ed., pp. 143–179. McKay, New York.

Neubert, P., Wrazidlo, W., Vielemeyer, H.P., Hundt, I., Gollmick, Fr., and Bergmann, W. 1970. "Tabellen zur Pflanzenanalyse—Erste orientierende Übersicht." Inst. fur Pflanzenernährung, Jena, der deutschen Akademie der Landwirtschaftswissenschaften zu Berlin. 69 Jena, Naumburger Strasse 98.

Norrish, K. 1975. *In* "Trace Elements in Soil-Plant-Animal Systems" (D.J.D. Nicholas and A.R. Egan, eds.), pp. 55–81. Academic Press, New York.

Nusbaum, C.J. 1947. *Phytopathology* **37**, 435.

Oertli, J.J., and Grgurevic, E. 1975. *Agron. J.* **67**, 278–280.

Ohki, K. 1973. *Agron. J.* **65**, 482–485.

Okazaki, E., and Chao, T.T. 1968. *Soil Sci.* **105**, 255–259.

Page, N.R., and Cooper, H.P. 1955. *J. Agric. Food Chem.* **3**, 222–225.

Parks, W.L., and White, J.L. 1952. *Soil Sci. Soc. Am. Proc.* **16**, 298–300.

Peterson, J.R., and MacGregor, J.M. 1966. *Agron. J.* **58**, 141–142.

Peterson, L.A., and Newman, R.C. 1976. *Soil Sci. Soc. Am. J.* **40**, 280–282.

Piper, C.S. 1940. *Emp. J. Exp. Agric.* **8**, 85–96.

Pollard, A.S., Parr, A.J., and Loughman, B.C. 1977. *J. Exp. Bot.* **28**, 831–841.

Prasad, M., and Byrne, E. 1975. *Agron. J.* **67**, 553–556.

Purves, D., and MacKenzie, E.J. 1973. *Plant Soil* **39**, 361–371.

Purvis, E.R., and Carolus, R.L. 1964. *In* "Hunger Signs in Crops. A symposium" (3rd ed., H.B. Sprague, ed.), pp. 245–286. McKay, New York.

Reeve, E., and Shive, J.W. 1943. *Better Crops Plant Food* **27**(4), 14–16, 45–48.

Reeve, E., and Shive, J.W. 1944. *Soil Sci.* **57**, 1–14.

Reisenauer, H.M., Walsh, L.M., and Hoeft, R.G. 1973. *In* "Soil Testing and Plant Analysis" (L.M. Walsh and J.D. Beaton, eds.), pp. 173–200. Soil Sci. Soc. Am., Madison, Wisconsin.

Robertson, G.A., and Loughman, B.C. 1974. *New Phytol.* **73**, 821–832.

Robertson, L.S., Knezek, B.D., and Belo, J.O. 1975. *Commun. Soil Sci. Plant Anal.* **6**, 359–373.

Sherstnev, E.A., and Kurilenok, G.V. 1964. *Bot. Zh.* (*Leningrad*) **49**, 699–702.

Shorrocks, V.M. 1974. "Boron Deficiency—Its Prevention and Cure." Borax Consolidated Ltd., London, p. 56.

Simojoki 1969. *Maaseudun Tulevaisuus Suppl. Koetoiminta ja Käytäntö* No. 1 (*Soils Fertil.* **37**, 123 (1974)).

Sims, J.R., and Bingham, F.T. 1968. *Soil Sci. Soc. Am. Proc.* **32**, 364–369.

Singh, D.V., Chauhan, R.P.S., and Charan, R. 1976. *Indian J. Agron.* **21**, 309–310.

Sippola, J., and Erviö, R. 1977. *Finn. Chem. Lett.* (4–5) 138–140.

Sisler, E.C., Dugger, W.M., Jr., and Gauch, H.G. 1956. *Plant Physiol.* **31**, 11–17.

Slack, C.R., and Whittington, W.J. 1964. *J. Exp. Bot.* **15**, 495–514.

Smilde, K.W., and Luit, B. Van. 1970. *Bedrijfsontwikkeling* **1**, 30–31.

Smithson, J.B., and Heathcote, R.G. 1976. *Samaru Agric. Newsl.* **18**(1), 59–63.

Sparr, M.C. 1970. *Commun. Soil Sci. Plant Anal.* **1**, 241–262.

Stoyanov, D.V. 1971. *Pochvozn. Agrokhim.* **6**, 99–107.

Tanaka, H. 1966. *Plant Soil* **25**, 425–434.

Tanaka, H. 1967. *Soil Sci. Plant Nutr.* **13**, 41–44.

Touchton, J.T., and Boswell, F.C. 1975. *Agron. J.* **67**, 197–200.

Turner, J.R. 1975. *Fert. Solutions* **19**, 72–76.

Ulrich, A., and Hills, F.J. 1967. *In* "Soil Testing and Plant Analysis Part II." SSSA Special Publ. Ser. No. 2, pp. 11–24. Soil Sci. Soc. Am., Madison, Wisconsin.

Van de Venter, H.A., and Currier, H.B. 1977. *Am. J. Bot.* **64**, 861–865.

Vlamis, J., and Ulrich, A. 1971. *J. Am. Soc. Sugarbeet Technol.* **16**, 428–439.

Vlamis, J., and Williams, D.E. 1970. *Plant Soil* **33**, 623-628.

Walker, J.C., McLean, J.G., and Jolivette, J.P. 1941. *J. Agric. Res.* **62**, 573-587.

Wall, J.R., and Andrus, C.F. 1962. *Am. J. Bot.* **49**, 758-762.

Wallace, T. 1951. "The Diagnosis of Mineral Deficiencies in Plants by Visual Symptoms." H.M. Statinery Office, London.

Warington, K. 1923. *Ann. Bot. (London)* **37**, 629-672.

Wear, J.I. 1965. *In* "Methods of Soil Analysis, Part 2, Agronomy 9" (C.A. Black, ed.), pp. 1059-1063. Am. Soc. Agron., Madison, Wisconsin.

Wear, J.I., and Patterson, R.M. 1962. *Soil Sci. Soc. Am. Proc.* **26**, 344-346.

Weir, C.C. 1970. *J. Sci. Food Agric.* **21**, 545-547.

Weiser, C.J., Blaney, L.T., and Li, P. 1964. *Physiol. Plant.* **17**, 589-599.

Whittington, W.J. 1959. *Rep. School. Agric. Univ. Nottingham* pp. 51-53.

Wilcox, L.V. 1960. *Agric. Info. Bull.* 211, A.R.S. U.S. Dept. Agric., Washington, D.C., pp. 3-7.

Williams, D.E., and Vlamis, J. 1970. *Soil Sci. Plant Anal.* **1**, 131-139.

Willis, A.L. 1970. *Soil Sci. Plant Anal.* **1**, 205-211.

Wolf, B. 1940. *Soil Sci.* **50**, 209-217.

Wolf, B. 1971. *Commun. Soil Sci. Plant Anal.* **2**, 363-374.

Wolf, B. 1974. *Commun. Soil Sci. Plant Anal.* **5**, 39-44.

Yamaguchi, M., Howard, F.D., and Minges, P.A. 1958. *Proc. Am. Soc. Hortic. Sci.* **71**, 455-467.

Zapata, R.M. 1973. *J. Agric. Univ. Puerto Rico* **57**, 9-23.

Subject Index